An Elementary Overview of Mathematical Structures

Algebra, Topology and Categories

An Elementary Overview of Mathematical Structures

Algebra, Topology and Categories

Marco Grandis
Università di Genova, Italy

World Scientific

NEW JERSEY • LONDON • SINGAPORE • BEIJING • SHANGHAI • HONG KONG • TAIPEI • CHENNAI • TOKYO

Published by

World Scientific Publishing Co. Pte. Ltd.
5 Toh Tuck Link, Singapore 596224
USA office: 27 Warren Street, Suite 401-402, Hackensack, NJ 07601
UK office: 57 Shelton Street, Covent Garden, London WC2H 9HE

Library of Congress Control Number: 2020032226

British Library Cataloguing-in-Publication Data
A catalogue record for this book is available from the British Library.

AN ELEMENTARY OVERVIEW OF MATHEMATICAL STRUCTURES
Algebra, Topology and Categories

ISBN 978-981-122-031-9 (hardcover)
ISBN 978-981-122-032-6 (ebook for institutions)
ISBN 978-981-122-033-3 (ebook for individuals)

For any available supplementary material, please visit
https://www.worldscientific.com/worldscibooks/10.1142/11828#t=suppl

Desk Editor: Soh Jing Wen

To my grandchildren

Alessandra, Elena, Luigi

Preface

Since the last century, a large part of Mathematics is concerned with the study of mathematical structures, from groups to fields and vector spaces, from lattices to Boolean algebras, from metric spaces to topological spaces, from topological groups to Banach spaces.

More recently, these structured sets and their transformations have been assembled in higher structures, called categories.

We want to give a structural overview of these topics, where the basic facts of the different theories are unified through the 'universal properties' that they satisfy, and their particularities stand out, perhaps even more.

This book can be used as a textbook for Undergraduate Studies and for self-study. It can provide students of Mathematics with a unified perspective of subjects which are often kept apart. It is also addressed to students and researchers of disciplines having strong interactions with Mathematics, like Physics, Chemistry, Statistics, Computer Science, Engineering.

Contents

Preface *page* vii
Introduction 1
0.1 Structures in Mathematics 1
0.2 A structural perspective 1
0.3 An inductive approach 2
0.4 Borders and links 3
0.5 Notation and conventions 3
0.6 Acknowledgements 4
1 Algebraic structures, I 5
1.1 Introducing fields and vector spaces 6
1.2 Sets and algebraic structures 17
1.3 Abelian groups 31
1.4 Preordered sets and lattices 44
1.5 Semigroups, monoids and groups 57
1.6 Complements on groups and semigroups 68
1.7 Complements on the real line and set theory 75
2 Algebraic structures, II 82
2.1 Rings and fields 82
2.2 Rings and fields, continued 90
2.3 Modules and vector spaces 97
2.4 Linear dependence and bases 105
2.5 Algebras of polynomials 111
2.6 Matrices, linear and affine spaces 120
2.7 Constructing the complex field 130
3 Topological structures, I 138
3.1 Introducing continuity 139

3.2 Topological spaces and continuous mappings 149
3.3 Sequences and countability axioms 158
3.4 Subspaces and quotients 163
3.5 Topological products and sums 171
3.6 Limits of mappings and Hausdorff spaces 177

4 Topological structures, II 182
4.1 Connected spaces 182
4.2 Compact spaces 189
4.3 Limits at infinity and compactifications 195
4.4 Projective spaces and compact surfaces 201
4.5 Metric spaces 209
4.6 From topological groups to topological vector spaces 221

5 Categories and functors 227
5.1 An overview of Category Theory 227
5.2 Categories 232
5.3 Functors and natural transformations 241
5.4 Universal arrows and representable functors 250
5.5 Monomorphisms and epimorphisms 259

6 Categorical limits and adjunctions 267
6.1 Basic limits and colimits 267
6.2 General limits and completeness 275
6.3 Adjoint functors 284
6.4 Adjunctions between ordered sets 294
6.5 Adjoints and categorical limits in Topology 297
6.6 *Complements 301

7 Solutions and hints 308
7.1 Exercises of Chapter 1 308
7.2 Exercises of Chapter 2 324
7.3 Exercises of Chapter 3 336
7.4 Exercises of Chapter 4 352
7.5 Exercises of Chapter 5 360
7.6 Exercises of Chapter 6 365

References 371

Index 375

Introduction

0.1 Structures in Mathematics

The focus of various sectors of Mathematics has shifted, in the last century, from equations to structures. The reader has likely encountered some of them, in Mathematics or related sciences.

Algebra denotes now the study of algebraic structures, like groups and fields, vector spaces, lattices and Boolean algebras: all these are sets equipped with some operations, satisfying some conditions.

Topology deals with the study of continuity, based on metric spaces, topological spaces, uniform spaces, etc. Algebro-topological structures, from topological groups to Banach and Hilbert spaces, are at the roots of Functional Analysis.

Furthermore, each of these structures has transformations that preserve the structure, like the *homomorphisms* of an algebraic structure or the *continuous mappings* of topological spaces. Collecting all the structures of a given type, with the appropriate transformations, we form a higher structure called a 'category'; for instance, the category Gp of all groups and their homomorphisms (with their composition law), or the category Top of all topological spaces and their continuous mappings.

Categories can now be investigated, along the same lines used for algebraic or topological structures; we arrive thus at a theory of mathematical structures.

0.2 A structural perspective

We want to present a general, elementary overview of these frameworks: the main algebraic structures in the first two chapters, then the main topological ones in Chapters 3 and 4, all of them organised in categories in Chapters 5 and 6.

In all these structures, the basic facts are *governed by universal properties* that follow and repeat the same pattern. Listing a few examples of the many cases dealt with, we have:

- the cartesian product of structures of the same kind (in 1.2.3, 1.3.4, 1.4.8, etc.),

- the free group on a set, or the free ring, or the free structure of a certain kind (in 1.6.3, 5.4.2, etc.),

- the field of fractions of an integral ring (in 2.2.4),

- the completion of a metric space (in 4.5.6),

- the Hausdorff space associated to a general topological space (in 6.5.1),

- the universal compactification of a topological space (in 6.6.5).

Loosely speaking, in each case we want to find the 'best solution' of a problem, with respect to the structures that we are considering and their transformations; this solution is determined up to 'isomorphism', i.e. an invertible transformation in the theory we are considering. Highlighting this fact should help the reader to have a better understanding of these constructions, and hopefully a deeper one.

A general formalisation of a universal property will be given within Category Theory, in Section 5.4, but the various instances we are interested in can be made precise from the start, as the reader can see in the examples listed above.

This perspective involves Category Theory itself: for instance, the cartesian product of categories (in 5.2.5) is governed by the same universal property as all cartesian products.

0.3 An inductive approach

This book draws on the author's experience, while teaching courses of Algebra, or Topology, or Category Theory, or Calculus.

Notions are presented in a concrete, 'inductive' way, starting from elementary examples. Then their theory is developed, with new examples and many exercises. Rich structures are often presented before the more general ones, formally simpler but didactically more abstract – in the same way as, in the historical construction of mathematics, the former often preceded the latter.

Whenever possible, the reader is guided to build the theory, through a series of exercises. The unfolding – and beauty – of mathematics combines and alternates 'natural parts', where everything seems to go on by itself, with sudden turns where new directions or unexpected results appear; the reader should learn to work out the natural developments, and to realise the power of real advances.

Each chapter and section has its own introduction; many references for further reading or study are given.

0.4 Borders and links

This book is reasonably self contained. We assume, from the beginning, the existence of the field $\mathbb{R}$ of real numbers, satisfying the axioms listed in Section 1.1; this framework, the heritage of centuries of thinking and research, is a source of inspiration and examples for our study of algebraic and topological structures.

Some basic issues of Set Theory are reviewed in Section 1.2. Hints at a foundational setting of Mathematics, where everything is constructed *ab ovo*, can be found in Subsection 4.6.7.

Calculus is not studied here. The basic transcendental functions – exponential and logarithm, sine and cosine – are only used in a marginal way, e.g. for linear groups and complex numbers.

However, the properties of continuous real functions are explored in Chapters 3 and 4. Limits of functions 'at infinity' are interpreted by compactifications of euclidean spaces, in Section 4.3.

Some unusual links with Calculus and Physics also appear in the first chapters on algebraic structures: for instance, the interpretation of a linear differential equation with constant coefficients as a linear equation, in a module of C^∞-functions over a polynomial ring (in Subsection 2.6.8), or the interpretation of physical dimensions as elements of a vector space over the rational field, written in multiplicative notation (in Remark 2.3.6(b)).

0.5 Notation and conventions

Weak inclusion of sets is denoted by the symbol $\subset$, instead of the more usual $\subseteq$, which is not used here. General notation for number sets and other issues can be found at the beginning of Section 1.1.

A part marked with * is out of the main line of exposition. It may refer to issues dealt with further down, or be addressed to readers with some knowledge of the subject which is being analysed, or give references for higher topics.

Most exercises have a solution or convenient hints. These are deferred to the last chapter, or can be found below the exercise when they are important for the sequel. Easy exercises or exercises marked with * can be left to the reader.

0.6 Acknowledgements

I developed the approach to teaching mentioned above at my Department, beginning in a climate of discussion and experimentation which was alive in the late 1960's and the 1970's.

I inherited in part this approach from Gabriele Darbo, my professor and dear friend, together with my interest in Algebraic Topology, Homological Algebra and Category Theory.

I discussed the didactical solutions of this book with many colleagues, in particular with Ettore Carletti and Giacomo Caviglia, receiving helpful suggestions on its organisation.

I would also like to thank Dr Lim Swee Cheng, Ms Tan Rok Ting, and Ms Soh Jing Wen, at World Scientific Publishing Co., for their kind, effective help in the publication of this book, together with Mr Loo Chuan Min, the author of the cover.

Diagrams and figures are composed with 'xy-pic', by K.H. Rose and R. Moore – a free package.

1
Algebraic structures, I

An algebraic structure is a set equipped with some operations satisfying some conditions – the axioms of the structure that we are considering.

A mapping $f: X \to Y$ between two structures of the same kind is called a 'homomorphism' when it preserves the operations of the structure, and an 'isomorphism' if – moreover – it is bijective. Then the inverse mapping is also a homomorphism, and the objects X, Y have, essentially, the same structure.

The algebraic properties of the set $\mathbb{R}$ of real numbers, with respect to the main operations (addition and multiplication), are well known: properties of associativity, commutativity, distributivity, etc. We begin by listing them in Section 1.1, exploring their consequences, and how they can be extended to other set of numbers, or other sets.

All the main algebraic structures get out of this analysis, from groups to fields and vector spaces; they are explored in this chapter and the next. Some elementary points of Set Theory are reviewed in Section 1.2; we also need something about ordered sets, dealt with in Section 1.4.

Historically, group theory begins with Évariste Galois, in the 1830's, and Arthur Cayley in the 1850's. Vector spaces on the real field were formally introduced by Giuseppe Peano in 1888, as 'linear systems' [Pe].

The real development of the theory of groups and rings began in the first decades of the last century, under the impulsion of Emmy Noether.

Van der Waerden's 'Moderne Algebra', in 1940, was a milestone in this process: the book had many translations and augmented editions, as 'Modern Algebra', and later as 'Algebra' [Wa], in 1991. The dropping of 'Modern', in the title, reflects the fact that 'Algebra' denotes now – in Mathematics – the study of all algebraic structures: algebraic equations, the origins of this discipline, are just a part of it.

Serge Lang's [La2] is also a general textbook on this discipline. Books on the main algebraic structures will be cited in the appropriate section.

1.1 Introducing fields and vector spaces

The set $\mathbb{R}$ of real numbers contains the set $\mathbb{N}$ of natural numbers $0, 1, 2, \dots$, $n, \dots$ used to count the elements of a finite set. It also contains the set $\mathbb{Z}$ of all integers

$$\dots -2,\ -1,\ 0,\ 1,\ 2,\ \dots$$

and the set $\mathbb{Q}$ of rational numbers, i.e. the quotients h/k of two integers, with $k \neq 0$.

We write as $\mathbb{N}^*$, $\mathbb{Z}^*$, $\mathbb{Q}^*$, $\mathbb{R}^*$ the same sets without 0. The letter n usually stands for a natural number; saying $n \geqslant 1$ we mean $n \in \mathbb{N}^*$.

The elementary construction of the set $\mathbb{C}$ of complex numbers will be dealt with in the first part of Section 2.7; it can also be read after the present section.

We use as of now some basic notation for sets, that will be reviewed in Section 1.2. Writing $A \subset X$ (or $X \supset A$) we mean that A is a subset of X, and X is a superset of A: in other words, every element of A belongs to X; the sets can be equal.

If A and B are subsets of X, $A \cup B$ denotes their union and $A \cap B$ denotes their intersection, while $A \setminus B$ is the set of elements of A which do not belong to B. The complement $X \setminus A$ is also written as $C_X A$.

The symbol $\{x_1, \dots, x_n\}$ denotes the set formed by the elements of this list. In particular, the set $\{x\}$ has a unique element, and is called a *singleton*; we can write $\{*\}$ to avoid choosing a name for the element, when this is not relevant.

1.1.1 The real field and other fields

The set $\mathbb{R}$ of real numbers, also called the *real line*, comes with two main operations, the addition, or sum, $x + y$ and the multiplication, or product, $x.y$ (often written as xy), which are defined for all $x, y \in \mathbb{R}$.

Their main properties can be listed as follows.

(A.1) (*Associativity of the sum*) For every $x, y, z \in \mathbb{R}$ we have: $x + (y + z) = (x + y) + z$.

The result of both procedures can be written as $x + y + z$. Similarly, a finite sum of real numbers $x_1 + x_2 + \dots + x_n$ has a precise meaning.

(A.2) (*Identity of the sum*) There is a real number 0 such that, for all $x \in \mathbb{R}$: $0 + x = x = x + 0$.

This element is uniquely determined and called the *identity* of the sum. In fact, if $0'$ also satisfies the same relations, we deduce that: $0' = 0 + 0' = 0$ (using first the fact that 0 is an identity, and then the fact that $0'$ is also).

(A.3) (*Opposite element*) For every $x \in \mathbb{R}$ there is some $x' \in \mathbb{R}$ such that $x + x' = 0 = x' + x$.

This element is determined by x and called the *opposite*, or *additive inverse*, of x; it is written as $-x$. In fact, if x'' also satisfies the same relations, we deduce that:

$$x' = 0 + x' = (x'' + x) + x' = x'' + (x + x') = x'' + 0 = x'',$$

using first the fact that 0 is an identity, then a property of x'', then associativity, then a property of x', then again a property of 0.

(A.4) (*Commutativity of the sum*) For every x, $y \in \mathbb{R}$ we have: $x + y = y + x$.

The reader will note that, taking this into account, one could write (A.2) and (A.3) in a simplified form. We prefer to avoid this shortcut, for future developments where commutativity is not assumed.

(A.5) (*Associativity of the product*) For every $x, y, z \in \mathbb{R}$ we have: $x(yz) = (xy)z$.

Again, this allows us to write iterated products, like xyz and $x_1x_2 \ldots x_n$, without parentheses.

(A.6) (*Distributivity of the product over the sum*) For every $x, y, z \in \mathbb{R}$ we have:

$$x(y + z) = xy + xz, \qquad (x + y)z = xz + yz.$$

As usual, it is understood that $xy + xz$ means $(xy) + (xz)$: by default, a product has priority over a sum.

(A.7) (*Identity of the product*) There is a real number 1 such that, for all $x \in \mathbb{R}$: $1.x = x = x.1$.

This element is uniquely determined and called the *identity* of the product, or also the *unit* of $\mathbb{R}$. The proof is the same as in (A.2), changing notation from sum to product: if $1'$ satisfies the same relations, we deduce that: $1' = 1.1' = 1$.

(A.8) (*Commutativity of the product*) For every $x, y \in \mathbb{R}$ we have: $xy = yx$.

Also here we could simplify the other axioms using this property; but we do not.

(A.9) (*Inverse element*) $1 \neq 0$, and for every $x \neq 0$ there is some y such that $xy = 1 = yx$.

This element y is determined by x; it is written as x^{-1} and called the *inverse* of x. The proof is the same as in (A.3), changing sum to product. The property $1 \neq 0$ cannot be deduced from the others, as the 'null ring' will show, below.

More generally, a set K equipped with two operations satisfying the previous properties is called a *field*. The properties (A.1–9) are called *the axioms of fields*. The set of real numbers, equipped with its sum and multiplication, is called the *real field*.

The set $\mathbb{Q}$ of rational numbers is also a field, with the same operations of real numbers (restricted to $\mathbb{Q}$): see Exercise 1.1.3(i). It is called the *rational field.*

*The field $\mathbb{C}$ of *complex numbers* is constructed in Section 2.7.*

1.1.2 The ring of integers and other rings

The set of integers $\mathbb{Z}$, with its addition and multiplication, satisfies all the axioms above, except (A.9); it is called the *ring of integers.*

More generally, a set R equipped with two operations that satisfy the axioms (A.1–6) is called a *ring*. It is a *unital ring* if also (A.7) holds true, and a *commutative ring* if (A.8) is satisfied.

$\mathbb{Z}$ is thus a commutative unital ring, and is not a field. Rings of square matrices will give examples of unital rings that are not commutative, in Section 2.6. (Note that, in a ring, the addition is always assumed to be commutative.)

The singleton $\{0\}$, with the operation $0 + 0 = 0 = 0.0$ (the only possible one), is a commutative unital ring, called the *trivial ring*, or the *null ring*. Note that the additive and multiplicative identities coincide – the only point of the axioms of fields that is not satisfied here. (The name of the unique element is of no relevance.)

In a unital ring R, an element x is said to be *invertible* if there is some $y \in R$ such that $xy = 1 = yx$. Then y is determined by x, and written as x^{-1}. These elements form the set $\mathrm{Inv}(R)$, analysed in Exercise 1.1.3(f) below.

For instance, $\mathrm{Inv}(\mathbb{Z}) = \{-1, 1\}$. In a ring, the set $R \setminus \{0\}$ is often written as R^*; thus, a non-trivial commutative unital ring R is a field if and only if $\mathrm{Inv}(R) = R^*$.

1.1.3 Exercises and complements

For a beginner, it is important to understand how the axioms (A.1–9) are indeed the foundation of our use of the basic operations of real numbers.

This can be done with the following exercises, stated for a ring R, under additional hypotheses when it is the case; all of them hold for a field. Solutions can be found below.

(a) In a ring R we have:

$$-0 = 0, \qquad -(-x) = x, \qquad -(x + y) = (-x) + (-y). \tag{1.1}$$

(b) (*Cancellation law of the sum*) In a ring R, from $x + y = x + z$ it follows that $y = z$.

(c) For every $x \in R$ we have: $x.0 = 0 = 0.x$. A unital ring where $0 = 1$ is trivial (a singleton). A unital ring where 0 is invertible is trivial.

(d) For every $x, y \in R$ we have:

$$x.(-y) = -xy = (-x).y, \qquad (-x).(-y) = xy. \tag{1.2}$$

(e) One defines the *difference* $x - y = x + (-y)$. Prove that the product distributes over this 'derived' operation.

(f) If R is unital, the set $\mathrm{Inv}(R)$ of invertible elements contains the unit 1. If x and y are invertible, also x^{-1} and xy are, and

$$1^{-1} = 1, \qquad (x^{-1})^{-1} = x, \qquad (xy)^{-1} = y^{-1}.x^{-1}. \tag{1.3}$$

Therefore the set $\mathrm{Inv}(R)$ inherits a multiplication from the ring; this operation is associative, has an identity (the unit of the ring), and every element has an inverse. *(This will be expressed saying that $\mathrm{Inv}(R)$ is a group.)*

(g) (*Cancellation laws of the product*) If $x \in \mathrm{Inv}(R)$, from $xy = xz$ it follows that $y = z$. Similarly, if $yx = zx$ then $y = z$.

Thus, in a field, this multiplicative cancellation holds for the elements $x \neq 0$. The same is true in any ring which is contained in a field, with the 'same' operations, like $\mathbb{Z} \subset \mathbb{Q}$. (But we will see rings where there exist elements $x, y \neq 0$ with $xy = 0$.)

(h) In a field K one defines the *quotient* $x/y = x.y^{-1}$, provided that $y \neq 0$. Prove that $(xz)/(yz) = x/y$, if $y, z \neq 0$.

(i) The set $\mathbb{Q}$ of rational numbers inherits from $\mathbb{R}$ the structure of a field.

(j) A field can be finite: for instance, one can form a field $\mathbb{F}_2 = \{0, 1\}$ consisting of two distinct elements, which can be viewed as 'even' and 'odd'.

Solutions. (a) Obviously 0 is the opposite of itself, x is the opposite of $-x$, and $(-x)+(-y)$ is the opposite of $x + y$. Let us note that, without the commutativity of the sum, in (A.4), we should compute the opposite of $x + y$ as $(-y) + (-x)$.

(b) Add $-x$ to each member of the equation.

(c) We have: $x.0 + 0 = x.0 = x.(0 + 0) = x.0 + x.0$; cancelling $x.0$ we get $0 = x.0$.

(d) We have: $x.y + x.(-y) = x(y + (-y)) = x.0 = 0$, which means that $x.(-y)$ is the opposite of xy. The rest is an obvious consequence.

(e) Applying previous results, we have:

$$x(y - z) = x(y + (-z)) = xy + x(-z) = xy + (-xz) = xy - xz,$$

and symmetrically.

(f) As in (a); here the product is not assumed to be commutative.

(g) One multiplies by x^{-1}, at the left or the right.

(h) In fact $(xz)/(yz) = x.z.y^{-1}.z^{-1} = x.y^{-1}$.

(i) The rational numbers are 'stable' in $\mathbb{R}$ under sum, product and all the derived items considered in properties (A.1–9), as shown by the following well-known formulas of the 'calculus of fractions', for $a, b, c, d \in \mathbb{Z}$ and $b, d \neq 0$

$$a/b + c/d = (ad + bc)/(bd), \qquad 0 = 0/1, \qquad -(a/b) = (-a)/b,$$
$$(a/b).(c/d) = (ac)/(bd), \qquad 1 = 1/1, \qquad (b/d)^{-1} = d/b.$$

(j) Following the interpretation of the elements $0, 1$ as 'even' and 'odd', we define the operations in $\mathbb{F}_2$ as:

$$0 + 0 = 1 + 1 = 0, \qquad 0 + 1 = 1 + 0 = 1,$$
$$0.0 = 0.1 = 1.0 = 0, \qquad 1.1 = 1.$$

(Note that, here, $1 + 1$ is not defined as in $\mathbb{Z}$ and $\mathbb{R}$.)

These operations satisfy the axioms of fields. A direct proof, here, would be tiresome and of little interest: the reader can wait until this result will follow from a general construction, that will give a finite field having any prime number p of elements (in 2.1.6(c)).

1.1.4 Subrings and subfields

A *subring* of a ring R is a subset R' that satisfies the following conditions:

(i) if $x, y \in R'$ then $x + y \in R'$,

(ii) $0 \in R'$,

(iii) if $x \in R'$ then $-x \in R'$,

(iv) if $x, y \in R'$ then $xy \in R'$.

Among them we always have the *null* subring $\{0\}$ and the *total* subring R. A subring of a ring is a ring, with the restricted operations.

For a unital ring, a *unital subring* is assumed to contain the unit of the ring.

In a field K, a *subfield* K' of K is a unital subring that also satisfies

(v) if $x \in K'$ and $x \neq 0$, then $x^{-1} \in K'$.

A subfield of a field is a field. $\mathbb{Q}$ is a subfield of $\mathbb{R}$, while $\mathbb{Z}$ is a unital subring of both, but not a subfield.

A subring or subfield is said to be *proper* if it is not the total one.

Exercises and complements. The solutions can be found in Section 7.1.

(a) Prove that, in the definition of a subring, conditions (ii) and (iii) cannot be left out.

(b) The set $2\mathbb{Z}$ of all even integers forms a non-unital subring of $\mathbb{Z}$. There are infinitely many others. The only unital subring of $\mathbb{Z}$ is the total one.

(c) The rational field $\mathbb{Q}$ has no proper subfields. Any subfield of $\mathbb{R}$ contains $\mathbb{Q}$, that is called the *minimal subfield* of $\mathbb{R}$. (This topic will be developed in 2.2.1.)

1.1.5 Homomorphisms of rings

Let R, S be rings. A mapping $f\colon R \to S$ is said to be a *homomorphism* (of rings) if it preserves sum and multiplication, in the sense that, for all $x, y \in R$

(i) $f(x+y) = f(x) + f(y)$,

(ii) $f(xy) = f(x).f(y)$.

One can write

$$f(x +_R y) = f(x) +_S f(y), \qquad f(x \cdot_R y) = f(x) \cdot_S f(y),$$

to distinguish the operations of our rings. This is only done when useful.

The homomorphism f also preserves the identity of the sum and all opposites.

In fact $f(0_R) = 0_S$, as follows from $f(0) = f(0+0) = f(0) + f(0)$, cancelling $f(0)$ in S. Moreover, for $x \in R$, $f(x) + f(-x) = f(x-x) = f(0) = 0$.

The identity mapping $R \to R$ of a ring is a homomorphism, written as $\mathrm{id}\, R$. Given two consecutive homomorphisms $f\colon R \to S$ and $g\colon S \to T$, the composed mapping $gf\colon R \to T$ is a homomorphism: it takes any $x \in R$ to the element $g(f(x))$ of T. It will also be written as $g.f$, when useful.

This *partial composition law* is associative (whenever legitimate): given a third consecutive homomorphism $h\colon T \to U$, we have

$$h(gf) = (hg)f,$$

as both composites take any $x \in R$ to the element $h(g(f(x)))$ of U. Moreover, an identity homomorphism acts as an identity, for every legitimate composition: for a homomorphism $f\colon R \to S$ we have

$$f.(\mathrm{id}\, R) = f = (\mathrm{id}\, S).f. \tag{1.4}$$

If R' is a subring of R, the inclusion $R' \to R$ is a homomorphism. We have already considered some of them

$$\{0\} \to 2\mathbb{Z} \to \mathbb{Z} \to \mathbb{Q} \to \mathbb{R}. \tag{1.5}$$

An *isomorphism* $f\colon R \to S$ of rings is a homomorphism that has an inverse: there is a homomorphism $g\colon S \to R$ such that $gf = \mathrm{id}\, R$ and $fg = \mathrm{id}\, S$. This is equivalent to saying that the homomorphism f is bijective, i.e. injective and surjective, and g is the inverse mapping (see Exercise 1.1.6(c)).

When this is the case, we say that the rings R and S are *isomorphic*, and write $R \cong S$. The inverse homomorphism is written as f^{-1}.

1.1.6 Exercises and complements

The following properties are important; their proof can be found below.

(a) For all rings R, S the constant mapping $R \to S$ at 0_S is a homomorphism, called the null homomorphism. If R, S are unital rings, a *unital homomorphism* $f\colon R \to S$, or *homomorphism of unital rings*, is a homomorphism that preserves the unit (and can only be null when S is the null ring). If K is a field, there are no unital homomorphisms $K \to \mathbb{Z}$.

(b) Given two fields K and K', a *homomorphism* (of fields) $f\colon K \to K'$ is defined as a homomorphism of unital rings. Prove that it preserves all inverses, and that it is necessarily injective.

(c) An isomorphism $f\colon R \to S$ is bijective. Conversely, if $f\colon R \to S$ is a bijective homomorphism of rings, the inverse mapping is a homomorphism. The same holds for homomorphisms of unital rings.

(d) Isomorphism is an equivalence relation between rings.

(e) An isomorphism $f\colon R \to S$ of rings, between unital rings, is necessarily unital; more generally, this holds for every surjective homomorphism of rings whose domain is a unital ring.

Solutions. (a) The first point is obvious. If $f\colon K \to \mathbb{Z}$ is a unital homomorphism, $f(1_K + 1_K) = 2$, therefore $1_K + 1_K \neq 0_K$, and 2 should be invertible in $\mathbb{Z}$.

(b) First, if $x \neq 0$ in K, then $f(x).f(x^{-1}) = f(x.x^{-1}) = f(1) = 1$. Secondly, suppose that $x, y \in K$ and $f(x) = f(y)$. Therefore $f(x - y) = 0$ is not invertible in K', and $x - y = 0$, which means that $x = y$.

(c) The inverse mapping g is also a homomorphism, as we can cancel f in the following equalities

$$f(g(x) + g(y)) = f(g(x)) + f(g(y)) = x + y = f(g(x + y)),$$
$$f(g(x).g(y)) = f(g(x)).f(g(y)) = x.y = f(g(x.y)).$$

In the unital case one cancels f in: $f(g(1_S)) = 1_S = f(1_R)$.

(d) Reflexivity comes from the identity mapping $\mathrm{id}\,R$ of any ring. Symmetry follows from the definition, and transitivity from composing two consecutive isomorphisms.

In fact, the composite gf of two consecutive isomorphisms $f\colon R \to S$ and $g\colon S \to T$ is an isomorphism, with inverse $f^{-1}.g^{-1}\colon T \to R$.

(e) Plainly, the element $f(1_R)$ is a unit for every element $f(x)$ of S.

1.1.7 Vector spaces on a field

The reader likely knows that many physical quantities – like velocity, acceleration, force – are expressed by a vector. Vectors can be added and multiplied by real numbers, forming a 'vector space', an algebraic structure which appears everywhere in Mathematics and Physics.

We begin by an important instance, likely known in some form to the reader: the set $F(T,\mathbb{R})$ of all functions $f\colon T \to \mathbb{R}$ defined on a set T, with values in the real field.

This set has two basic operations (for $f, g \in F(T,\mathbb{R})$ and $\lambda \in \mathbb{R}$):

$$\begin{aligned} (f+g)(t) &= f(t) + g(t), \\ (\lambda f)(t) &= \lambda.f(t). \end{aligned} \tag{1.6}$$

The first operation acts on two functions $f, g\colon T \to \mathbb{R}$, yielding their *sum* $f+g$; this is computed pointwise, at each $t \in T$, by a sum $f(t) + g(t)$ of real numbers. The second operation acts on a number $\lambda \in \mathbb{R}$ and a function $f\colon T \to \mathbb{R}$, yielding their *scalar multiplication* λf (also written as $\lambda.f$); again, this is computed pointwise, by products $\lambda.f(t)$ in $\mathbb{R}$.

In this context a function $f \in F(T,\mathbb{R})$ is called a *vector*, a number $\lambda \in \mathbb{R}$ is called a *scalar* and the set $F(T,\mathbb{R})$ is called a *vector space*, or a *linear space*, on the real field.

More generally, we can replace the real field with any field K. A *vector space on the field* K is a set X equipped with two operations: the *sum* $x+y$ (for $x, y \in X$) and the *scalar multiplication* λx (for $x \in X$ and $\lambda \in K$). In both cases the result is an element of X; these elements are called *vectors*, while the elements of K are called *scalars*. The following axioms must be satisfied.

(VS.1) (*Associativity of the sum*) For every $x, y, z \in X$ we have: $x + (y+z) = (x+y) + z$.

(VS.2) (*Identity of the sum*) There is a vector $0 \in X$ such that, for all $x \in X$: $0 + x = x = x + 0$.

(VS.3) (*Opposite element*) For every $x \in X$ there is some $x' \in X$ such that $x + x' = 0 = x' + x$.

(VS.4) (*Commutativity of the sum*) For every $x, y \in X$ we have: $x + y = y + x$.

(VS.5) (*Distributive property, I*) For every $x, y \in X$ and $\lambda \in K$ we have: $\lambda(x+y) = \lambda x + \lambda y$.

(VS.6) (*Distributive property, II*) For every $x \in X$ and $\lambda, \mu \in K$ we have: $(\lambda + \mu).x = \lambda x + \mu x$.

(VS.7) (*Compatibility*) For every $x \in X$ and $\lambda, \mu \in K$ we have: $(\lambda\mu).x = \lambda.(\mu x)$.

(VS.8) (*Unitarity*) For every $x \in X$ we have: $1.x = x$.

The reader will note that the axioms (VS.1–4) coincide with the axioms

(A.1–4), for the sum in a field. (In both cases, as we will see in Section 1.3, we are saying that our structure is a *commutative group*, with respect to the sum.) Also here the *null vector* 0 is uniquely determined; the vector opposite to a vector x is determined by the latter, and written as $-x$; the cancellation law of the sum holds true.

The axioms (VS.5) and (VS.6) are distributive properties of the scalar multiplication, with respect to the sum of vectors or scalars. Usually the context is sufficient to distinguish the null vector from the null scalar; otherwise, the null vector can be written as $\underline{0}$, or 0_X.

(In Physics and Mathematical Physics, vectors are often distinguished by special characters, either underlined, or boldface, or marked with an arrow.)

The structure of vector spaces will be studied in Section 2.3, but a reader can find attractive (and certainly useful) to explore as of now their notions of homomorphism, isomorphism and substructure, with respect to a fixed field of scalars.

Exercises and complements. (a) For every set T, the set $F(T, \mathbb{R})$ equipped with the operations defined above is indeed a real vector space. More generally, any field K gives a vector space $F(T, K)$ of functions $f: T \to K$. *(There is also a pointwise product of functions $(fg)(t) = f(t).g(t)$, which makes $F(T, \mathbb{R})$ into a K-*algebra*, see 2.5.1.)*

(b) A singleton $\{x\}$ has a unique structure of vector space on K, with $x + x = x$ and $\lambda x = x$. We have thus the *trivial*, or *null*, vector space on the field K, often written as $\{0\}$.

The vector space $F(T, K)$ is trivial when $T = \emptyset$: it has one element, the unique mapping $\emptyset \to K$ (see 1.2.1).

(c) For a positive integer n, the set $K^n = K \times ... \times K$ of n-tuples $x = (x_1, ..., x_n)$ of elements of K is a vector space on the field K, with the following operations

$$\begin{aligned} (x_1, ..., x_n) + (y_1, ..., y_n) &= (x_1 + y_1, ..., x_n + y_n), \\ \lambda(x_1, ..., x_n) &= (\lambda x_1, ..., \lambda x_n). \end{aligned} \qquad (1.7)$$

For $n = 1$ we get the set K as a vector space on itself.

In particular, $\mathbb{R}^n$ is a vector space on the real field $\mathbb{R}$, formed of all the n-tuples $x = (x_1, ..., x_n)$ of real numbers. We assume that the reader is familiar with the representation of $\mathbb{R}$ as a line, of $\mathbb{R}^2$ as a plane and $\mathbb{R}^3$ as the three-dimensional space (after a system of cartesian coordinates is fixed in each of these geometrical structures).

(d) For the finite set $T = \{1, ..., n\}$, the vector space $F(T, K)$ can be identified with K^n.

(e) For every vector x in a vector space X we have: $0.x = \underline{0}$ and $(-1).x = -x$.

(f) (*Cancellation law of the scalar multiplication*) If $\lambda \neq 0$, from $\lambda x = \lambda y$ it follows that $x = y$.

(g) An expression $\lambda x + \mu y$ is called a *linear combination of the vectors x, y with scalar coefficients λ, μ*. More generally we have linear combinations

$$\textstyle\sum_i \lambda_i x_i = \lambda_1 x_1 + ... + \lambda_n x_n \qquad (\lambda_i \in K,\ x_i \in X). \qquad (1.8)$$

1.1.8 The natural order of the real field

We come back to examining the set $\mathbb{R}$ of real numbers. After addition and sum, governed by the axioms (A.1–9), in 1.1.1, the set $\mathbb{R}$ is equipped with a binary relation $x \leqslant y$, called the *natural order of real numbers.*

The main new properties are listed below, as (A.10–16).

(A.10) (*Reflexivity*) For every $x \in \mathbb{R}$ we have: $x \leqslant x$.

(A.11) (*Transitivity*) For every $x, y, z \in \mathbb{R}$, if $x \leqslant y$ and $y \leqslant z$ then $x \leqslant z$.

(A.12) (*Anti-symmetry*) For every $x, y \in \mathbb{R}$, if $x \leqslant y$ and $y \leqslant x$ then $x = y$.

(A.13) (*Totality*) For every $x, y \in \mathbb{R}$ we have $x \leqslant y$ or $y \leqslant x$.

This group of axioms only deals with the order relation; it says that $\mathbb{R}$, equipped with the relation $x \leqslant y$, is a *totally ordered set.* The relation $x \leqslant y$ is also written as $y \geqslant x$, while $x < y$ (and $y > x$) means that $x \leqslant y$ and $x \neq y$.

One can rewrite the axioms (A.10–15) using the relation $x < y$ (see Exercise 1.1.9(a)), but this is not convenient in the general theory of ordered sets.

(A.14) (*Addition and order*) For every $x, y, z \in \mathbb{R}$, if $x \leqslant y$ then $x + z \leqslant y + z$.

(A.15) (*Multiplication and order*) For every $x, y, z \in \mathbb{R}$, if $x \leqslant y$ and $z \geqslant 0$ then $xz \leqslant yz$.

These two axioms state the compatibility of the order relation with the main operations. Globally, the axioms (A.1–15) say that $\mathbb{R}$, equipped with addition, multiplication and order, is a *totally ordered field.* Note that we have written the compatibility conditions (A.14–15) in a form that takes advantage of the commutativity of both operations.

The rational field $\mathbb{Q}$, with the natural order, is also a totally ordered field.

(A.16) (*Completeness*) Every subset A of $\mathbb{R}$ which is non-empty and *upper bounded* has a *least upper bound*, written as $\sup A$.

Now, the axioms (A.1–16) say that $\mathbb{R}$ is a *complete totally ordered field.* We will see that this list of axioms determines the real field, up to isomorphism, i.e. up to a bijection that preserves addition, multiplication and order (see 2.2.7).

To make sense of the last axiom, we examine now various notions for a subset A of $\mathbb{R}$, related to the ordering. We take advantage of the reversion symmetry $r(x) = -x$, which reverses the order of $\mathbb{R}$ (by Exercise 1.1.9(b)); the image of A under this symmetry is written as:

$$-A = \{-x \mid x \in A\} = \{x \in \mathbb{R} \mid -x \in A\}. \tag{1.9}$$

The subset A is said to be *upper bounded* if it has an upper bound in $\mathbb{R}$, i.e. there is some $k \in \mathbb{R}$ such that $x \leqslant k$, for all $x \in A$. Symmetrically, A is said to be *lower bounded* if there is some lower bound $h \in \mathbb{R}$, with $h \leqslant x$ for all $x \in A$; it is said to be *bounded* if it satisfies both conditions. Plainly, A is lower bounded if and only if $-A$ is upper bounded.

We write as $\max A$ the *greatest element*, or *maximum*, of A, if it exists: it is a real number a such that:

$$a \in A, \qquad \text{for every } x \in A,\ x \leqslant a. \tag{1.10}$$

It is uniquely determined, because of the anti-symmetry property of the order. Symmetrically, $\min A$ denotes the *least element*, or *minimum*, of A, if it exists; then $-\min A = \max(-A)$.

We write as $\sup A$ the *least upper bound* of A, if it exists

$$\sup A = \min\{k \in \mathbb{R} \mid \text{for every } x \in A,\ x \leqslant k\}. \tag{1.11}$$

It is characterised as the real number α such that:

(i) for every $x \in A$, $x \leqslant \alpha$,

(ii) if $k \in \mathbb{R}$ and for every $x \in A$, $x \leqslant k$, then $\alpha \leqslant k$.

Plainly, A has a maximum if and only if the supremum of A exists and belongs to A; then $\max A = \sup A$.

Symmetrically, we write as $\inf A$ the *greatest lower bound* of A, if it exists; then $-\inf A = \sup(-A)$. The completeness axiom is equivalent to saying that every subset of $\mathbb{R}$ which is non-empty and lower bounded has a greatest lower bound.

1.1.9 Exercises and complements

(a) Write a set of axioms for the relation $x < y$, equivalent to the previous (A.10–15).

(b) Consider the *reversion symmetry* $r\colon \mathbb{R} \to \mathbb{R}$ defined by $r(x) = -x$, and note that this mapping is *involutive*, i.e. inverse to itself: $rr = \mathrm{id}\,\mathbb{R}$ (and therefore bijective).

Prove that r reverses the order relation: if $x \leqslant y$ then $-x \geqslant -y$. Prove also that $x \leqslant y$ and $z \leqslant 0$ imply $xz \geqslant yz$.

(c) Prove that $x^2 \geqslant 0$, for every $x \in \mathbb{R}$; this implies that $0 < 1$ (an 'obvious' fact, but also a consequence of the axioms) and $x < x + 1$, for every x.

(d) (*Modulus*) The *modulus*, or *absolute value*, of the real number x is defined as:

$$|x| = x, \text{ for } x \geqslant 0, \qquad |x| = -x, \text{ for } x \leqslant 0, \tag{1.12}$$

so that, for every $x \in \mathbb{R}$, $|x| \geqslant 0$, and $|x| = 0$ if and only if $x = 0$. Note that $x \leqslant |x|$, $|-x| = |x|$ and $|x|^2 = x^2$. There are other important properties, for $x, y, z \in \mathbb{R}$

$$\begin{aligned} |x+y| &\leqslant |x|+|y| && \textit{(subadditive property)}, \\ |xy| &= |x|.|y| && \textit{(multiplicative property)}, \\ |x-y|+|y-z| &\geqslant |x-z| && \textit{(triangle inequality)}. \end{aligned} \quad (1.13)$$

(e) For the empty subset, every real number is (trivially) a lower and upper bound; there is no inf nor sup. The total subset has no lower nor upper bound in $\mathbb{R}$. The subset $A = \{x \in \mathbb{R} \mid x > 0\}$ of all positive real numbers has $\inf A = 0$, no minimum and no upper bound.

(f) (*The integral part*) Every $x \in \mathbb{R}$ has an *integral part* in $\mathbb{Z}$

$$[x] = \max\{k \in \mathbb{Z} \mid k \leqslant x\}, \quad (1.14)$$

that satisfies the following inequalities

$$[x] \leqslant x < [x] + 1. \quad (1.15)$$

In particular, for every $x \in \mathbb{R}$, there is an integer $> x$.

1.2 Sets and algebraic structures

Mathematics is built with some *primitive* items: these are not defined but their use has to respect some rules. At an informal level, these primitive terms have a concrete meaning, which guides our use.

The foundation commonly used is Set Theory. Here we only give a brief, informal review of some basic notions about sets, that will be used throughout the book. This approach will necessarily leave some points undefined, without affecting our use.

Formal Set Theory is a complex subject, outside of our scopes. An interested reader is referred to [Ha, Kap, Je, Fk]; the first two books are more elementary.

1.2.0 Sets and elements

We think of a *set* X as a 'collection of elements'. The expression $x \in X$ is read as *x is an element of X*, or *x belongs to X*, or *x is in X*.

The sets X, Y are *equal* (written as $X = Y$) if and only if they have the same elements.

The relation $X \subset Y$, read as *X is contained in Y*, or *X is a subset of Y*, means that every element of X also belongs to Y. Equivalently we write

$Y \supset X$, read as Y *contains* X, or Y *is a superset of* X. Thus $X = Y$ is equivalent to the conjunction: $X \subset Y$ and $Y \subset X$.

X is a *proper subset* of Y if $X \subset Y$ and $X \neq Y$, a derived notion of marginal importance.

The *empty set* $\emptyset$ is defined by having no elements, and is contained in any set. A *singleton* $\{x\}$ is defined by having a unique element, namely x. One writes as $\{x_1, x_2, ..., x_n\}$ the set whose elements are specified in the list (and nothing else).

Curly brackets are also used to denote the subset of a set formed by the elements satisfying a certain property, as in the following examples:

$$\{x \in \mathbb{N} \mid x^2 = x\} = \{0, 1\}, \qquad \{x \in \mathbb{N} \mid x \text{ is even}\}. \tag{1.16}$$

Remarks. (a) Different letters or symbols can denote the same thing. Thus, if $n \geqslant 1$, the set $\{x_1, x_2, ..., x_n\}$ has at least one element and at most n; it has precisely n elements if and only if all x_i are different. Even when we speak of "*two elements* x, y", the common use in mathematics does not assume that they are different: this should be explicitly said, if needed.

(b) The expression "x *is an element of* X" is a relation between sets, and does not mean that x has a different status. This relation is subject to various axioms, which imply that $x \in x$ cannot happen. It is also well known that one cannot form 'the set of all sets'.

(c) The procedure exemplified in (1.16) describes *a subset of a given set.* The reader probably knows that an illegitimate use, like $S = \{x \mid x \notin x\}$ leads to a contradiction, *Russell's paradox*: $S \in S$ implies $S \notin S$, and conversely.

An expression $\{x \mid p(x)\}$, where $p(x)$ is some property in the variable x, is only acceptable when we are leaving understood that x is required to belong to some set, specified by the context.

1.2.1 Mappings and cardinals

A *mapping* $f\colon X \to Y$ *from the set* X *to the set* Y is defined by a formula $f(x)$ that transforms every element x of the set X into a unique element $f(x)$ of the set Y, read as "f of x". The notation $x \mapsto f(x)$ denotes the action of f on an element of X.

The mapping f is said to be defined on $X = \operatorname{Dom} f$, the *domain* of f, and to take values in $Y = \operatorname{Cod} f$, the *codomain* of f. All the mappings $X \to Y$ are elements of a set, written as $\operatorname{Map}(X, Y)$, or also as Y^X (a 'cartesian power': see (1.30)).

We also speak of a *function* $f\colon X \to Y$; this term is commonly used when the codomain Y is the real field, as in 1.1.7, where the set $\operatorname{Map}(T, \mathbb{R})$ is written as $F(T, \mathbb{R})$.

Given two *consecutive mappings* $f\colon X \to Y$ and $g\colon Y \to Z$ (where the codomain of f coincides with the domain of g), the *composed mapping* is written as gf, or $g.f$, and defined as:

$$gf\colon X \to Z, \qquad (gf)(x) = g(f(x)) \qquad \text{(for } x \in X). \qquad (1.17)$$

Composition is associative (when legitimate): given a third mapping $h\colon Z \to T$ we have: $h(gf) = (hg)f$. Moreover, every set X has an *identity mapping*, written as $\mathrm{id}\, X$ or 1_X

$$\mathrm{id}\, X\colon X \to X, \qquad (\mathrm{id}\, X)(x) = x \qquad \text{(for } x \in X), \qquad (1.18)$$

which acts as an identity for legitimate compositions: $f.(\mathrm{id}\, X) = f$ and $(\mathrm{id}\, Y).f = f$, for $f\colon X \to Y$.

A mapping $f\colon X \to Y$ is *injective* if, for all $x, x' \in X$, the relation $f(x) = f(x')$ implies $x = x'$. It is *surjective* if, for every $y \in X$, there exists some $x \in X$ such that $f(x) = y$.

The mapping $f\colon X \to Y$ is *bijective*, or a *bijection*, or a *bijective correspondence*, if it is injective and surjective; equivalently, this means that for every $y \in X$ there exists a unique $x \in X$ such that $f(x) = y$. We can then construct a mapping $g\colon Y \to X$, backwards, letting

$$g(y) = x \ \text{ if and only if } \ f(x) = y \qquad \text{(for } y \in Y,\ x \in X), \qquad (1.19)$$

and we say that the sets X, Y are *equipotent.*

A mapping $f\colon X \to Y$ is *invertible* if there is a mapping $g\colon Y \to X$ such that $gf = \mathrm{id}\, X$ and $fg = \mathrm{id}\, Y$. Plainly, this is the case if and only if f is bijective. The inverse function g, constructed as in (1.19), is determined by f, and can be written as f^{-1}.

An *indexed family* $x = (x_i)_{i \in I}$ of elements of X is a mapping $x\colon I \to X$, written in *index notation*; the domain I is then called *the set of indices* of the family.

For each set X there is a unique mapping $\emptyset \to X$, and therefore a unique *empty family* $(x_i)_{i \in \emptyset}$ of elements of X.

For a finite set, $\sharp X$ will denote the (natural) number of its elements. More generally, each set X has an equipotent *cardinal set* $\sharp X$, and two sets X, Y have the same cardinal if and only if there exists a bijection $X \to Y$. The smallest infinite cardinal is $\aleph_0 = \sharp\mathbb{N}$, read as 'aleph-zero'. (Something more on cardinals will be said in Subsection 1.7.8.)

Plainly, a subset of a finite set X with the same cardinal must be the total one. The reader may know that this fact is no longer true for an infinite set X: see Exercise (e) below.

Exercises and complements. (a) (*Commutative diagrams*) Mappings between sets (or structured sets) can be represented by vertices and arrows in a *diagram*, as in the examples below, to make evident their relationship and which compositions are legitimate

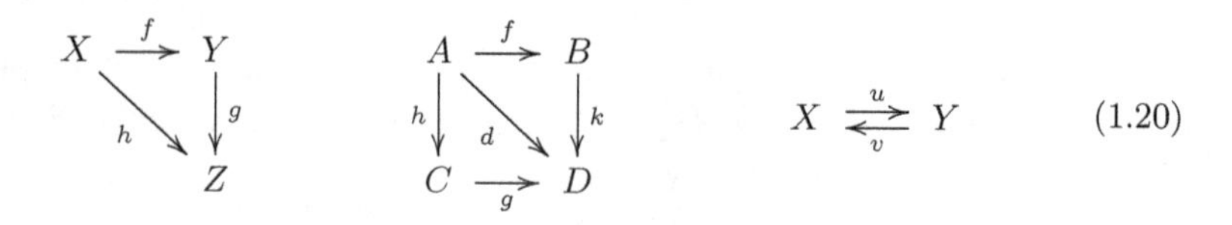

(1.20)

As an important property, we say that such a diagram is *commutative* if:

- whenever we have two 'paths' of consecutive arrows, from a certain object to another, the two composed mappings are the same,

- whenever we have a 'loop' of consecutive arrows, from an object to itself, then the composed mapping is the identity of that object.

Thus, the first diagram above is commutative if and only if $gf = h$. For the second, commutativity means that $kf = d = gh$. For the third, it means that $vu = \mathrm{id}\, X$ and $uv = \mathrm{id}\, Y$ (so that these mappings are inverse to each other).

(b) If X and Y are finite sets, with m and n elements, then the set $Y^X = \mathrm{Map}(X, Y)$ has n^m elements.

(c) In particular, for $X = \emptyset$, the set $Y^{\emptyset} = \mathrm{Map}(\emptyset, Y)$ is a singleton (also when $Y = \emptyset$), and $n^0 = 1$. *In the context of natural numbers*, 0^0 is defined and equal to 1. (The reader likely knows that, in the context of real numbers, the expression 0^0 is preferably left undefined.)

(d) For consecutive mappings $f\colon X \to Y$ and $g\colon Y \to Z$

- if f and g are injective (resp. surjective, bijective), so is the composed mapping gf,
- if gf is injective, then f is also; if gf is surjective, then g is also.

(e) The set $2\mathbb{N}$ of even natural numbers has the same cardinal as $\mathbb{N}$, as the mapping $f\colon \mathbb{N} \to 2\mathbb{N}$ defined by the formula $f(n) = 2n$ is bijective.

(f) Let us note that the mapping $\mathbb{N} \to \mathbb{N}$ defined by the same formula is not surjective. Injectivity and surjectivity of a mapping $f\colon X \to Y$ only make sense *with respect to assigned sets*, as a domain and a codomain.

*(g) As another example, the mapping $f\colon \mathbb{R} \to \mathbb{R}$, $f(x) = x^2$, is neither injective nor surjective, as $f(-1) = f(1)$ and we have seen that $f(x) \geqslant 0$, for all $x \in \mathbb{R}$.

But the reader likely knows (and will also find in 1.7.4) that, taking restrictions to the interval $J = \{x \in \mathbb{R} \mid x \geqslant 0\}$:

- the mapping $g\colon J \to \mathbb{R}$ defined by the same formula is injective, not surjective,

- the mapping $h\colon \mathbb{R} \to J$ defined by the same formula is surjective, not injective,

- the mapping $k\colon J \to J$ defined by the same formula is bijective, and its inverse is the *main* square root $\sqrt{-}\colon J \to J$.

We also note that an expression like 'the square roots of the number x' (or the formula $\pm\sqrt{x}$) does not define a mapping $J \to \mathbb{R}$, as it takes *two* values on each positive number. (Multi-valued relations *can* be considered, with due care, but are not mappings.)

1.2.2 The power set

Every set X has a *power set* $\mathcal{P}X$, whose elements are the subsets of X:

$$A \in \mathcal{P}X \quad \Leftrightarrow \quad A \subset X, \tag{1.21}$$

where the symbol $\Leftrightarrow$ stays for 'if and only if' (also written as 'iff').

The relation of (weak) inclusion $A \subset B$ in $\mathcal{P}X$ is an order relation (see Section 1.4). In other words, for all $A, B, C \in \mathcal{P}X$ we have

$$\begin{aligned}
&A \subset A && (\textit{reflexivity}),\\
&A \subset B \subset C \Rightarrow A \subset C && (\textit{transitivity}),\\
&A \subset B \subset A \Rightarrow A = B && (\textit{anti-symmetry}),
\end{aligned} \tag{1.22}$$

where the symbol $\Rightarrow$ stays for 'implies'.

The least element of $\mathcal{P}X$ is the *empty subset* $\emptyset$; the greatest element is the *total subset* X.

The set $\mathcal{P}X$ has two main operations, called *union* and *intersection*

$$\begin{aligned}
A \cup B &= \{x \in X \mid x \in A \text{ or } x \in B\},\\
A \cap B &= \{x \in X \mid x \in A \text{ and } x \in B\}.
\end{aligned} \tag{1.23}$$

Let us note that 'or' is (always) meant in the inclusive sense, which admits that both conditions can hold: $A \cap B \subset A \cup B$. The algebraic properties of these operations will be examined in Section 1.4. Two subsets A, B are said to be *disjoint* when $A \cap B = \emptyset$; otherwise, we say that A *meets* B. The set $A \cap B$ is also called the *trace* of A on B, or the *trace* of B on A.

More generally, we can start from a family $(A_i)_{i \in I}$ of subsets of X, indexed by a set I (possibly infinite), and consider their union and intersection:

$$\begin{aligned}
\textstyle\bigcup_i A_i &= \{x \in X \mid \text{ there exists } i \in I \text{ such that } x \in A_i\},\\
\textstyle\bigcap_i A_i &= \{x \in X \mid \text{ for all } i \in I,\ x \in A_i\}.
\end{aligned} \tag{1.24}$$

The empty family of subsets has union $\emptyset$ and intersection X. We also recall that $A \setminus B$ is the set of elements of A which do not belong to B.

A mapping $f \colon X \to Y$ induces two mappings

$$f_* \colon \mathcal{P}X \to \mathcal{P}Y, \qquad f^* \colon \mathcal{P}Y \to \mathcal{P}X, \tag{1.25}$$

where f_* takes a subset $A \subset X$ to its *image* $f(A) \subset Y$, while f^* takes a subset $B \subset X$ to its *preimage* $f^{-1}(B) \subset X$

$$\begin{aligned}
f(A) &= \{y \in Y \mid \text{ there exists } x \in A \text{ such that } y = f(x)\},\\
f^{-1}(B) &= \{x \in X \mid f(x) \in B\}.
\end{aligned} \tag{1.26}$$

In particular, the subset $\operatorname{Im} f = f(X) \subset Y$ is said to be the *image of the mapping* f; the latter is surjective if and only if $\operatorname{Im} f = Y$. One often writes $f(A)$ in the shortened form $\{f(x) \mid x \in A\}$. Note also that, for a bijective mapping f, the preimage $f^{-1}(B)$ is the same as the image of B with respect to the inverse mapping f^{-1}, and there is no conflict of notation.

This topic, the *transfer of subsets along a mapping*, will be further examined in 6.4.3.

Exercises and complements. (a) Given a mapping of sets $f: X \to Y$, the mapping $f_*: \mathcal{P}X \to \mathcal{P}Y$ of direct images preserves the unions, but need not preserve intersections, including the empty one.

On the other hand, the preimage-mapping $f^*: \mathcal{P}Y \to \mathcal{P}X$ preserves all unions and intersections.

(b) If I is a set, and for each $i \in I$ we have a set A_i (defined by some 'well-formed formula'), one assumes the existence of a set X that contains all A_i. Thus all A_i belong to $\mathcal{P}X$, and form an indexed family $(A_i)_{i \in I}$ in the latter.

The formulas (1.24), for union and intersection, make sense also in this case, as their result does not depend on the superset X we are using, *with one exception*: the intersection of the empty family is only defined for a specified superset X.

(c) For each set X, there is a canonical bijection

$$\chi: \mathcal{P}X \to \operatorname{Map}(X, \{0, 1\}), \tag{1.27}$$

that takes a subset $A \subset X$ to its *characteristic function* $\chi_A: X \to \{0, 1\}$.

The latter is defined on each $x \in X$ as

$$\chi_A(x) = 1, \quad \text{if } x \in A, \qquad\qquad \chi_A(x) = 0, \ \text{otherwise}. \tag{1.28}$$

Loosely speaking, the term 'canonical' highlights the fact that χ is defined by an explicit formula, not depending on choice.

(d) Therefore, if X is finite, with $n \geqslant 0$ elements, the power set $\mathcal{P}X$ has 2^n elements. From combinatorics, we know that X has $\binom{n}{k}$ subsets of k elements, for $0 \leqslant k \leqslant n$; this gives again $\sharp(\mathcal{P}X) = 2^n$.

(e) For every set X, $\sharp X \leqslant \sharp\mathcal{P}X$. (Set theory proves that $\sharp X < \sharp\mathcal{P}X$, see 1.7.8.)

1.2.3 Cartesian products and their universal property

Let $(A_i)_{i \in I}$ be a family of sets, indexed by a set I. As we have seen (in 1.2.2(b)), there is some set X such that $A_i \subset X$ for all indices i, and we can take $X = \bigcup_i A_i$. The *cartesian product* $A = \prod_{i \in I} A_i$ is defined as a subset of the set $\operatorname{Map}(I, X)$

$$A = \{x: I \to X \mid x(i) \in A_i, \ \text{for all } i \in I\}. \tag{1.29}$$

An element x is generally written as an indexed family $(x_i)_{i \in I}$, or simply as (x_i).

In particular, if all the factors A_i are the same set X, we have a *cartesian power*

$$X^I = \textstyle\prod_{i \in I} X = \operatorname{Map}(I, X). \tag{1.30}$$

Coming back to the general case, the cartesian product comes with a family of *cartesian projections*

$$p_i \colon A \to A_i, \qquad\qquad p_i((x_i)_{i\in I}) = x_i, \tag{1.31}$$

which allows us to formalise our construction, in a way that can be adapted to any kind of structured sets we will consider – a unifying approach already stressed in the general Introduction, in Section 0.2.

The cartesian product of a family of sets $(A_i)_{i\in I}$ can be viewed as a set A provided with a family of mappings $p_i \colon A \to A_i$ (for $i \in I$), satisfying the following *universal property of the product* (of sets):

(i) for every similar pair $(B, (f_i \colon B \to A_i)_{i\in I})$ formed of set B and a family of mappings $f_i \colon B \to A_i$, there exists precisely one mapping $f \colon B \to A$ such that

$$\begin{array}{ccc} B & \overset{f}{\dashrightarrow} & A \\ & \underset{f_i}{\searrow} & \downarrow{\scriptstyle p_i} \\ & & A_i \end{array} \qquad\qquad p_i f = f_i \ \ (\text{for } i \in I). \tag{1.32}$$

In fact, all these triangles commute (as defined in 1.2.1(a)) if and only if the mapping $f \colon B \to A$ is defined as $f(y) = (f_i(y))_{i\in I}$, for all $y \in B$.

It is crucial to note that the universal property determines its solution *up to a unique bijection.* In fact, if the pair $(A', (q_i)_{i\in I})$ is also a solution of (i), we have two (well determined) mappings

$$\begin{array}{lll} f \colon A' \to A, & p_i f = q_i & (\text{for } i \in I), \\ g \colon A \to A', & q_i g = p_i & (\text{for } i \in I), \end{array} \tag{1.33}$$

and they are inverse to each other. This comes out of the fact that

$$q_i.(gf) = p_i f = q_i = q_i.\mathrm{id}\, A',$$

for all indices i, so that $gf = \mathrm{id}\, A'$; similarly, $fg = \mathrm{id}\, A$.

A binary product is written as $X_1 \times X_2$, and an element is written as an (ordered) *pair* (x_1, x_2), with $x_1 \in X_1$ and $x_2 \in X_2$. This is an indexed family (on the set $I = \{1, 2\} \subset \mathbb{N}$), and determines its *first term* x_1 and its *second term* x_2; therefore $(x_1, x_2) = (y_1, y_2)$ if and only if $x_1 = y_1$ and $x_2 = y_2$.

Similarly, in a finite product $X_1 \times X_2 \times ... \times X_n$, an element is written as an (ordered) *n-tuple* $(x_1, x_2, ..., x_n)$, with $x_i \in X_i$.

Exercises and complements. (a) If in formula (1.29) we replace the set $X = \bigcup A_i$ with any set X' which contains all the sets A_i, we get a set $A' \subset \mathrm{Map}(I, X')$ related to A by a canonical bijection.

(b) The reader is warned that the cartesian projections $p_i \colon A \to A_i$ of a product are *not always* surjective. This fails whenever some factor A_i is empty (so that the product is empty) and some other factor is not. Outside of this situation, the axiom of choice allows us to conclude that all projections are surjective (see 1.7.6).

(c) A unary product, of a family (A) consisting of a single term, is the set A with its identity projection. The product of the empty family $(A_i)_{i \in \emptyset}$ of sets has one element, the empty mapping $\emptyset \to X$ (however we choose the superset X), with no projection.

(d) (*Disjoint unions*) Given a family $(A_i)_{i \in I}$ of sets, we construct their 'disjoint union', namely the set

$$A = \bigcup_i A_i \times \{i\}, \tag{1.34}$$

where we have replaced the original A_i with a set $B_i = A_i \times \{i\}$ in obvious bijection with the former, so that the new sets are pairwise disjoint: if $i \neq j$ then $B_i \cap B_j = \emptyset$.

This set comes equipped with a family of mappings

$$u_i \colon A_i \to A, \qquad\qquad u_i(x) = (x, i), \tag{1.35}$$

which satisfies the following *universal property of the sum* (of sets):

(ii) for every similar pair $(B, (f_i \colon A_i \to B)_{i \in I})$ formed of set B and a family of mappings $f_i : A_i \to B$, there exists precisely one mapping $f : A \to B$ such that

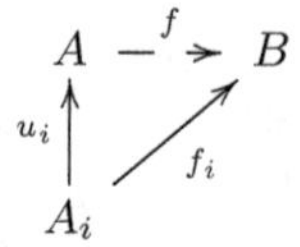

$$\begin{array}{ccc} A & \overset{f}{\dashrightarrow} & B \\ {\scriptstyle u_i}\uparrow & \nearrow_{f_i} & \\ A_i & & \end{array} \qquad\qquad f u_i = f_i \ \text{ (for } i \in I). \tag{1.36}$$

Also here the solution of the universal property is 'essentially unique', *up to a unique bijection.* Note also that this property is 'dual' to the universal property of the product, in the sense that any of them is turned into the other 'by reversing the arrow of each mapping'; all this will be made precise within category theory, in Chapter 5.

(e) If, in the previous point, all A_i coincide with a set X, their disjoint union (1.34) is the cartesian product $X \times I$.

1.2.4 Equivalence relations and quotient sets

A *relation* R in a set X is a subset $R \subset X \times X$. When $(x, y) \in R$, we say that x is *R-related* to y; this is often written as $x\, R\, y$.

The relation R is said to be:

(i) *reflexive* if, for all $x \in X$, we have $x\, R\, x$,

(ii) *symmetric* if, for all $x, y \in X$, $x\, R\, y$ implies $y\, R\, x$,

(iii) *transitive* if, for all $x, y, z \in X$, $x\, R\, y$ and $y\, R\, z$ imply $x\, R\, z$.

We say that R is an *equivalence relation* when these three properties are satisfied. Then, for each $x \in X$, the *equivalence class* of x (with respect to R) is the subset

$$[x] = \{x' \in X \mid x\,R\,x'\} \subset X, \tag{1.37}$$

also written as $[x]_R$ or $\overline{x}$. The element x is said to be a *representative* of the class $[x]$; the element x' is also if and only if $x\,R\,x'$.

The *quotient of the set X modulo R*, written as X/R, is the set of all equivalence classes of X.

Formally, X/R is a subset of $\mathcal{P}X$; but we generally think of X/R in a more intuitive way, as if we had 'identified' all the elements of X which lie in the same equivalence class. Thus, the equivalence relation $x = \pm x'$ (more formally: $x = x'$ *or* $x = -x'$) in the real line can be described as: 'the relation that identifies each number with the opposite one'.

The *canonical projection*

$$p\colon X \to X/R, \qquad p(x) = [x] \quad (\text{for } x \in X), \tag{1.38}$$

is always surjective.

The equivalence relations of a set X are ordered by inclusion $R \subset R'$ (as subsets of $X \times X$). The *finest*, or smallest, is the equality relation $x = y$ in X, determined by the *diagonal* of the product $X \times X$

$$\Delta_X = \{(x, y) \in X \times X \mid x = y\}. \tag{1.39}$$

The *coarsest*, or largest, is the relation $x, y \in X$, determined by the total subset $X \times X$.

A mapping $f\colon X \to Y$ has an *associated equivalence relation* R_f on X:

$$xR_fx' \text{ if and only if } f(x) = f(x'). \tag{1.40}$$

There is a unique mapping $g\colon X/R_f \to Y$ such that

$$\begin{array}{ccc} X & \xrightarrow{f} & Y \\ {\scriptstyle p}\downarrow & \nearrow{\scriptstyle g} & \\ X/R & & \end{array} \qquad f = gp. \tag{1.41}$$

In fact, we can (and must) define $g([x]) = f(x)$, for all $x \in X$. The mapping f is injective if and only if R_f is the equality relation.

Exercises and complements. (a) The quotient of the set X modulo the finest equivalence relation Δ_X is in canonical bijection with X itself. The quotient of the set X modulo the coarsest equivalence relation is a singleton if $X \neq \emptyset$, and is empty otherwise.

(b) In the set of all straight lines of the euclidean plane (or the euclidean 3-dimensional space), *parallelism* is an equivalence relation. The quotient set can be interpreted as the set of *directions* of the plane (or the space).

(c) (*Partitions*) A *partition* of a set X is a family of disjoint subsets $(A_i)_{i \in I}$ that *cover* X

$$X = \bigcup_{i \in I} A_i, \qquad A_i \cap A_j = \emptyset \quad (\text{for } i \neq j \text{ in } I). \tag{1.42}$$

This amounts to giving an equivalence relation R in the set X.

(d) Every binary relation R on the set X *generates* an equivalence relation E, the least equivalence relation of X containing R.

1.2.5 The canonical factorisation

A mapping $f \colon X \to Y$ between sets has a *canonical factorisation*

$$\begin{array}{ccc} X & \xrightarrow{\ f\ } & Y \\ {\scriptstyle p}\downarrow & & \uparrow{\scriptstyle m} \\ X/R_f & \xrightarrow[g]{} & \operatorname{Im} f \end{array} \qquad f = mgp, \tag{1.43}$$

where:

- the (surjective) mapping p is the canonical projection of the domain X onto its quotient modulo the associated equivalence relation R_f,

- the (injective) mapping m is the inclusion of the image $\operatorname{Im} f$ into the codomain Y,

- the (bijective) mapping $g \colon X/R_f \to \operatorname{Im} f$ is defined by $g([x]) = f(x)$ (for $x \in X$), and is the only mapping such that $f = mgp$.

From (1.43), we can deduce a factorisation of f formed of a surjective and an injective mapping

$$f = (mg).p, \tag{1.44}$$

which is 'essentially unique', as made precise in the exercise below.

(Strictly speaking, there are infinitely many such factorisations, including $f = m.(gp)$.)

Exercises and complements. (a) Prove the following property of 'essential uniqueness', up to a determined bijection.

Given two factorisations $f = mp = m'p'$, where the mappings p, p' are surjective and m, m' are injective, there is precisely one bijection i such that the following diagram commutes

$$\begin{array}{ccccc} X & \xrightarrow{\ p\ } & A & \xrightarrow{\ m\ } & Y \\ \| & & \downarrow{\scriptstyle i} & & \| \\ X & \xrightarrow[p']{} & A' & \xrightarrow[m']{} & Y \end{array} \qquad p' = ip, \quad m'i = m. \tag{1.45}$$

1.2.6 Induction

Let $A \subset \mathbb{N}$. If the following conditions are satisfied

(*i*) $0 \in A$ (*initial step*),

(*ii*) for every $n \in A$, $n+1 \in A$ (*inductive step*),

we conclude that $A = \mathbb{N}$. (A well-known procedure, called *a proof by induction.*)

In fact, if there is some natural number that is not in A, we can let m be the least of them. But $m > 0$, by (i), and therefore $m-1$ must be in A. Applying (ii) we get $m \in A$, a contradiction.

If we replace the initial step with $n_0 \in A$, the conclusion says that A contains all the natural numbers $\geqslant n_0$.

Exercises and complements. (a) Prove by induction that the sum $s_n = 0+1+2+ ... +n$ (of integers) can be expressed as $n(n+1)/2$, for all $n \in \mathbb{N}$.

(b) (*Complete induction*) Replacing the inductive step (ii) by the following (apparently) weaker assumption:

(ii′) for every $n \in \mathbb{N}^*$, if $\{0, 1, ..., n-1\} \subset A$ then $n \in A$,

the conclusion still holds: $A = \mathbb{N}$. The procedure is now called *a proof by complete induction.*

(c) (*Prime factor decomposition*) Let us recall that a natural number $p > 1$ is said to be *prime* if it has no proper divisor in $\mathbb{N}$: if $p = ab$ then $a = 1$ or $a = p$. Prove that any natural number $n > 1$ is a product of prime numbers. (One can add 1, as the product of the empty family of prime numbers.)

1.2.7 Structures and categories

A set can carry structures of various kinds. We have seen various examples in Section 1.1:

- algebraic structures, defined by operations, like fields, rings, and so on,
- order structures, defined by a relation, like ordered and preordered sets,
- ordered algebraic structures, like ordered fields,

and will see other kinds later, like the topological and the algebro-topological structures.

In each of these kinds, there are ‘privileged mappings’, or *morphisms*, that preserve the structure in a specified sense, like homomorphisms of fields, or order preserving mappings between ordered sets, or order preserving homomorphisms between ordered fields. In each kind, the morphisms are closed under composition, and include the identity of each object. This partial composition law is associative (whenever legitimate), and any identity morphism acts as an identity for every legitimate composition.

These objects and morphisms form thus a *category of structured sets.* This topic will be investigated in Chapter 5, but it will be useful to present now – informally – its basic elements.

In each category, a morphism $f\colon X \to Y$ is said to be an *isomorphism* if it admits an inverse, i.e. a backward morphism $g : Y \to X$ (of the category that we are considering) such that $gf = \mathrm{id}\, X$ and $fg = \mathrm{id}\, Y$. By the usual proof (as in Section 1.1), the morphism g is determined by f, and can be written as f^{-1}. The isomorphism relation $X \cong Y$, meaning that there exists an isomorphism $X \to Y$ (in the category that we are considering) is an equivalence relation.

A morphism $X \to X$ is called an *endomorphism* of X, and an *automorphism* if it is invertible.

In any category of structured sets, an isomorphism is necessarily a bijective mapping. The converse need not be true. For instance the identity mapping $\mathrm{id}\, \mathbb{R}$ gives an order-preserving mapping

$$f\colon (\mathbb{R}, =) \to (\mathbb{R}, \leqslant), \qquad (1.46)$$

from the set $\mathbb{R}$ equipped with the discrete order, to the same set equipped with the natural order; this is not an isomorphism of ordered sets, because the inverse mapping (of sets) is not order-preserving, and does not belong to the category that we are considering. The same example works in the category of ordered fields, and similar ones will be given for topological spaces.

However, it is important to remark once for all that in a 'category of (pure) algebraic structures' every bijective morphism is an isomorphism: in each case this can be proved as in Exercise 1.1.6(c), for rings.

The *transport of a structure*, along a bijection, is also a useful tool. For instance, suppose that K is an ordered field, A is a set and $f\colon K \to A$ is a bijective mapping. Then there is one and only one structure of ordered field on A that makes f into an isomorphism (of ordered fields): in fact each element of A can be written in a unique way as $f(x)$ (with $x \in K$), and we can (and must) let, for all $x, y \in K$:

$$\begin{gathered} f(x) + f(y) = f(x+y), \qquad f(x).f(y) = f(x.y), \\ f(x) \leqslant f(y) \quad \Leftrightarrow \quad x \leqslant y. \end{gathered} \qquad (1.47)$$

We have already seen some instance of a *universal property.* It is an important issue, that will be developed in various forms, in all the structures we will consider: either algebraic, or order-like, or topological, or some combination of the previous kinds. A general definition can be given within category theory, as we will see in Section 5.4 and exploit thereafter.

1.2.8 Algebraic structures and equational algebras

Let us reconsider, more formally, the algebraic structure of rings.

A ring R is usually presented as a set equipped with two binary operations (in additive and multiplicative notation, respectively)

$$\begin{aligned} &\sigma_R \colon R^2 \to R, \qquad &\sigma_R(x, y) = x + y, \\ &\mu_R \colon R^2 \to R, \qquad &\mu_R(x, y) = xy, \end{aligned} \tag{1.48}$$

satisfying the axioms (A.1–6).

It can also be presented as a set R equipped with four operations, adding to the previous ones a *unary* operation and a *constant*, or *zero-ary* operation

$$\begin{aligned} &\omega_R \colon R \to R, \qquad &\omega_R(x) = -x, \\ &\zeta_R \colon R^0 \to R, \qquad &\zeta_R(*) = 0, \end{aligned} \tag{1.49}$$

defined, respectively, on $R^1 = R$ and the singleton $R^0 = \{*\}$.

The second presentation, if more complex, has a crucial advantage: now the axioms of the structure can be written in *equational form*, only depending on the universal quantifier *for all*, applied to all the elements of R. In the present case, we require that, *for all* $x, y, z \in R$

$$\begin{gathered} x + (y + z) = (x + y) + z, \qquad x + 0 = x = 0 + x, \\ x + (-x) = 0 = (-x) + x, \qquad x + y = y + x, \\ x(yz) = (xy)z, \\ x(y + z) = xy + xz, \qquad (x + y)z = xz + yz. \end{gathered} \tag{1.50}$$

An algebraic structure which can be presented in such a form will be called an *equational algebraic structure*, or an *equational algebra*, and their complex will be called a *variety of algebras*; the homomorphisms are always defined as the mappings that preserve all the operations. (The study of varieties of algebras is the subject of Universal Algebra [Gr1, Coh]. A brief presentation of this discipline can be found in [G4].)

Other equational algebraic structures, to be studied later, include: semigroups, monoids, groups, commutative groups, unital rings, modules on a ring, vector spaces on a field, lattices, boolean algebras, etc.

Automatically, a variety of algebras $\mathcal{V}$ has important properties: let us simply mention here that the cartesian product $A \times B$ of two algebras in $\mathcal{V}$, equipped with the componentwise extension of all the operations of the structure, is again an algebra in $\mathcal{V}$.

Fields are an important example of an algebraic structure *which is not equational*, as readily detected by the fact that the cartesian product $K \times K'$ of two fields is a unital ring, but not a field: the element $(1, 0)$ cannot have an inverse, as $(1, 0).(x, y) = (x, 0) \neq (1, 1)$.

Concretely, we cannot require $0 \neq 1$ in equational form, and we cannot transform the existence of inverses into a global unary operation $x \mapsto x'$ satisfying global equational conditions. *Not because we have not yet been able*, but because this possibility is contradicted by the previous remark on cartesian products.

1.2.9 Commments and remarks

(a) To describe the structure of an $\mathbb{R}$-vector space X as an equational algebra, one uses an infinite set of operations, comprising the addition $\sigma_X : X^2 \to X$, the opposite $\omega_X : X \to X$, the additive identity $\zeta_X : X^0 \to X$, and – for each scalar $\lambda \in \mathbb{R}$ – the *multiplication by* λ as a unary operation

$$\lambda_X : X \to X, \qquad \lambda_X(x) = \lambda x. \tag{1.51}$$

Then the distributivity axioms are written in the (universally quantified) form

$$\lambda_X(x+y) = \lambda_X(x) + \lambda_X(y), \quad (\lambda+\mu)_X(x) = \lambda_X(x) + \mu_X(x). \tag{1.52}$$

(There are more economical presentations; for instance, one can get the opposite by $(-1)_X$. But this is not really important.)

(b) (*Horror vacui*) We will not require that an algebra be non-empty, as is often done in Universal Algebra (e.g. in [Gr1], but not in [Coh]).

If the structure we are considering contains some zero-ary operation, this condition is automatically satisfied (and no problem arises): for instance, a ring always contains its additive identity.

But consider now a structure without zero-ary operations, like that of a semigroup: this is a set X equipped with a binary operation that is required to be associative. A subsemigroup is a subset stable under this operation. If we require all semigroups to be non-empty, the intersection of two disjoint subsemigroups will not be a subsemigroup and we lose an important property of substructures. More important properties would also be lost, as we will see later (in 6.6.4(e)).

Disjoint subsemigroups can occur: for instance, in the multiplicative semigroup $\mathbb{N}$, we have $\{1\}$ and $2\mathbb{N} = \{2n \mid n \in \mathbb{N}\}$.

(c) If A is a structured set, for instance a commutative group, it can be useful to denote as $|A|$ the *underlying set*, deprived of its structure. For instance, we will see in the next section that a commutative group A has a ring of endomorphisms $\mathrm{End}(A)$, which should not be confused with the semigroup $\mathrm{End}(|A|)$ of all endomappings of the underlying set.

(d) Sets are equational algebras, with no operations and no axioms (and for sure we cannot forbear the empty set). The homomorphisms of this structure are just mappings.

(e) A *pointed set* is a pair (X, x_0) consisting of a set X and a *base*-element $x_0 \in X$, while a *pointed mapping* $f\colon (X, x_0) \to (Y, y_0)$ is a mapping $f\colon X \to Y$ such that $f(x_0) = y_0$. This structure is again a very simple equational algebra, defined by a zero-ary operation $X^0 \to X$, under no axioms.

1.3 Abelian groups

We begin our study of algebraic structures with commutative groups, usually called abelian groups.

More elementary structures, from semigroups to monoids and (general) groups, will be studied in Section 1.5. Being more general, they are often treated before abelian groups, but our approach is aimed at 'concreteness', as discussed in the general Introduction.

Abelian groups are named after Niels Henrik Abel, a 19th century mathematician. Their theory goes far beyond the basic aspects examined here. An interested reader can see [Fu].

1.3.0 Main definitions

An *abelian group*, or *commutative group*, is a set A equipped with a binary operation $x + y$ which satisfies the axioms (A.1–4) of 1.1.1: associativity, existence of the identity (written as 0), existence of the opposite (or additive inverse) of any element x (written as $-x$), and commutativity.

The *cancellation law* holds: from $x + y = x + z$ it follows that $y = z$ (as in Exercise 1.1.3(b)). Also here, we define the *difference* $x - y = x + (-y)$.

If A is finite, the number of its elements is called the *order* of the group.

While developing the theory of abelian groups, the additive notation is generally preferred, but an example will carry its 'natural' (or historical) notation, be it additive, or multiplicative, or of a different kind. When a multiplicative notation is used, like xy, or $x.y$, or $x * y$, the identity is written as 1, while the 'opposite' of an element x is called an inverse, and written as x^{-1}.

Examples, exercises and complements. (a) The singleton $\{0\}$, with the only possible operation $0 + 0 = 0$, is an abelian group, the *trivial* or *null* group.

(b) Each ring R is an abelian group, with respect to its addition; if necessary, we can write $(R, +)$ to make clear that we are only considering this operation.

(c) In a commutative unital ring R, the set $\mathrm{Inv}(R)$ of invertible elements inherits a multiplication xy from the ring, and forms a commutative group, as we have seen in Exercise 1.1.3(f). Of course this group is still written in multiplicative notation. For a field K, the group $\mathrm{Inv}(K) = K^*$ is formed of all non-zero elements of K.

(d) The sets $\mathbb{Z}$, $\mathbb{Q}$, $\mathbb{R}$ are abelian groups, with respect to their (usual) addition.

The sets $\mathbb{Q}^*$, $\mathbb{R}^*$ are commutative groups, with respect to multiplication. The set $\mathrm{Inv}(\mathbb{Z}) = \{-1, 1\}$ is also a commutative group.

(e) Prove that the power set $\mathcal{P}X$ of any set X is an abelian group with respect to the *symmetric difference*

$$A\Delta B = (A \cup B) \setminus (A \cap B) = (A \setminus B) \cup (B \setminus A). \tag{1.53}$$

In this abelian group *each element is opposite to itself.*

(f) Adding binary intersection, $(\mathcal{P}X, \Delta, \cap)$ is a commutative unital ring.

If $X = \emptyset$, $\mathcal{P}X$ is the null ring. If X is a singleton, the ring $\mathcal{P}X$ has two elements, $\emptyset$ and X, and is isomorphic to the two-element field $\mathbb{F}_2$ of Exercise 1.1.3(j).

(g) In an abelian group, the sum $\sum_{i\in I} x_i$ of a finite family of elements makes sense without any ordering on the set of indices I, because the operation is associative and commutative. The sum of the empty family is defined to be 0.

It will be useful to extend this notation to an *essentially finite sum* $\sum_{i\in I} x_i$, where the set of indices I is arbitrary but the family (x_i) is *quasi null*: this means that its *support* $J = \{i \in I \mid x_i \neq 0\}$ is a finite subset of I. Then we let $\sum_{i\in I} x_i = \sum_{i\in J} x_i$.

(h) Abelian groups from a variety of algebras, in the sense of 1.2.8.

1.3.1 Homomorphisms, subgroups and kernels

A *homomorphism* $f\colon A \to B$ *of abelian groups* is a mapping that preserves the operation:

$$f(x + y) = f(x) + f(y) \qquad (\text{for } x, y \in A). \tag{1.54}$$

(One can write $f(x +_A y) = f(x) +_B f(y)$, when useful to distinguish the operations.) It follows that f preserves the identity of the sum, all opposites and all differences, as we have verified in 1.1.5.

As examined in 1.2.7 for any algebraic structure, two consecutive homomorphisms of abelian groups, $f\colon A \to B$ and $g\colon B \to C$, give a composed homomorphism $gf\colon A \to C$. This partial composition law is associative (whenever legitimate), and any identity homomorphism $\mathrm{id}\, A\colon A \to A$ acts as a unit for every legitimate composition.

An *isomorphism* $f\colon A \to B$ of abelian groups is a homomorphism that has an inverse homomorphism: there is a homomorphism $g\colon B \to A$ such that $gf = \mathrm{id}\, A$ and $fg = \mathrm{id}\, B$. This happens if and only if the homomorphism f is a bijective mapping, and $g = f^{-1}$ is the inverse homomorphism. Then the abelian groups A and B are said to be *isomorphic*, and we write $A \cong B$, an equivalence relation between abelian groups.

A *subgroup* of an abelian group A is a subset $H \subset A$ such that:

(i) H is stable in A under the addition: if $x, y \in H$ then $x + y \in H$,

(ii) H contains the identity 0 of A,

(iii) H is stable in A under opposites: if $x \in H$ then $-x \in H$.

Then H is an abelian group, under the restricted operation, and the inclusion $H \to A$ is a homomorphism. There is always the null subgroup $\{0\}$ and the total subgroup A.

A homomorphism $f\colon A \to B$ of abelian groups determines a subset of its domain, called the *kernel*, and a subset of its codomain, called the *image*

$$\begin{aligned} \operatorname{Ker} f &= \{x \in A \mid f(x) = 0\} \subset A, \\ \operatorname{Im} f &= \{y \in B \mid \text{there exists } x \in A \text{ such that } f(x) = y\} \subset B, \end{aligned} \tag{1.55}$$

which are proved below to be subgroups.

1.3.2 *Exercises and complements*

The following exercises are important, and easy.

(a) Each of the conditions (i), (ii), (iii) of 1.3.1 is independent of the other two. One can replace (ii) with the condition: A is not empty.

(b) The subsets $\operatorname{Ker} f$ and $\operatorname{Im} f$, in (1.55), are subgroups of A and B, respectively.

(c) (*The lattice of subgroups*) For an abelian group A, we write as $\operatorname{Sub}(A)$ the set of all subgroups of A, ordered by inclusion. This ordered set has a minimum $\{0\}$ and a maximum A.

Moreover, two subgroups H and K have a *meet* (the greatest lower bound) and a *join* (the least upper bound), with respect to inclusion in $\operatorname{Sub}(A)$

$$\begin{aligned} H \wedge K &= \inf\{H, K\} = H \cap K, \\ H \vee K &= \sup\{H, K\} = H + K \\ &= \{a \in A \mid \text{there are } h \in H,\, k \in K \text{ such that } a = h + k\}. \end{aligned} \tag{1.56}$$

All this will be expressed saying that the ordered subset $\operatorname{Sub}(A)$ is a *lattice*, in Section 1.4.

(d) A homomorphism $f\colon A \to B$ is injective if and only if $\operatorname{Ker} f = \{0\}$, the null subgroup of its domain. On the other hand, of course, $f\colon A \to B$ is surjective if and only if $\operatorname{Im} f = B$, the total subgroup of its codomain.

(e) (*The group of homomorphisms*) For two abelian groups A, B, the set

Hom(A, B) of all homomorphisms from A to B is an abelian group, when equipped with the *pointwise sum* of $f, g \in \text{Hom}(A, B)$

$$(f+g)(x) = f(x) + g(x) \qquad (\text{for } x \in A). \tag{1.57}$$

The identity of this operation is the *zero homomorphism* from A to B

$$0_{AB} \colon A \to B, \qquad 0_{AB}(x) = 0_B, \tag{1.58}$$

and the opposite of $f \in \text{Hom}(A, B)$ is the *opposite homomorphism*, also computed pointwise:

$$(-f)(x) = -f(x). \tag{1.59}$$

(f) For any abelian group A, the group Hom$(\mathbb{Z}, A)$ is canonically isomorphic to A.

(g) (*The ring of endomorphisms*) For an abelian group A, the set End(A) = Hom(A, A) of all *endomorphisms* of A is a unital ring, when equipped with the previous sum and the composition law $(f, g) \mapsto gf$. This ring is not commutative, generally. It is a multiplicative subsemigroup of the semigroup End$(|A|)$ of all endomappings of the underlying *set* $|A|$, and of course we should not confuse these items.

1.3.3 Multiples and linear combinations

In an abelian group A we can write any finite sum $x_1 + x_2 + ... + x_n$ of elements without parentheses. In particular, for every $x \in A$ and every integer $n \geqslant 0$ we have the *multiple element* $nx = x + x + ... + x$ (a sum of n terms), inductively defined by:

$$0.x = 0_A, \qquad (n+1)x = nx + x \qquad (n \geqslant 0). \tag{1.60}$$

Moreover, for a negative integer $k = -n < 0$, we let

$$kx = n(-x). \tag{1.61}$$

Multiples have the following properties, for $x, y \in A$ and $h, k \in \mathbb{Z}$

$$\begin{array}{lll} (i)\ \ hx + kx = (h+k)x, & 0_{\mathbb{Z}}.x = 0_A, & (-h)x = -(hx), \\ (ii)\ \ h(kx) = (hk)x, & 1_{\mathbb{Z}}.x = x, & \\ (iii)\ hx + hy = h(x+y), & h.0_A = 0_A, & h(-x) = -(hx). \end{array}$$

This will be proved in Section 1.5, working in multiplicative notation, where the multiple kx becomes the power x^k. More precisely, we will see in Exercises 1.5.1(d) and 1.5.2(c) that these properties hold in a commutative semigroup for positive integers h, k; in a commutative unital semigroup for

$h, k \geqslant 0$; in a commutative group for $h, k \in \mathbb{Z}$. In fact, commutativity will be replaced by asking that the given elements x, y commute.

Complements. (a) In an abelian group A, we can consider any (finite) *linear combination*

$$\textstyle\sum_i \lambda_i x_i = \lambda_1 x_1 + \lambda_2 x_2 + \ldots + \lambda_n x_n, \tag{1.62}$$

of elements $x_i \in A$, with integral coefficients $\lambda_i \in \mathbb{Z}$. For a subset $X \subset A$, the set of all linear combinations of elements of X forms a subgroup, and actually the least subgroup of A containing the subset X. It is called the subgroup of A *generated by* X, and written as $\langle X \rangle$. We say that A is *finitely generated* if it has a finite set of generators.

(b) In particular, we have subgroups

$$\langle x \rangle = \{kx \mid k \in \mathbb{Z}\} \subset A, \tag{1.63}$$

generated by a single element, including $\{0\} = \langle 0 \rangle$. In the group of integers, such a subgroup is more often written as

$$(n) = n\mathbb{Z} = \{kn \mid k \in \mathbb{Z}\} \subset \mathbb{Z}, \tag{1.64}$$

for reasons related to the theory of rings (see 2.1.6). We generally use a generator $n \geqslant 0$, but of course $(-n) = (n)$.

(c) A *cyclic group* is an abelian group A which has one generator, i.e. one element x such that $A = \langle x \rangle$. For instance, $\mathbb{Z}$ is cyclic, with generator 1 (or also -1), and we will soon see that there is a finite cyclic group of any order $n \geqslant 0$ (in 1.3.6).

(One simply speaks of a 'cyclic group' because every group with one generator is necessarily commutative; this rather obvious fact can be found in 1.5.2(c).)

(d) One can easily verify that a homomorphism $f \colon A \to B$ of abelian groups preserves multiples: $f(kx) = kf(x)$, working by induction on $k \geqslant 0$ and using opposites. More generally, f preserves all linear combinations with integral coefficients

$$f(\textstyle\sum_i \lambda_i x_i) = \sum_i \lambda_i f(x_i) \qquad (\lambda_i \in \mathbb{Z},\ x_i \in A). \tag{1.65}$$

1.3.4 Cartesian products

For two abelian groups A, B, the *cartesian product* $A \times B$ is an abelian group, when equipped with the *componentwise sum*

$$(x, y) + (x', y') = (x +_A x', y +_B y') \qquad (x, x' \in A,\ y, y' \in B). \tag{1.66}$$

The verifications are straightforward; the identity is the pair $(0_A, 0_B)$, and the opposite of (x, y) is the pair $(-x, -y)$.

More generally, for a family of abelian groups $(A_i)_{i \in I}$ indexed by a set I, the cartesian product $A = \prod_{i \in I} A_i$ is an abelian group under the componentwise sum

$$(x_i) + (y_i) = (x_i + y_i) \qquad (x_i, y_i \in A_i), \tag{1.67}$$

where it is understood that each sum $x_i + y_i$ is computed in A_i.

The cartesian product comes with a family of *cartesian projections*

$$p_i \colon A \to A_i, \qquad p_i((x_i)_{i \in I}) = x_i, \tag{1.68}$$

which are homomorphisms.

As for sets, in 1.2.3(i), we have a *universal property of the product* (of abelian groups). Given a family of abelian groups $(A_i)_{i \in I}$, their cartesian product can be viewed as an abelian group A provided with a family of homomorphisms $p_i \colon A \to A_i$ (for $i \in I$), such that:

(i) for every similar pair $(B, (f_i \colon B \to A_i)_{i \in I})$ formed of an abelian group B and a family of homomorphisms $f_i \colon B \to A_i$, there is precisely one homomorphism $f \colon B \to A$ such that

$$\begin{array}{ccc} B & \xrightarrow{f} & A \\ & {\scriptstyle f_i}\searrow & \downarrow {\scriptstyle p_i} \\ & & A_i \end{array} \qquad p_i f = f_i \quad (i \in I). \tag{1.69}$$

In fact all these triangles commute if and only if $f(y) = (f_i(y))_{i \in I}$, for all $y \in B$. Moreover, defining the mapping $f \colon B \to A$ in this way we do get a homomorphism.

Again, the universal property determines its solution *up to canonical isomorphism*: if the pair $(A', (q_i)_{i \in I})$ is also a solution of property (i), we have two (well determined) homomorphisms

$$\begin{array}{lll} f \colon A' \to A, & p_i f = q_i & (\text{for } i \in I), \\ g \colon A \to A', & q_i g = p_i & (\text{for } i \in I), \end{array} \tag{1.70}$$

and they are inverse to each other (as in 1.2.3(i)).

1.3.5 Quotients

Let A be an abelian group and E an equivalence relation in A. We say that E is a *congruence* (of abelian groups) if it is consistent with the operation in A. Precisely, this means that:

$$\text{if } xEx' \text{ and } yEy' \text{ then } (x + y)\, E\, (x' + y'). \tag{1.71}$$

One denotes as A/E the *quotient* of the set A modulo the equivalence relation E. By definition, its elements are the equivalence classes of E

$$[x] = \{x' \in A \mid x'Ex\}, \tag{1.72}$$

which form a partition of the set A (by Exercise 1.2.4(c)).

Condition (1.71) allows us to define an induced operation on the set A/E

$$[x] + [y] = [x + y], \tag{1.73}$$

which is obviously associative, has identity $[0]$ and opposites $-[x] = [-x]$.

A/E becomes thus an abelian group, called the *quotient* of the abelian group A modulo the congruence E. The canonical projection on the quotient set

$$p \colon A \to A/E, \qquad p(x) = [x], \tag{1.74}$$

is a homomorphism, and the structure we have put on A/E is the only one having this outcome.

For a fixed abelian group A, there is a natural bijection *between congruences and subgroups*

$$\begin{aligned} E &\mapsto \{x \in A \mid x\, E\, 0_A\} = \operatorname{Ker} p, \\ H &\mapsto \equiv_H, \qquad x \equiv_H x' \;\Leftrightarrow\; x - x' \in H. \end{aligned} \tag{1.75}$$

The reader can easily prove this fact (or see the solution of Exercise 1.3.9(a)).

The quotient of A modulo the associated congruence $\equiv_H$ is denoted as A/H and read as X *modulo* H. In this quotient, the equivalence class of an element x is determined as

$$[x] = x + H = \{x + h \mid h \in H\}, \tag{1.76}$$

and called a *coset* of H (with respect to the element x).

The null subgroup $\{0\}$ determines the discrete congruence $x = y$, so that $A/\{0\}$ can (and will) be identified with A. The total subgroup A determines the indiscrete congruence $x, y \in A$, so that A/A is a null group, and can be written as $\{0\}$.

Since a subgroup is more elementary notion than a congruence, the quotient of abelian groups are often presented in the form A/H. Yet, the notion of congruence in an object A makes sense for any equational algebraic structure, and in various cases cannot be expressed by means of a substructure of A (see 1.5.7).

Exercises and complements. (a) For a homomorphism $f \colon A \to B$, the equivalence relation R_f coincides with the congruence of A associated to the subgroup $\operatorname{Ker} f$, defined (for $x, y \in A$) by

$$f(x) = f(y) \;\Leftrightarrow\; f(x - y) = 0 \;\Leftrightarrow\; (x - y) \in \operatorname{Ker} f. \tag{1.77}$$

(b) A congruence E of A is always a subgroup of $A \times A$.

1.3.6 Exercises and complements (Modular arithmetic)

Everyone is familiar with adding integers *modulo 7*, when we want to know the day of the week in (say) 15 days; or *modulo 12*, when we want to know the time in (say) 15 hours; or *modulo 9*, when we check an addition of integers by 'casting out nines'.

Each of these 'modular arithmetics' is based on a quotient of the abelian group $\mathbb{Z}$, as the following list of exercises will show. Because of their importance, the proof can be found below. The argument is closed by a warning against a frequent error of beginners, in (g).

(a) The basic tool we are to use is the well-known *euclidean division*, or *division with remainder*. For two integers x, n, with $n > 0$, there are unique integers q, r (the *quotient* and the *remainder*) such that:

$$x = nq + r, \qquad 0 \leqslant r < n. \tag{1.78}$$

(b) (*The subgroups of the integers*) We already know that, for every natural number $n \in \mathbb{N}$, the set of its integral multiples

$$n\mathbb{Z} = \{nh \mid h \in \mathbb{Z}\}, \tag{1.79}$$

is the subgroup (n) of $\mathbb{Z}$ generated by n (see 1.3.3(b)). We can now prove that, in this way, we get all the subgroups of $\mathbb{Z}$ (and each of them only once). In particular $0\mathbb{Z} = \{0\}$ is the null subgroup, $1\mathbb{Z} = \mathbb{Z}$ is the total one and $2\mathbb{Z}$ is the subgroup of even integers.

(c) Each quotient of $\mathbb{Z}$ is thus of the form

$$\mathbb{Z}/(n\mathbb{Z}) \qquad (n \geqslant 0), \tag{1.80}$$

also written as $\mathbb{Z}/n$. The case $n = 0$ gives back the group $\mathbb{Z}$.

(We avoid the notation $\mathbb{Z}_n$, which is used for this and a different structure.)

(d) If $n > 0$, the quotient $\mathbb{Z}/n$ is finite and has n elements, namely the cosets

$$[0], \ [1], ..., \ [n-1].$$

For every $x \in \mathbb{Z}$, formula (1.78) gives $[x] = [r]$. In other words, the *distinguished representative* of the class $[x]$ is the remainder r of the division of x by n.

Adding, or subtracting, integers modulo n simply means to compute in the abelian group $\mathbb{Z}/n$, as a quotient of $\mathbb{Z}$. We will see that this practice also works for multiplication (in Section 2.1).

(e) When 'casting out nines', the class modulo 9 of the integer $k = 30421$ (written in decimal notation) is computed as $[k] = [1+2+4+0+3] = [1]$. Why is this correct?

(f) Every finite non-empty set can be given a structure of abelian group. *(The same is true of infinite sets: see Exercise 2.1.6(d).)*

(g) (*An important warning*) We have seen that each quotient $\mathbb{Z}/n\mathbb{Z}$ has n elements, and two of them are never isomorphic, even though it is easy to prove that all the groups $n\mathbb{Z}$ are isomorphic to $\mathbb{Z}$, for $n > 0$. The relevant point is that all of them are distinct *subgroups* of $\mathbb{Z}$.

A quotient of the abelian group A is determined by a *subgroup* of A, not by a *group*. To get the same quotient, up to isomorphism, one can change *both* numerator and denominator, in a coherent way; changing one of them gives a different result, generally. For instance, $\mathbb{Z}/2\mathbb{Z}$ is isomorphic to $2\mathbb{Z}/4\mathbb{Z}$ and $3\mathbb{Z}/6\mathbb{Z}$, not to $\mathbb{Z}/4\mathbb{Z}$.

Solutions. (a) Take $x \geqslant 0$, and let $x = nq + r$, with $q, r \geqslant 0$; this is possible, as we can always take $q = 0$ and $r = x$. If $r < n$ we are done. Otherwise $r \geqslant n$; we take $q' = q + 1$ and $r' = r - n$, so that $x = nq' + r'$ still holds, with $q', r' \geqslant 0$ and $r' < r$. We go on this way, stopping (after a finite number of steps) when we get a remainder $< n$.

The case $x < 0$ can be reduced to the previous one, applied to $-x - 1 \geqslant 0$. Letting $-x - 1 = nq + r$, with $0 \leqslant r < n$, we have:

$$x = -nq - r - 1 = n(-q - 1) + (n - r - 1),$$

with $0 < n - r \leqslant n$ and $0 \leqslant n - r - 1 < n$.

(We have used the fact that $x \leqslant y$ in $\mathbb{R}$ implies $-y \leqslant -x$, see 1.1.9(b).)

(b) Let H be a non-null subgroup of $\mathbb{Z}$. Then H contains some $x \neq 0$, and also $-x$; one of them is positive. The subset $\{x \in H \mid x > 0\} \subset \mathbb{N}$, being non-empty, has a least element n. It follows that $n\mathbb{Z} \subset H$. On the other hand, any $x \in H$ can be written in the form $x = nq + r$, with $0 \leqslant r < n$; then $r = x - nq \in H$ must be 0, and $x \in n\mathbb{Z}$. We have also seen that the subset $n\mathbb{Z}$ determines n, as its least positive element.

(c) An obvious consequence, by 1.3.5.

(d) Any $x \in \mathbb{Z}$ can be written as $x = nq + r$, with $0 \leqslant r < n$, so that $[x] = [r]$ in $\mathbb{Z}/n$. The classes $[0], [1], ..., [n - 1]$ are distinct: if $0 \leqslant r \leqslant r' < n$ and $[r] = [r']$, then $r' - r \in n\mathbb{Z}$ and $0 \leqslant r' - r < n$, which implies $r' - r = 0$ (because n is the least positive element of $n\mathbb{Z}$).

(e) The given integer k can be rewritten as

$$k = 30421 = 1 + 2.10 + 4.10^2 + 0.10^3 + 3.10^4.$$

But $10 = 9 + 1$, $10^2 = 99 + 1$, $10^3 = 999 + 1$ and $10^4 = 9999 + 1$; all of them have remainder 1 modulo 9. Since the projection $p\colon \mathbb{Z} \to \mathbb{Z}/9$ preserves linear combinations with integral coefficients, we conclude that $[k] = [1 + 2 + 4 + 0 + 3] = [10] = [1]$. Similarly for all integral numbers in decimal notation. Let us note that:

- one can check addition of integers in any quotient $\mathbb{Z}/n$,

- this will leave undetected any error by a multiple of n,

- working in $\mathbb{Z}/9$ is effective, because we have a quick way of computing remainders modulo 9; this comes out of the decimal notation, in base $10 = 9 + 1$.

Since $10^n = (11 - 1)^n = m + (-1)^n$, where m is a multiple of 11, there is also a quick way of computing remainders modulo 11. For instance, the class modulo

11 of the same $k = 30421$ can be computed as: $[1 - 2 + 4 - 0 + 3]_{11} = [6]_{11}$. Starting from the digit of units and working backwards is now relevant.

(f) If the set X has a finite number $n > 0$ of elements, any bijection $f\colon \mathbb{Z}/n \to X$ defines a structure of abelian group on X, by transport of structure (see 1.2.7).

(g) There is an obvious isomorphism $f\colon \mathbb{Z} \to n\mathbb{Z}$, $f(k) = nk$.

1.3.7 The canonical factorisation and Noether isomorphisms

The canonical factorisation of a mapping has basic consequences for abelian groups: the three Isomorphism Theorems, also named Noether's Isomorphism Theorems, after Emmy Noether, a founder of the theory of algebraic structures in the first half of the 20th century [No2].

(a) (*First isomorphism theorem*) For a homomorphism $f\colon A \to B$ of abelian groups, the canonical factorisation of the mapping f which we have seen in 1.2.5 can be rewritten in the following form

$$\begin{array}{ccc} A & \xrightarrow{\;f\;} & B \\ {\scriptstyle p}\downarrow & & \uparrow{\scriptstyle m} \\ A/\operatorname{Ker} f & \xrightarrow[g]{} & \operatorname{Im} f \end{array} \qquad f = mgp, \tag{1.81}$$

where each mapping is a homomorphism of abelian groups: a surjective homomorphism p, followed by an isomorphism g and an injective homomorphism m. In fact:

- the relation R_f coincides with the congruence of A modulo $\operatorname{Ker} f$ (see (1.77)), and the canonical projection p is a surjective homomorphism,

- the subset $\operatorname{Im} f \subset B$ is a subgroup (see (1.54)), and the inclusion m is an injective homomorphism,

- the unique mapping $g\colon A/\operatorname{Ker} f \to \operatorname{Im} f$ such that $f = mgp$ is a bijective homomorphism, and therefore an isomorphism of abelian groups. It is called the *isomorphism induced by f*.

We only have to verify that g preserves the addition, which is straightforward: for $x, y \in A$, we have the following equality, where m is injective

$$\begin{aligned} m(g([x] + [y])) &= m(g([x + y])) = mgp(x + y) = f(x + y) = f(x) + f(y) \\ &= mgp(x) + mgp(y) = mg[x] + mg[y] = m(g[x] + g[y]). \end{aligned}$$

In particular, a surjective homomorphism f induces an isomorphism $A/\operatorname{Ker} f \to B$.

(b) (*Second isomorphism theorem*) Let A be an abelian group, and let H

and K be subgroups of A. Then the homomorphism

$$f\colon H \to H + K \to (H+K)/K, \qquad f(x) = [x] = x + K,$$

(the composite of an inclusion and a canonical projection) induces an isomorphism

$$g\colon H/(H \cap K) \to (H+K)/K. \tag{1.82}$$

This is an obvious consequence of the canonical factorisation of f, because:

$$\operatorname{Ker} f = H \cap K, \qquad \operatorname{Im} f = (H+K)/K.$$

(c) (*Third isomorphism theorem*) Let A be an abelian group, and let $K \subset H$ be subgroups of A. Then the surjective homomorphism

$$f\colon A/K \to A/H, \qquad f(x+K) = x + H,$$

induces an isomorphism

$$(A/K)/(H/K) \to A/H. \tag{1.83}$$

In fact, f is obviously well defined: if $x - y \in K$ then $x + H = y + H$. It is a surjective homomorphism, with $\operatorname{Ker} f = H/K$; its canonical factorisation gives the isomorphism (1.83).

Note. All equational algebras have a version of these theorems; in the other structures examined below we will only insist on the canonical factorisation.

1.3.8 Direct sums

For a family of abelian groups $(A_i)_{i \in I}$ indexed by a set I, we are now interested in the *support* of an element $x = (x_i) \in \prod_{i \in I} A_i$

$$\operatorname{supp}(x) = \{i \in I \mid x_i \neq 0\} \subset I, \tag{1.84}$$

(extending a notion already considered in 1.3.0(g), where all x_i belong to a given group).

We say that an element x *has finite support*, or is *quasi null*, if $\operatorname{supp}(x)$ is a finite set. All these families form a subgroup

$$\bigoplus_{i \in I} A_i = \{x \in \prod_{i \in I} A_i \mid \operatorname{supp}(x) \text{ is finite}\},$$

$$\operatorname{supp}(x+y) \subset \operatorname{supp}(x) \cup \operatorname{supp}(y), \qquad \operatorname{supp}(0) = \emptyset,$$
$$\operatorname{supp}(-x) = \operatorname{supp}(x).$$

This abelian group $A = \bigoplus_{i \in I} A_i$ is called the *direct sum* of the family

$(A_i)_{i\in I}$. It comes equipped with a family of *canonical injections* (which are injective homomorphisms)

$$u_i\colon A_i \to A, \tag{1.85}$$

where u_i sends an element $x \in A_i$ to the family $(x_j)_{j\in I}$ such that $x_i = x$ and $x_j = 0$ for all $j \neq i$.

The important fact is that any element $x = (x_i) \in A$ can be expressed as an essentially finite sum in A (as considered in 1.3.0(g)):

$$x = \textstyle\sum_{i\in I} u_i(x_i), \tag{1.86}$$

so that A is the subgroup of $\prod_i A_i$ generated by the subset $\bigcup_i u_i(A_i)$.

Even more importantly, the abelian group A, equipped with the family $(u_i)_{i\in I}$ of homomorphisms, satisfies the *universal property of the sum* (of abelian groups), analogous to that of sets (in Exercise 1.2.3(d))

- for every similar pair $(B, (f_i\colon A_i \to B)_{i\in I})$ formed of an abelian group B and a family of homomorphisms $f_i\colon A_i \to B$, there exists a unique homomorphism $f\colon A \to B$ such that

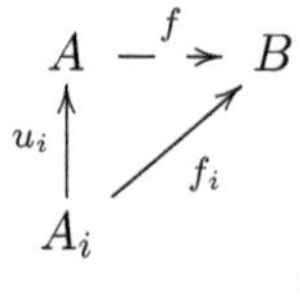

$$\begin{array}{ccc} A & \overset{f}{\dashrightarrow} & B \\ {\scriptstyle u_i}\uparrow & \nearrow{\scriptstyle f_i} & \\ A_i & & \end{array} \qquad fu_i = f_i \ \ (\text{for } i \in I). \tag{1.87}$$

In fact, on an element $x = \sum_{i\in I} u_i(x_i)$, we must take

$$f(x) = \textstyle\sum_{i\in I} fu_i(x_i) = \sum_{i\in I} f_i(x_i),$$

an essentially finite sum in B; conversely, defining f in this way, we do get a homomorphism, that satisfies condition (1.87). Also here the solution of the universal property is essentially unique, up to a determined isomorphism.

For a finite set I, $\bigoplus_{i\in I} A_i$ coincides with the cartesian product. This abelian group satisfies thus the universal property of the product (when equipped with the canonical projections p_i) *and* the universal property of the sum (when equipped with the canonical injections u_i).

We also write $A_1 \oplus A_2 \oplus ... \oplus A_n$; for abelian groups, this notation is often preferred to $A_1 \times A_2 \times ... \times A_n$.

1.3.9 Exercises and complements

(a) (*Congruences*) Prove that, as already stated in (1.75), for any abelian group A there is a natural bijection between congruences and subgroups

$$\begin{aligned} E &\mapsto \{x \in A \mid x\, E\, 0_A\} = \mathrm{Ker}\,(A \to A/E), \\ H &\mapsto \equiv_H, \qquad x \equiv_H x' \;\Leftrightarrow\; x - x' \in H. \end{aligned} \tag{1.88}$$

(b) For a subgroup $H \subset A$, all the equivalence classes $[x] = x + H$ have the same cardinal. If A is finite

$$\sharp A = (\sharp H).(\sharp(A/H)), \tag{1.89}$$

and the order of H divides the order of A. (This fact holds for all groups, see 1.5.9.)

The relation (1.89) also holds for infinite cardinals.

(c) (*The transfer of subgroups*) A homomorphism $f\colon A \to B$ of abelian groups induces two mappings, called *image* and *preimage* along f

$$f_*\colon \mathrm{Sub}(A) \to \mathrm{Sub}(B), \qquad f^*\colon \mathrm{Sub}(B) \to \mathrm{Sub}(A), \tag{1.90}$$

which are restrictions of the transfer of subsets (in 1.2.2).

(d) (*Cyclic groups, I*) Every infinite cyclic group is isomorphic to the additive group $\mathbb{Z}$. Every cyclic group of finite order n is isomorphic to the group $\mathbb{Z}/n$, examined in Exercises 1.3.6(c), (d).

(e) (*Cyclic groups, II*) Every subgroup and every quotient of a cyclic group is cyclic.

(f) (*Cyclic groups, III*) Every finite abelian group of prime order p is cyclic and generated by every element $x \neq 0$. It is thus isomorphic to the group $\mathbb{Z}/p$.

Therefore, on a set of p elements there is *essentially* one structure of abelian group (up to isomorphism, i.e. up to renaming elements).

(g) The direct sum $\mathbb{Z}/2 \oplus \mathbb{Z}/3$ is cyclic, while $\mathbb{Z}/2 \oplus \mathbb{Z}/2$ is not.

We have thus a non-cyclic group of order four, called the *Klein four-group*, after the mathematician C. Felix Klein.

Because of the previous point, four is the smallest cardinal admitting two essentially different structures of abelian group, the cyclic one and Klein's; we will see that there are no others, in Exercise 1.5.9(a).

We will also see that the group $\mathbb{Z}/n \oplus \mathbb{Z}/m$ is cyclic if and only if m and n are coprime, i.e. have no common proper divisors (in Exercise 1.4.5(e)).

(h) (*Torsion elements*) We know that an element x of an abelian group A generates the cyclic subgroup $\langle x \rangle = \{kx \mid k \in \mathbb{Z}\}$. The *order* of x is defined as the order of this group, either infinite or a positive integer. In the second case, x is said to be a *torsion element* of A, and its order is the least positive integer n such that $nx = 0$. Prove that the torsion elements of A form a subgroup $\mathrm{t}A$, characterised by an evident universal property.

An abelian group is said to be a *torsion* group if $\mathrm{t}A = A$, and to be *torsion free* if $\mathrm{t}A = \{0\}$. (The trivial group is both.) Every finite abelian group is torsion.

*(i) A reader with a basic knowledge of the complex field $\mathbb{C}$ can be interested in examining now the torsion subgroup of the multiplicative group $\mathbb{C}^*$ (to be explored in Section 2.7.)

1.4 Preordered sets and lattices

We begin by reviewing and completing the terminology for preordered and ordered sets. Then we introduce various order-theoretical structures, like lattices, modular lattices, distributive lattices and boolean algebras, that can equivalently be presented as equational algebras.

The power set $\mathcal{P}X$ of any set is a boolean algebra. The ordered set $\mathrm{Sub}(A)$ of subgroups of an abelian group is a modular lattice, useful to study A. More generally, the substructures of any equational algebra form a lattice.

A reader interested in this beautiful domain will find pleasure in browsing, or studying, the classical texts of Garrett Birkhoff and George Grätzer [Bi, Gr2].

1.4.1 Preordered and ordered sets

We use the following terminology for orderings, partially outlined in 1.1.9.

A *preordered set* X is a set equipped with a preorder relation $x \prec x'$ (read as x *precedes* x'), which is assumed to be reflexive and transitive.

In an *ordered set*, the relation is also *anti-symmetric*: if $x \prec x'$ and $x' \prec x$, then $x = x'$. An order relation is more often written as $x \leqslant x'$, or $x \leqslant_X x'$ when useful. It is a *total order relation* if for all x, x' we have $x \leqslant x'$ or $x' \leqslant x$.

An ordered set is also called (redundantly) a 'partially ordered set', abbreviated to *poset*, to mean that totality is not assumed (but not excluded).

A symmetric preorder relation is the same as an equivalence relation.

Every set X has two trivial preorder relations: the equality relation $x = y$ (in X), also called the *discrete order* (or the *discrete equivalence relation*), and the chaotic relation $x, y \in X$, also called the *indiscrete preorder* (or the *indiscrete equivalence relation*).

A preordered set X has an associated equivalence relation $x \sim x'$ defined by the conjunction: $x \prec x'$ and $x' \prec x$. The quotient $X/{\sim}$ has an induced order:

$$[x] \leqslant [x'] \;\Leftrightarrow\; x \prec x'. \tag{1.91}$$

If X is a preordered set, X^{op} is the *opposite* one, or *dual* one, with reversed preorder: $x \prec^{\text{op}} x'$ if $x' \prec_X x$. Every aspect of the theory of preordered sets can thus be *dualised.*

The *minimum* $\min X$ is an element that precedes all the elements of X (and can exist or not, of course); the *maximum* $\max X$ is an element preceded by all the elements of X. They are determined up to the associated equivalence relation in X, and uniquely determined if X is ordered. They are also written as $\bot$ and $\top$ (*bottom* and *top*).

If the preordered set X has a minimum $\bot$, an element $p \in X$ is said to be an *atom* if

$$\begin{gathered} p \nsim \bot, \\ x \prec p \text{ implies } (x \sim \bot \text{ or } x \sim p). \end{gathered} \tag{1.92}$$

The set $\mathbb{R}$ equipped with the natural order $x \leqslant y$ is called the *ordered line.* We have already seen, in 1.1.8, that it is a totally ordered set, with no minimum nor maximum.

For every set S, the power set $X = \mathcal{P}S$ is ordered by inclusion, with minimum $\emptyset$ and maximum X. This order is not total, as soon as S has more than 1 element. The atoms of $\mathcal{P}S$ are the singletons $\{p\}$, for $p \in S$ (if any).

1.4.2 Monotone mappings

A mapping $f\colon X \to Y$ between preordered sets is said to be *monotone,* or *preorder-preserving,* or *increasing,* if for all $x, x' \in X$

$$x \prec_X x' \text{ implies } f(x) \prec_Y f(x'). \tag{1.93}$$

(There is some difference with the terminology used in Calculus, see Remark (a), below.) Monotone mappings compose (in an associative way), and the identity mapping $\mathrm{id}\, X$ of any preordered set is monotone.

An *isomorphism* $f\colon X \to Y$ of preordered sets is a monotone mapping that has a monotone inverse $g\colon Y \to X$. Equivalently, it is a bijective monotone mapping whose inverse mapping is also monotone (we have already seen, in (1.46), that the last condition is not redundant). Or also, a bijective mapping that preserves and reflects the preorder relation: $x \prec_X x'$ if and only if $f(x) \prec_Y f(x')$.

In this case, X and Y are said to be *isomorphic* preordered sets, written as $X \cong Y$.

A mapping $f\colon X \dashrightarrow Y$ is *anti-monotone* (or *preorder-reversing*, or *decreasing*) if it is monotone from X^{op} to Y (or equivalently from X to Y^{op}). Note that the composition of two anti-monotone mappings is monotone.

A preordered set X is said to be *self-dual* if it is isomorphic to X^{op}. This is often realised by an *involution* $r\colon X \dashrightarrow X$, i.e. a preorder-reversing mapping which is involutive (i.e. inverse to itself): $rr = \mathrm{id}\, X$.

A *preordered subset* X' of a preordered set X is a subset equipped with the restricted preordering $x \prec x'$ (for $x, x' \in X'$). This is the greatest preorder relation on X' which makes the inclusion mapping $X' \to X$ monotone.

A quotient X/R of a preordered set modulo an equivalence relation can be equipped with the least preorder relation that makes the projection $p\colon X \to X/R$ monotone. This can be obtained by considering the relation $R' = (R \cup \prec) \subset X \times X$, then its transitive extension R'' (described in the solution of 1.2.4(d)), and finally

$$[x] \prec [x'] \text{ in } X/R \quad \Leftrightarrow \quad x\, R'' x' \text{ in } X. \tag{1.94}$$

Note that, if $\prec_X$ is anti-symmetric, the relation (1.94) need not be.

A *quotient of ordered sets* is realised in a more complex way: first we take the quotient X/R *as a preordered set*, with the preorder relation $[x] \prec [x']$ defined previously; then we take the associated ordered set $(X/R)/\!\sim$, with the induced order relation $[[x]] \leqslant [[x']]$ (see (1.91)). An example is given below, in Exercise (f).

Remarks, exercises and complements. (a) For a function $f\colon A \to \mathbb{R}$, defined on a subset $A \subset \mathbb{R}$, we will always use a terminology consistent with the previous, general one: the function f is said to be (weakly) *increasing* if $x \leqslant x'$ (in A) implies $f(x) \leqslant f(x')$.

We also say that f is *strictly increasing* if $x < x'$ implies $f(x) < f(x')$, a stronger property, which also makes sense for mappings of arbitrary ordered sets.

The reader may have seen a different terminology in Calculus: the second case can be called 'increasing', and the first 'non-decreasing'. Moreover, 'monotone' is generally used to mean what here would be called 'monotone *or* anti-monotone'.

Clashes of terminology, in different domains of Mathematics, do occur. (But let us remark that 'non-decreasing' is a dangerous term, leading to confusion with 'not decreasing'; it can be replaced with 'weakly increasing', in any context.)

(b) The ordered line $\mathbb{R}$ is self-dual. The power set $\mathcal{P}X$ of any set is also.

(c) (*Divisibility for natural numbers*) We write as $\mathbb{N}_D$ the set of natural numbers equipped with the *divisibility* relation

$$m|n \quad \Leftrightarrow \quad (\text{there exists } k \in \mathbb{N} \text{ such that } km = n) \quad \Leftrightarrow \quad n \in m\mathbb{Z}. \tag{1.95}$$

This is an order relation. It is not total (of course), and it is not self-dual. For this relation, $\min \mathbb{N}_D = 1$ and $\max \mathbb{N}_D = 0$. An atom p of the ordered set $\mathbb{N}_D$, defined (in (1.92)) by the property

$$p \neq 1, \qquad \text{if } n|p \text{ in } \mathbb{N}, \text{ then } n = 1 \text{ or } n = p, \tag{1.96}$$

is the same as a prime number p (see 1.2.6(c)).

If we restrict divisibility to $\mathbb{N}^*$, then $m|n$ implies $m \leqslant n$ (and there is no greatest element).

(d) Consider also the divisibility relation in $\mathbb{Z}$. (Divisibility in commutative rings will be briefly examined in Exercises 2.1.8(c), (d).)

(e) Define the canonical factorisation of a monotone mapping $f\colon X \to Y$ of preordered sets.

(f) On the ordered interval $X = [0, 1]$ of the real line, we take the equivalence relation that identifies 0 and 1. Consider the quotient X/R as a preordered set, and the associated ordered set.

1.4.3 Infima and suprema

Let X be a preordered set. Extending the terminology of 1.1.8 for the ordered line, the minimum of a preordered subset $A \subset X$ can be generalised by the greatest lower bound of A in X, a relative notion (while $\min A$ is an absolute one, independent of X). Symmetrically, the maximum will be generalised by the least upper bound.

For $a \in X$ and $A \subset X$, the sets of their *lower bounds* and *upper bounds* in X will be denoted as

$$\begin{aligned} \downarrow a &= \{x \in X \mid x \prec a\}, \quad L(A) = \{x \in X \mid x \prec a, \text{ for all } a \in A\}, \\ \uparrow a &= \{x \in X \mid a \prec x\}, \quad U(A) = \{x \in X \mid a \prec x, \text{ for all } a \in A\}. \end{aligned} \tag{1.97}$$

We define the *infimum* of A in X (or *greatest lower bound*, or *meet*) as

$$\inf_X A = \max(L(A)), \tag{1.98}$$

also written as $\inf A$, or $\wedge A$. Dually, the *supremum* of A in X (or *least upper bound*, or *join*) is defined as

$$\sup_X A = \min(U(A)) = \inf_{X^{\mathrm{op}}} A, \tag{1.99}$$

and also written as $\sup A$, or $\vee A$. Again, these outcomes can exist or not, and are determined by A up to the associated equivalence relation in X.

If A has a minimum, its meet also exists and $\inf_X A = \min A$; if A has a maximum, $\sup_X A = \max A$. Every element of X is (trivially) a lower bound and an upper bound of the empty subset, so that

$$\inf_X \emptyset = \max X = \sup_X X, \qquad \sup_X \emptyset = \min X = \inf_X X.$$

For an indexed family $(a_i)_{i\in I}$ of elements of X, we use the same terminology referred to the subset $A = \{a_i \mid i \in I\} \subset X$

$$\inf a_i = \wedge\, a_i = \inf_X A, \qquad \sup a_i = \vee\, a_i = \sup_X A.$$

(Any repetition in the family has no influence on these outcomes.)

Examples and exercises. (a) As we have seen in 1.1.8, in the ordered line $\mathbb{R}$ every subset A which is non-empty and upper bounded (resp. lower bounded) has $\sup A$ (resp. $\inf A$). The existence of $\max A$ or $\min A$ is a stronger fact.

(b) In a power set $\mathcal{P}S$, a subset $\mathcal{A} \subset \mathcal{P}S$ is more easily understood as an indexed family $(A_i)_{i\in I}$ of subsets of S. A lower bound is any subset contained in all of them, and the greatest is $\bigcap A_i$. Symmetrically, the least upper bound is $\bigcup A_i$.

Finally, the ordered set $\mathcal{P}S$ has all meets and joins:

$$\wedge\, A_i = \bigcap A_i, \qquad \vee\, A_i = \bigcup A_i.$$

(c) The euclidean plane $X = \mathbb{R}^2$, (partially) ordered by the relation

$$(x, y) \leqslant (x', y') \;\Leftrightarrow\; (x \leqslant x' \text{ and } y \leqslant y'),$$

is another instance where the previous notions can be usefully explored. (It is a cartesian power of the ordered line $\mathbb{R}$, see 1.4.8.)

Below, the subset $A \subset \mathbb{R}^2$ is a solid square (including its edges); it has a minimum $p = \min A\, (= \inf A)$, and a maximum $q = \max A\, (= \sup A)$

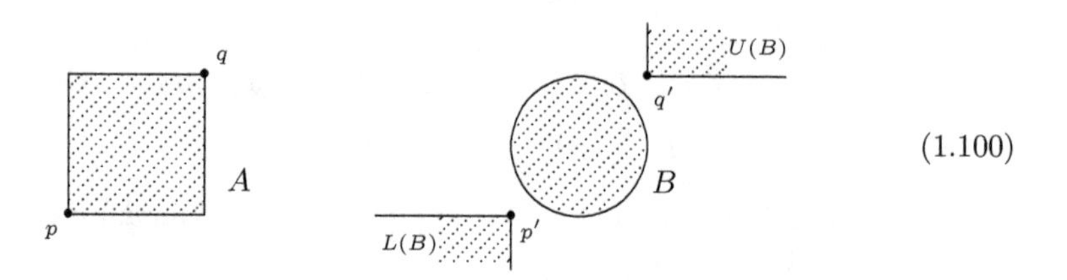

(1.100)

The subset B is a disc, and has no minimum nor maximum in this ordering; we have represented the sets $L(B)$ and $U(B)$ of its lower bounds and upper bounds, together with the points

$$p' = \max L(B) = \inf_X B, \qquad q' = \min U(B) = \sup_X B.$$

If A' denotes the previous square, *deprived* of the four edges, it is easy to see that the new subset has no minimum nor maximum, but $p = \inf_X A'$ and $q = \sup_X A'$ are still true. *(Topology will allow us to describe A' as the interior of A in the euclidean plane, in 3.1.6.)*

*(d) One can consider a similar order relation in $\mathbb{R}^n$ and prove that a non-empty lower bounded subset A has a greatest lower bound, whose i-th coordinate is the greatest lower bound of the projection of A on the i-th axis.

1.4.4 Lattices

A *lattice* is an ordered set X such that every pair of elements x, x' has a *join* $x \vee x' = \sup\{x, x'\}$ (the least element of X greater than both) and a *meet* $x \wedge x' = \inf\{x, x'\}$ (the greatest element of X smaller than both).

Then the order relation $x \leqslant x'$ is determined by the join-operation, as the condition $x \vee x' = x'$, and by the meet-operation, as the condition $x \wedge x' = x$. (We will see that a lattice can be described as an equational algebra, in 1.4.6.)

Every totally ordered set is a lattice, with $x \vee x' = \max\{x, x'\}$ and $x \wedge x' = \min\{x, x'\}$. Every power set $\mathcal{P}X$ is a lattice, with $A \vee B = A \cup B$ and $A \wedge B = A \cap B$.

A *bounded lattice* is a lattice X which has a minimum and a maximum, generally written as

$$0 = \min X = \sup_X \emptyset, \qquad 1 = \max X = \inf_X \emptyset, \tag{1.101}$$

or as $\bot$ and $\top$ (*bottom* and *top*), when the previous notation would be confusing. These bounds are equal in the one-point lattice $\{*\}$, and only there.

A *lattice homomorphism* $f\colon X \to Y$ has to preserve all binary joins and meets; as a consequence, it is monotone. These mappings compose (in an associative way), and the identity mapping $\mathrm{id}\,X$ of any lattice is a homomorphism.

An *isomorphism* $f\colon X \to Y$ of lattices is a homomorphism that has an inverse homomorphism $g\colon Y \to X$. This is equivalent to a bijective homomorphism, where the inverse mapping $g\colon Y \to X$ automatically preserves joins and meets (as we have seen for rings, or abelian groups, or general algebraic structures).

Moreover, if $f\colon X \to Y$ is an isomorphism of ordered sets and X (resp. Y) is a lattice, so is Y (resp. X), and f is an isomorphism of lattices. In fact f preserves and reflects the order relation, and therefore preserves and reflects all the existing joins and meets.

A *homomorphism of bounded lattices* is also assumed to preserve minimum and maximum.

A *sublattice* $X' \subset X$ of a lattice is a subset which is stable in X under the binary operations of join and meet. It is a lattice with the restricted operations, and the inclusion $X' \to X$ is a lattice homomorphism.

Remarks. (a) We are not excluding the empty lattice, as many authors do: see 1.2.9(b). Note that, in any lattice, all singletons are sublattices and the intersection of any (distinct) two of them is empty. A bounded lattice cannot be empty, of course.

(b) In a lattice X every finite non-empty subset A has $\inf A$ and $\sup A$, as is easily proved by induction on $\sharp A$.

(c) A bounded lattice is the same as an ordered set X where every finite subset $A \subset X$ has $\inf A$ and $\sup A$ (including the case $A = \emptyset$). Every finite non-empty lattice is bounded.

(d) In some contexts, for instance in Category Theory, bounded lattices are the important structure and are just called 'lattices'.

1.4.5 Exercises and complements

We can now study the lattices of subsets of a set, and the lattices of subgroups of an abelian group.

(a) (*Complete lattices*) Prove that a preordered set has all infima (of its subsets) if and only if it has all suprema. In this case – if it is an ordered set – it is called a *complete lattice.*

A *homomorphism of complete lattices* is assumed to preserve arbitrary meets and joins: one of these conditions is not sufficient to ensure the other, as the next point shows.

Let us note that $\mathbb{R}$, a 'complete totally ordered field', is a lattice, but not a complete lattice; one can embed it in the extended real line $\overline{\mathbb{R}}$ (see 1.7.2), which is a complete lattice but is not a field.

One says that $\mathbb{R}$ is a *conditionally complete lattice*, to mean that every non-empty upper-bounded subset has a join. This is a self-dual condition, as above.

(b) We have already seen that a power set $\mathcal{P}X$ is a complete lattice, where meets and joins are intersections and unions of subsets. Every element $A \in \mathcal{P}X$ is a join of atoms of $\mathcal{P}X$, its singletons (the empty join for $A = \emptyset$).

Given a mapping of sets $f\colon X \to Y$, we have already seen (in 1.2.2(a)) that the mapping $f_*\colon \mathcal{P}X \to \mathcal{P}Y$ of direct images preserves arbitrary unions, but need not preserve binary intersections, nor the maximum, while the preimage-mapping $f^*\colon \mathcal{P}Y \to \mathcal{P}X$ preserves arbitrary meets and joins.

(c) The ordered set $\mathrm{Sub}(A)$ of subgroups of an abelian group A is a complete lattice. (This ordered set need not have any atom: see the next exercise.)

(d) The ordered set $\mathbb{N}_D$ (of natural numbers with the divisibility relation (1.95)) is anti-isomorphic to $\mathrm{Sub}(\mathbb{Z})$, and is a complete lattice as well. In particular

$$(m) \cap (n) = (m \vee_D n), \qquad (m) + (n) = (m \wedge_D n), \tag{1.102}$$

where $m \vee_D n = \mathrm{lcm}(m, n)$ is the least common multiple of m, n (in $\mathbb{N}$), and $m \wedge_D n = \gcd(m, n)$ is their greatest common divisor. The numbers

m, n are said to be *coprime* when $\gcd(m, n) = 1$, which means that they have no common factor except 1, or equivalently that $(m) + (n) = \mathbb{Z}$.

We already know that the atoms of $\mathbb{N}_D$ are the prime natural numbers (Exercise 1.4.2(c)); the number 6 is a join of atoms, while 4 is not. On the other hand, the (anti-isomorphic) ordered set $\mathrm{Sub}(\mathbb{Z})$ has no atoms.

Let us note that the usual definition of $\mathrm{lcm}(m, n)$ and $\gcd(m, n)$ is based on the prime factor decomposition of natural numbers (in 1.2.6). Here we are deducing their existence from Exercise 1.3.2(c), on the lattice of subgroups, and Exercise 1.3.6(b), on the subgroups of $\mathbb{Z}$.

(e) The group $A = \mathbb{Z}/m \oplus \mathbb{Z}/n$ is cyclic if and only if the positive integers m, n are coprime, and then $A \cong \mathbb{Z}/mn$.

*(f) Frames are particular complete lattices, related with topological spaces: see 6.5.6.

1.4.6 Exercises and complements (Lattices as algebras)

The following exercises are important. The solution is written below, and should at least be read.

(a) A lattice can be equivalently presented as an equational algebra, namely a set X equipped with two operations, $x \vee y$ and $x \wedge y$, called *join* and *meet*, that satisfy the following (redundant) axioms, for all $x, y, z \in X$:

$$\begin{array}{lll} (L.1) & (\textit{Associativity}) & x \vee (y \vee z) = (x \vee y) \vee z, \quad x \wedge (y \wedge z) = (x \wedge y) \wedge z, \\ (L.2) & (\textit{Commutativity}) & x \vee y = y \vee x, \quad x \wedge y = y \wedge x, \\ (L.3) & (\textit{Idempotence}) & x \vee x = x = x \wedge x, \\ (L.4) & (\textit{Absorption}) & x \vee (x \wedge y) = x = x \wedge (x \vee y). \end{array}$$

Given this presentation, one defines the ordering by letting $x \leqslant y$ if $x \vee y = y$, or equivalently $x \wedge y = x$. The opposite lattice X^{op} is now defined by interchanging join and meet.

Note that the homomorphisms of lattices are indeed the homomorphisms of this algebraic structure.

(b) For a bounded lattice, we further require the existence of an identity for each operation, written as 0 and 1, so that:

(L.5) (*Identities*) $\quad x \vee 0 = x = x \wedge 1.$

Let us recall that one can have $0 = 1$, and then all the elements coincide.

(c) A *join-semilattice* X is an ordered set where every pair x, x' of elements has a join $x \vee x'$. Describe this structure as an equational algebra.

(d) Dually, a *meet-semilattice* X is an ordered set where every pair x, x' of elements has a meet $x \wedge x'$, which is equivalent to saying that the opposite ordered set X^{op} is a join-semilattice. The reader will note that this structure, as an equational algebra, has the same description as the previous one, up to a different notation of the operation.

Solutions. (a), (b) The proof is straightforward, and an easy, useful exercise.

If the ordered set X is a lattice, as defined in 1.4.4, the properties (L.1–4) are easily proved. For instance, both $x \vee (y \vee z)$ and $(x \vee y) \vee z$ coincide with $\sup\{x, y, z\}$.

Conversely, suppose that the algebraic structure $(X, \vee, \wedge)$ satisfies the axioms (L.1–4). By (L.4), the condition $x \vee y = y$ implies $x \wedge y = x \wedge (x \vee y) = x$, while $x \wedge y = x$ implies $x \vee y = y$.

We define $x \leqslant y$ by these two equivalent conditions, which plainly give an order relation. Then $x \vee y$ is indeed the least upper bound of x and y, as:

- $x \vee (x \vee y) = (x \vee x) \vee y = x \vee y, \qquad y \vee (x \vee y) = y \vee (y \vee x) = y \vee x,$
- if $x \leqslant z$ and $y \leqslant z$ then: $z \vee (x \vee y) = (z \vee x) \vee y = z \vee y = z$.

Similarly $x \wedge y$ is indeed the greatest lower bound of x and y. The bounded case is obvious.

(c) Plainly, a join-semilattice X gives a commutative idempotent semigroup $(X, \vee)$. Conversely, starting from such a structure, we define $x \leqslant y$ if $x \vee y = y$, and proceed as in (a).

(d) A meet-semilattice X can also be viewed as a commutative idempotent semigroup $(X, \wedge)$. The structure is the same, but we interpret the associated order in the opposite way: $x \leqslant y$ if $x \wedge y = x$.

1.4.7 Distributive and modular lattices

A lattice X is said to be *distributive* if the meet operation distributes over the join operation (for $x, y, z \in X$)

(D) (*Distributivity*) $\qquad (x \vee y) \wedge z = (x \wedge z) \vee (y \wedge z).$

This is equivalent to saying that the join distributes over the meet. In fact, assuming (D) we have:

$$\begin{aligned}(x \vee y) \wedge (x \vee z) &= ((x \vee y) \wedge x) \vee ((x \vee y) \wedge z) \\ &= x \vee (x \wedge z) \vee (y \wedge z) = x \vee (y \wedge z),\end{aligned} \tag{1.103}$$

and the converse holds by duality.

A *boolean algebra* X (named after George Boole, who began the algebrisation of mathematical logic in the 1850's) is a distributive bounded lattice where every element x has a *complement* x^*, defined by the properties:

(C) $\quad x \wedge x^* = 0, \qquad x \vee x^* = 1.$

The complement is determined by x (see Exercise (b), below). A boolean algebra X is said to be *complete* if it is a complete lattice; to be *atomic* if, for every element x, there is some atom $p \leqslant x$ (as defined in (1.92)). These properties are studied in Exercise (f).

A *homomorphism of boolean algebras* $f\colon X \to Y$ is a homomorphism of bounded lattices between boolean algebras; then it necessarily preserves complements. A *homomorphism of complete boolean algebras* preserves arbitrary joins and meets (and therefore the whole structure).

The subsets of a set S form the 'classical' boolean algebra $\mathcal{P}S$, with the usual set-theoretical complement $A^* = S \setminus A = C_S A$. This algebra is complete and atomic. We have already seen, in Exercise 1.4.5(b), that a mapping $f\colon S \to T$ gives a homomorphism of complete boolean algebras $f^*\colon \mathcal{P}T \to \mathcal{P}S$, by preimages of subsets of T.

More generally, a lattice is said to be *modular* if it satisfies the following self-dual property (for all elements x, y, z)

(M) (*Modularity*) if $x \leqslant z$ then $(x \vee y) \wedge z = x \vee (y \wedge z)$,

which obviously holds in a distributive lattice, where $(x \wedge z) \vee (y \wedge z) = x \vee (y \wedge z)$ when $x \leqslant z$. There are useful, equivalent formulations:

(M′) if $x \leqslant z$ then $(x \vee y) \wedge z \leqslant x \vee (y \wedge z)$,

(M″) $((x \wedge z) \vee y) \wedge z = (x \wedge z) \vee (y \wedge z)$.

For (M′) note that the other inequality holds in any lattice (for $x \leqslant z$). For (M″), note that an element $x \leqslant z$ can always be written as $x \wedge z$. We also remark that (M′) is simpler to verify than (M), while (M″) has the interest of being an equational axiom, universally quantified without restrictions.

We verify below that the (complete) lattice Sub$A$ of subgroups of an abelian group is always modular, but need not be distributive.

Exercises and complements. (a) Every totally ordered set is a distributive lattice.

(b) In a boolean algebra X, the complement x^* is determined by x. The mapping

$$(-)^*\colon X \dashrightarrow X, \tag{1.104}$$

is an involutive anti-isomorphism of ordered sets, and therefore of bounded lattices. As a consequence, we have the *De Morgan laws* in X

$$(x \vee y)^* = x^* \wedge y^*, \qquad (x \wedge y)^* = x^* \vee y^* \qquad (\text{for } x, y \in X), \tag{1.105}$$

named after Augustus De Morgan, a mathematician and logician of the 19th century.

(c) Distributive lattices are equational algebras. The same is true of modular lattices, and of boolean algebras.

(d) The lattice Sub(A) of subgroups of an abelian group is modular.

(e) The lattice of subgroups of the group $A = \mathbb{Z} \oplus \mathbb{Z}$ is not distributive. The same holds for any power B^2 of a non-trivial abelian group B.

(f) Let X be an atomic boolean algebra and A the subset of its atoms. Then the mapping

$$f\colon X \to \mathcal{P}A, \qquad f(x) = \{p \in A \mid p \leqslant x\}, \tag{1.106}$$

is an injective homomorphism of boolean algebras, which allows us to view X as an 'algebra of sets'.

The mapping f is an isomorphism if and only if X is complete. The power sets are thus characterised as the complete atomic boolean algebras (up to isomorphism, of course).

(g) Every finite boolean algebra is isomorphic to the power set of its set of atoms. Its cardinal is a power 2^n of natural numbers. All finite boolean algebras are thus characterised. Moreover, two finite boolean algebras with the same number of elements are necessarily isomorphic.

*(h) Let S be an infinite set, and consider the set $\mathcal{X} \subset \mathcal{P}S$ formed by the finite subsets of S and their complements in S, the *cofinite subsets* of S. Then $\mathcal{X}$ is a *boolean subalgebra* of $\mathcal{P}S$ (in the obvious sense, to be made precise); it is atomic and not complete.

*(i) By Birkhoff's representation theorem ([Bi] III.5, Theorem 5) the free distributive lattice on n generators is finite and isomorphic to a lattice of subsets.

The reader may also be interested to know that the free modular lattice on three elements is finite and (obviously!) not distributive (see [Bi], III.6, Fig. 10), while four generators already give an infinite free modular lattice (see the final Remark in [Bi], III.6).

(The general notion of free structure is outlined in 1.6.1.)

1.4.8 Cartesian products of ordered structures

(a) We begin by considering preordered and ordered sets.

The *cartesian product* $X \times Y$ of two preordered sets is the cartesian product of their underlying sets, equipped with the componentwise relation

$$(x, y) \prec (x', y') \;\Leftrightarrow\; (x \prec x' \text{ in } X \text{ and } y \prec y' \text{ in } Y), \qquad (1.107)$$

which is obviously a preordering. If X and Y are ordered sets, the product $X \times Y$ is also.

The *cartesian projections*

$$p \colon X \times Y \to X, \qquad q \colon X \times Y \to Y,$$

are monotone mappings of preordered sets (or ordered sets), and the relation we have defined on $X \times Y$ is the coarsest relation (i.e. the biggest) with this effect.

Similarly, for a family $(X_i)_{i \in I}$ of preordered sets (or ordered sets), the cartesian product $X = \prod_i X_i$ is the product of the underlying sets, with the componentwise relation

$$(x_i)_{i \in I} \prec (y_i)_{i \in I} \;\Leftrightarrow\; (\text{for all } i \in I,\, x_i \prec y_i \text{ in } X_i). \qquad (1.108)$$

The product has a family of projections $p_i \colon X \to X_i$ ($i \in I$), which are monotone mappings.

(*Universal property of the product*) More formally, a product $\prod_i X_i$ of preordered sets is determined up to isomorphism as a preordered set X equipped with a family of monotone mappings $p_i \colon X \to X_i$ such that:

- for every preordered set Y and every family of monotone mappings $f_i \colon Y \to X_i$ there is a unique monotone mapping $f \colon Y \to X$ such that $p_i f = f_i$ (for all $i \in I$).

In fact, if we take X and p_i as defined above, we must let $f(x) = (f_i(x))_{i \in I}$ (for $x \in X$); on the other hand, defining $f \colon Y \to X$ in this way we do get a monotone mapping such that $p_i f = f_i$ (for $i \in I$). Finally, if the preordered set Y with the family $(f_i \colon Y \to X_i)$ also satisfies our property, we get a backward monotone mapping $g \colon X \to Y$ such that $f_i g = p_i$ (for all $i \in I$), and we conclude that f and g are inverse to each other, as in 1.3.4.

This universal property holds in the same way for ordered sets.

(b) All this can be extended to lattices. In fact, for a family $(X_i)_{i \in I}$ of lattices, the ordered set $X = \prod_i X_i$ defined above is a lattice, with

$$(x_i) \vee (y_i) = (x_i \vee y_i), \qquad (x_i) \wedge (y_i) = (x_i \wedge y_i). \tag{1.109}$$

The cartesian projections are homomorphisms of lattices, and satisfy the obvious universal property, in the domain of lattices and their homomorphisms. In the perspective of equational algebras, presented in 1.4.6, we can note that a product $\prod_i X_i$ of lattices has the unique structure that makes all cartesian projections into lattice homomorphisms.

(c) Similar facts holds for: bounded lattices, distributive lattices, modular lattices, boolean algebras. Indeed all these structures are equational algebras.

(d) The ordered line, as a totally ordered set, is a distributive lattice. Therefore the ordered set $\mathbb{R}^n$ is also a distributive lattice, with

$$(x_i) \vee (y_i) = (\max\{x_i, y_i\}), \qquad (x_i) \wedge (y_i) = (\min\{x_i, y_i\}). \tag{1.110}$$

(Joins and meets of 'big' subsets in $\mathbb{R}^n$ have been computed in 1.4.3(c), (d).)

This order of $\mathbb{R}^n$ is not total, for $n \geqslant 2$. The following drawing shows meet and join of a pair x, y of non-comparable points in $\mathbb{R}^2$

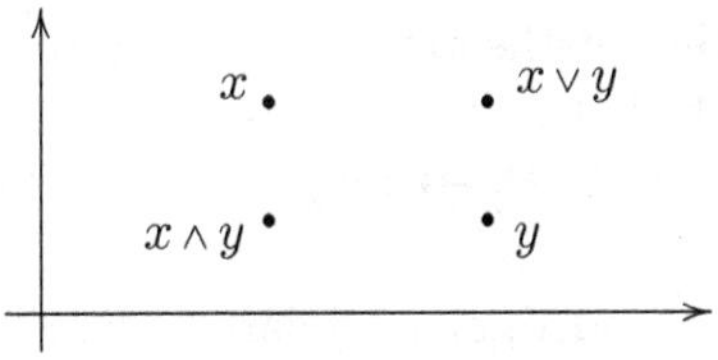

$x \bullet \qquad \bullet\, x \vee y$

$x \wedge y \bullet \qquad \bullet\, y$ (1.111)

(e) As a consequence, a totally ordered set cannot be reformulated as an equational algebra. But it can be viewed as a non-equational algebraic structure, namely a lattice X satisfying the axiom:

- for every $x, y \in X$, $x \vee y = x$ *or* $x \vee y = y$,

which is not a 'universally quantified equation'.

It follows that a bijective monotone mapping $f\colon X \to Y$ between totally ordered sets is always an isomorphism of ordered sets.

This can also be easily checked 'analytically', verifying that f reflects the order. Assuming that $f(x) \leqslant f(x')$ in Y, we have that $x \leqslant x'$ or $x' \leqslant x$ in X; in the first case we are done, in the second $f(x) = f(x')$ and then $x = x'$.

*(f) Strictly increasing mappings, defined in 1.4.2(a), are certainly important within real functions. Yet, in the general context of ordered sets, they give a poor framework without even binary products: the cartesian projections we have considered above are not strictly increasing, in general.

1.4.9 Quotients of lattices

Let X be a lattice. A *congruence* (of lattices) in X is an equivalence relation E consistent with the operations. This means that the equivalence relation E is a sublattice of the product $X \times X$:

- if xEx' and yEy', then $(x \vee y)E(x' \vee y')$ and $(x \wedge y)E(x' \wedge y')$.

The *quotient* X/E of the set X modulo the equivalence relation E is a lattice, under the operations induced by the operations of X

$$[x] \vee [y] = [x \vee y], \qquad [x] \wedge [y] = [x \wedge y]. \tag{1.112}$$

The canonical projection on the quotient

$$p\colon X \to X/E, \qquad p(x) = [x], \tag{1.113}$$

is a homomorphism, and the structure we have put on X/E is the only one giving this outcome.

The *canonical factorisation* of a homomorphism $f\colon X \to Y$ of lattices is easily deduced form the set-theoretical factorisation $f = mgp$ which we have seen in (1.43). In fact, the latter gives a quotient $p\colon X \to X/R_f$ of lattices, an isomorphism $g\colon X/R_f \to \operatorname{Im} f$ and an inclusion $m\colon \operatorname{Im} f \to Y$ of a sublattice.

All this carries on to bounded, or modular, or distributive lattices.

Exercises and complements. (a) Any binary relation R in a lattice *generates* a congruence E, the least congruence containing R.

1.5 Semigroups, monoids and groups

Semigroups, monoids and groups are equational algebras, more general than abelian groups. Their theory can be perceived as more abstract, which is why we introduce it after that of abelian groups.

Semigroup Theory and Group Theory are worlds, explored in many books, like [Ho, ClP, Hu, Ro].

Group theory governs symmetries, and intervenes in Physics and Chemistry. A famous Noether's Theorem, published in 1918 [No1], links symmetries and invariants in the laws of Physics. For applications of Group Theory in Physics and Crystallography one can see [Cor, Wo].

1.5.1 Semigroups and monoids

A *semigroup* is a set S equipped with a binary operation, which is associative. If we write it in multiplicative notation, as xy (or $x.y$, when convenient), we are only assuming that:

(i) for every $x, y, z \in S$ we have: $x(yz) = (xy)z$.

One can write any finite product $x_1x_2 \dots x_n$ of elements of S without parentheses. In particular, for every $x \in S$ and every integer $n \geqslant 1$ we have the *power* $x^n = xx \dots x$ (a product of n factors), inductively defined by:

$$x^1 = x, \qquad x^{n+1} = x^n.x \qquad (n \geqslant 1). \tag{1.114}$$

When it is useful to distinguish the semigroup S from the mere set of its elements, the latter is called the *underlying set*, and written as $|S|$.

We say that S is a *unital semigroup*, or a *monoid*, if it has a *unit*:

(ii) there exists an $e \in S$ such that $ex = x = xe$.

The element e is uniquely determined (as we have already seen), and can also be written as 1 (or e_S, or 1_S, if useful). Now a power x^n is defined for $n \in \mathbb{N}$, letting $x^0 = e$.

A semigroup (or monoid) X has an *opposite* semigroup (or monoid) X^{op}, with the same elements and *opposite* multiplication, obtained by reversing the original operation

$$x * y = y.x. \tag{1.115}$$

A semigroup, or a monoid, is said to be *commutative*, if its operation is, or equivalently if it coincides with the opposite semigroup. The set $\mathbb{N}$ of natural numbers has two 'natural' structures of commutative monoid, defined by 'natural' operations: $(\mathbb{N}, +)$ and $(\mathbb{N}, .)$.

A commutative monoid is more often called an *abelian monoid* when we are using the additive notation. Powers are now replaced by multiples nx ($n \in \mathbb{N}$), and we can use linear combinations $\Sigma_i \lambda_i x_i$ with natural coefficients $\lambda_i \in \mathbb{N}$.

Speaking of the abelian monoid $\mathbb{N}$ we always mean $(\mathbb{N}, +)$, with the 'natural' sum. As the reader can easily see, it is the free abelian monoid generated by a single element, 1; this will also follow from a general construction, in 1.6.2(b).

Exercises and complements. (a) The empty set $\emptyset$ has a trivial structure of semigroup. The singleton $\{x\}$ also has a unique structure of semigroup, with $x.x = x$; it is a commutative monoid.

(b) If X is a set, the set $\mathrm{End}(X)$ of all endomappings of X is a monoid with respect to the composition law, with unit $\mathrm{id}\, X$. This monoid is non-commutative, if X has at least two elements.

(c) Every set X has a structure of semigroup defined by $xy = x$, and another structure (the opposite one) defined by $xy = y$. These structures are different and non-commutative, as soon as X has two distinct elements.

(d) (*Properties of powers*) Prove the following properties in a semigroup S. It is understood that $x, y \in S$ and the integers m, n are positive; they can belong to $\mathbb{N}$ if S is a monoid

$$x^m.x^n = x^{m+n}, \tag{1.116}$$

$$e^n = e, \qquad (x^m)^n = x^{mn}, \tag{1.117}$$

$$\text{if } xy = yx \text{ then } x^m.y^n = y^n.x^m \text{ and } x^n.y^n = (xy)^n. \tag{1.118}$$

Note that all the powers of a given element commute: $x^m.x^n = x^n.x^m$.

(e) In a semigroup S, an element z is said to be *absorbing* if $zx = z = xz$, for all $x \in S$. Such an element is at most unique.

For instance, 0 is absorbing in the multiplicative monoid $\mathbb{N}$, and in the multiplicative semigroup of any ring. The additive monoid $\mathbb{N}$ has no absorbing element.

(f) Semigroups form a variety of algebras, in the sense of 1.2.8. The same is true of monoids.

1.5.2 Groups

A *group* G is a monoid where every element has an *inverse*:

(iii) for every $x \in G$ there is some $y \in G$ such that $xy = e = yx$.

The inverse y is determined by x (as we have already seen), and is written as x^{-1}. We have:

$$e^{-1} = e, \qquad (x^{-1})^{-1} = x, \qquad (xy)^{-1} = y^{-1}.x^{-1}. \tag{1.119}$$

Now we can define $x^{-n} = (x^{-1})^n$, for $n > 0$, and the power x^h is defined for all $h \in \mathbb{Z}$.

The trivial group is the singleton $\{*\}$, and is commutative.

Exercises and complements. (a) (*Cancellation laws*) We say that a semigroup S *satisfies the cancellation laws* if, for all $x, y, z \in S$:

$$xy = xz \;\Rightarrow\; y = z, \qquad\qquad yx = zx \;\Rightarrow\; y = z. \tag{1.120}$$

These laws hold in every group. They also hold in the additive monoid $\mathbb{N}$ of natural numbers, and in the multiplicative monoid $\mathbb{Z}^*$ of non-zero integers.

A semigroup with cancellation laws cannot have an absorbing element, unless it is the trivial group.

(b) In a group G, the relation $xy = e$ is sufficient to conclude that x and y are inverse to each other.

(c) (*Properties of powers*) In a group G, for $x, y \in G$ and $h, k \in \mathbb{Z}$, we have:

$$x^h.x^k = x^{h+k}, \tag{1.121}$$

$$e^h = e, \qquad (x^h)^k = x^{hk}, \tag{1.122}$$

$$\text{if } \; xy = yx \; \text{ then } \; x^h.y^k = y^k.x^h \; \text{ and } \; x^h.y^h = (xy)^h. \tag{1.123}$$

Again, all the powers x^h of a given element commute.

(d) Groups form a variety of algebras, in the sense of 1.2.8.

(e) The additive group $(\mathbb{Z}, +)$ can be embedded as a subgroup of the multiplicative group $(\mathbb{Q}^*, .)$. There are infinitely many ways of doing that.

*(f) The additive group $\mathbb{R}$ of real numbers is isomorphic to the multiplicative group $\mathbb{R}^*_+$ of positive real numbers.

1.5.3 Homomorphisms and substructures

(a) A *homomorphism* $f\colon S \to T$ *of semigroups* is a mapping that preserves the operation: $f(xy) = f(x)f(y)$, for all $x, y \in S$.

A *homomorphism* $f\colon S \to T$ *of monoids* is also assumed to preserve the unit: $f(e_S) = e_T$. This is not a consequence: see Exercise 1.5.4(b).

For a *homomorphism* $f\colon S \to T$ *of groups* one simply assumes that it preserves the main operation; in fact, here one easily proves that f preserves unit and inverses: see Exercise 1.5.4(c).

In each case, a composition gf of homomorphism gives a homomorphism. The identity $\mathrm{id}\, S$ of a semigroup (or a monoid, or a group) is a homomorphism. An *isomorphism* $f\colon S \to T$ (of semigroups, or monoids, or groups) is a homomorphism which has an inverse homomorphism $T \to S$; this is equivalent to saying that f is a bijective homomorphism: see 1.5.4(d).

Each group G is canonically isomorphic to the opposite group G^{op}, with reversed operation, by the isomorphism

$$\rho\colon G \to G^{\mathrm{op}}, \qquad \rho(x) = x^{-1}. \tag{1.124}$$

This will often be viewed as a bijection $|G| \to |G|$ of the underlying set.
If G is a group, any element $g \in G$ defines a homomorphism

$$\varphi_g \colon G \to G, \qquad \varphi_g(x) = g^{-1}xg, \tag{1.125}$$

which has an inverse of the same kind, produced by g^{-1}. It is called the *inner automorphism* associated to g.

(b) A *subsemigroup* of a semigroup S is a subset $S' \subset S$ that is closed under the operation of S: if $x, y \in S'$ then $xy \in S'$. S' is thus a semigroup with the restricted operation (still associative, of course). The inclusion $S' \to S$ is an injective homomorphism.

A *submonoid* of a monoid S is a subsemigroup $S' \subset S$ which contains the unit e_S. S' is thus a monoid, with the restricted operation, and the inclusion $S' \to S$ is an injective homomorphism of monoids.

A *subgroup* of a group S is a submonoid $S' \subset S$ which is closed under inverses: if $x \in S'$ then $x^{-1} \in S'$. S' is thus a group, with the restricted operation, and the inclusion $S' \to S$ is an injective homomorphism of monoids.

A subgroup $H \subset G$ is said to be *invariant*, or *normal*, in G if it is invariant under all the inner automorphisms of G

$$\text{for every } g \in G, \;\; \varphi_g(H) \subset H. \tag{1.126}$$

This condition, usually written as $H \triangleleft G$, has many equivalent expressions: see Exercise 1.5.5(b). Examples of invariant and non-invariant subgroups can be found in 1.6.7.

In a commutative group G these aspects are trivial: the only inner automorphism is $\operatorname{id} G$, and all subgroups are normal.

(c) If $f \colon S \to T$ is a homomorphism of semigroups (resp. of monoids, of groups), its *image*

$$\operatorname{Im} f = f(S) = \{y \in T \mid y = f(x), \text{ for some } x \in S\}, \tag{1.127}$$

is a subsemigroup of T (resp. a submonoid, a subgroup).

For a homomorphism $f \colon G \to G'$ of groups we also consider

$$\operatorname{Ker} f = f^{-1}\{1\} = \{x \in G \mid f(x) = 1\}. \tag{1.128}$$

This is (plainly) a subgroup of G. It is important to note that it is an invariant subgroup: if $x \in \operatorname{Ker} f$ and $g \in G$, then

$$f(g^{-1}xg) = f(g)^{-1}.f(g) = 1.$$

We also note that f is injective if and only if $\operatorname{Ker} f = \{1\}$, the trivial subgroup of G: in fact, if this is the case and $f(x) = f(y)$, then $f(xy^{-1}) = 1_{G'}$ and $xy^{-1} = 1_G$; the converse is obvious. (This argument would fail for a homomorphism of monoids, which is why kernels are much less important in this theory.)

We will see, in 1.5.8, that every normal subgroup of G is the kernel of some homomorphism of groups defined on G.

1.5.4 Exercises and complements

The following exercises are easy, but important for a reader who is beginning to study 'abstract' algebraic structures. The non-obvious solutions can be found below.

(a) Let $x \in S$. If S is a semigroup, the set of all powers x^n (for $n > 0$) is the subsemigroup generated by x, i.e. the least subsemigroup containing this element. If S is a monoid, the set of all powers x^n (for $n \in \mathbb{N}$) is the submonoid generated by x.

Finally, if S is a group, the set of all powers x^n (for $n \in \mathbb{Z}$) is the subgroup generated by x. The *order* of an element x in a group is defined as in 1.3.9(h).

All these structures, generated by an element, are commutative. In particular, a *cyclic group*, i.e. a group generated by one element, is necessarily commutative.

(b) Find a homomorphism of semigroups between two monoids, which does not preserve the unit.

(c) Prove that a homomorphism $f\colon S \to T$ of groups preserves unit and inverses.

(d) Let $f\colon S \to T$ be a homomorphism of semigroups (resp. of monoids, of groups). Prove that, if f is bijective, then the inverse mapping is a homomorphism $f^{-1}\colon T \to S$.

(e) (*Invertible elements*) In a monoid S, an element x is said to be *invertible* if there is some $y \in G$ such that $xy = e = yx$. Again, the inverse of x is determined by x and written as x^{-1}. The subset of these elements will be written as $\mathrm{Inv}(S)$. Prove the following properties:

$$\text{if } \; x, y \in \mathrm{Inv}(S), \; \text{ then } \; xy \in \mathrm{Inv}(S) \text{ and } (xy)^{-1} = y^{-1}.x^{-1}, \qquad (1.129)$$

$$e \in \mathrm{Inv}(S), \qquad e^{-1} = e, \qquad (1.130)$$

$$\text{if } \; x \in \mathrm{Inv}(S), \; \text{ then } \; x^{-1} \in \mathrm{Inv}(S) \text{ and } (x^{-1})^{-1} = x. \qquad (1.131)$$

Therefore $\mathrm{Inv}(S)$ is a submonoid of S, and a group in its own right.

(f) (*Universal property*) The inclusion $\varepsilon\colon \mathrm{Inv}(S) \to S$ is a homomorphism of monoids, defined on a group. Every homomorphism of monoids $f\colon G \to S$ defined on a group factorises uniquely as $\varepsilon f'$, for a unique homomorphism of groups $f'\colon G \to \mathrm{Inv}(S)$

$$\begin{array}{ccc} G & \xrightarrow{f} & S \\ & \underset{f'}{\searrow} & \uparrow{\scriptstyle \varepsilon} \\ & & \mathrm{Inv}(S) \end{array} \tag{1.132}$$

This property determines $\mathrm{Inv}(S)$, up to isomorphism of groups.

Solutions. (b) Take a set X. A mapping $\varphi\colon \{*\} \to \mathrm{End}(X)$ is determined by a function $\varphi(*) = f\colon X \to X$; φ is a homomorphism of semigroups if (and only if) f is idempotent: $ff = f$. If X has at least two elements, each constant function $f\colon X \to X$ is idempotent and is not the unit $\mathrm{id}\, X$ of $\mathrm{End}(X)$.

(c) The relation $f(e_S).f(x) = f(e_S x) = f(x) = e_T.f(x)$ gives $f(e_S) = e_T$, by cancellation in T. The relation $f(x^{-1}).f(x) = f(x^{-1}x) = f(e_S) = e_T$ proves that $f(x^{-1})$ is inverse to $f(x)$.

(d) This is a general fact for algebraic structures, as we have remarked in 1.2.7. Here, if $f\colon S \to T$ is a bijective homomorphism of semigroups (or groups), the relation

$$f(g(x).g(y)) = f(g(x)).f(g(y)) = xy = f(g(xy)),$$

proves that $g(x).g(y) = g(xy)$. For monoids we note that $f(g(e_T)) = e_T = f(e_S)$, whence $g(e_T) = e_S$.

(f) The only claim needing a proof is the last. Supposing we have two solutions $i'\colon G' \to S$ and $i''\colon G'' \to S$, each of them factorises through the other, by unique homomorphisms of groups

$$h\colon G' \rightleftarrows G'' : k, \qquad i' = i''h, \quad i'' = i'k,$$

and $kh\colon G' \to G'$ must be the identity, because $i'.(kh) = i''h = i' = i'.\mathrm{id}\, G'$. Similarly $hk = \mathrm{id}\, G''$, whence h and k are isomorphisms of groups, inverse to each other.

1.5.5 Exercises and complements, II

(a) Each semigroup S has a monoid $\mathrm{End}(S)$ of endomorphisms $S \to S$, and a group

$$\mathrm{Aut}(S) = \mathrm{Inv}(\mathrm{End}(S)), \tag{1.133}$$

of *automorphisms* $S \to S$, namely the invertible endomorphisms. If $X = |S|$ is the underlying set, $\mathrm{End}(S)$ is a submonoid of the monoid $\mathrm{End}(X)$ of endomappings of X – and of course these items should not be confused.

If G is a group, the mapping of inner automorphisms (see (1.125))

$$\begin{aligned} &\varphi\colon G \to \mathrm{Aut}(G), \\ &\varphi(g) = \varphi_g\colon G \to G, \qquad \varphi_g(x) = g^{-1}xg, \end{aligned} \tag{1.134}$$

is a homomorphism, whose image is the subgroup of inner automorphisms and whose kernel is the *centre* of G, formed by the elements which commute with each other

$$\mathrm{Cnt}(G) = \{g \in G \mid gx = xg, \text{ for every } x \in G\}. \tag{1.135}$$

Let us recall that, if A is a *commutative* group (in additive notation), then $\mathrm{End}(A)$ is a unital ring, under pointwise addition and composition (Exercise 1.3.2(g)). Moreover, all inner automorphisms are the identity.

(b) Let H be a subgroup of G and $g \in G$. The following subsets of G

$$gH = \{gh \mid h \in H\}, \qquad Hg = \{hg \mid h \in H\}, \tag{1.136}$$

are called *left cosets* and *right cosets* of H, respectively. We also write $\varphi_g(H) = g^{-1}Hg$.

The invariance condition $H \triangleleft G$ can be expressed in many equivalent ways as:

- for every $g \in G$, $g^{-1}Hg \subset H$,
- for every $g \in G$, $g^{-1}Hg = H$,
- for every $g \in G$, $gH \subset Hg$,
- for every $g \in G$, $Hg \subset gH$,
- for every $g \in G$, $gH = Hg$.

(c) For every set X, the group $\mathrm{Sym}(X) = \mathrm{Inv}(\mathrm{End}(X))$ consists of all the *permutations* of X, i.e. all the mappings $f\colon X \to X$ that are invertible with respect to composition, namely the bijective ones. It is called the *symmetric group* on the set X.

In particular, if X is the finite set $\{1, ..., n\}$ we get the group $\underline{S}_n$ of its permutations, called the *symmetric group* on n objects. It will be analysed in 1.6.7, showing that it is not commutative for $n \geqslant 3$.

(d) (*Cayley Theorem*) Every group G is isomorphic to a subgroup of the symmetric group $\mathrm{Sym}(X)$ on the underlying set $X = |G|$.

Again, it is useful to distinguish the group G from its underlying set X, because the group $\mathrm{Sym}(X)$ does not depend on the algebraic structure of G, and contains the group $\mathrm{Aut}(G)$ of automorphisms of G, as a subgroup.

Historically, Cayley Theorem was important in the process of abstracting the notion of a group from the groups of invertible transformations.

(e) Let S be a semigroup with multiplication $xy = x$, as in Exercise 1.5.1(c). Prove that S is isomorphic to a subsemigroup of End(X), where $X = |S|$ is the underlying set.

(f) In a group G, the torsion elements (i.e. those of finite order) form a normal subgroup $\mathrm{t}G$.

(g) In a group G, the *subgroup of commutators* $[G, G]$ is the subgroup generated by all *commutators* $[x, y] = xyx^{-1}y^{-1}$, for $x, y \in G$. Plainly, G is commutative if and only if this subgroup is trivial. Prove that this subgroup is always invariant in G.

1.5.6 Products

The *cartesian product* $S{\times}T$ of two semigroups is the cartesian product of their underlying sets, equipped with the componentwise operation, that combines the operations of S and T

$$(x, y).(x', y') = (xx', yy'), \qquad (x, x' \in S,\ y, y' \in T), \tag{1.137}$$

and is plainly associative.

If S and T are monoids, the product $S{\times}T$ is also, with unit (e_S, e_T). If S and T are groups, the product $S{\times}T$ is also, with $(x, y)^{-1} = (x^{-1}, y^{-1})$.

The *cartesian projections* $p\colon S{\times}T \to S$ and $q\colon S{\times}T \to T$ are homomorphisms of semigroups (or monoids, or groups) and the operation we have defined on $S{\times}T$ is the only one with this effect.

Similarly, for a family $(S_i)_{i\in I}$ of semigroups (or monoids, or groups), the cartesian product $S = \prod_i S_i$ is the product of the underlying sets, with the componentwise operation

$$(x_i)_{i\in I}.(y_i)_{i\in I} = (x_iy_i)_{i\in I}. \qquad \text{for } x_i, y_i \in S_i \quad (i \in I). \tag{1.138}$$

The product has a family of projections $p_i\colon S \to S_i$ $(i \in I)$, which are homomorphisms.

(*Universal property of the product*) More formally, a product $\prod_i S_i$ of semigroups is determined up to isomorphism as a semigroup S equipped with a family of homomorphisms $p_i\colon S \to S_i$ such that:

- for every semigroup X and every family of homomorphisms $f_i\colon X \to S_i$ there is a unique homomorphism $f\colon X \to S$ such that $p_if = f_i$ (for all $i \in I$).

In fact, if we take S and p_i as defined above, we must let $f(x) = (f_i(x))_{i\in I}$ (for $x \in X$); on the other hand, defining $f\colon X \to S$ in this way we do get a homomorphism such that $p_if = f_i$ (for $i \in I$). Finally, if the semigroup X

with the family $(f_i \colon X \to S_i)$ also satisfies our property, we get a backward homomorphism $g \colon S \to X$ such that $f_i g = p_i$ (for all $i \in I$), and we conclude that f and g are inverse to each other, as in 1.3.4.

The universal property of the product holds in the same way for monoids, and similarly for groups.

Remarks. (a) For a semigroup S, the multiplication mapping

$$m \colon S \times S \to S, \qquad m(x, y) = xy, \tag{1.139}$$

is a homomorphism (for the product structure of $S \times S$) if and only if we have

$$xx'yy' = xyx'y', \qquad \text{(for } x, x', y, y' \in S). \tag{1.140}$$

This is obviously true when S is commutative; but it is also true for the two structures considered in 1.5.1(c), where $xy = x$, or $xy = y$ (for all x, y).

(b) If S is a group, the previous property is equivalent to the commutativity of S, by cancellation in (1.140).

1.5.7 Quotients of semigroups

Let S be a semigroup. A *congruence* (of semigroups) in S is an equivalence relation E consistent with the operation (as in 1.3.5, for abelian groups). Precisely, this means that:

$$\text{if } \ xEx' \text{ and } yEy' \ \text{ then } \ (xy)E(x'y'). \tag{1.141}$$

The quotient S/E of the set S modulo the equivalence relation E consists of all the equivalence classes $[x] = \{x' \in S \mid x' \, E \, x\}$ of E (which form a partition of S). It is a semigroup, under the operation induced by the operation of S

$$[x].[y] = [xy], \tag{1.142}$$

called the *quotient* of the semigroup S modulo the congruence E. The canonical projection on the quotient set

$$p \colon S \to S/E, \qquad p(x) = [x], \tag{1.143}$$

is a homomorphism, and the structure we have put on S/E is the only one with this effect.

If S is a monoid, so is S/E, with identity $[1_S]$. If S is a group, so is S/E, with inverses $[x]^{-1} = [x^{-1}]$. In both cases the canonical projection is a homomorphism of the structure we are considering. In the case of groups, the kernel $\mathrm{Ker}\, p$ of the canonical projection determines the congruence, as we will see in the next subsection.

For a homomorphism $f \colon S \to T$ of semigroups (resp. monoids), the canonical factorisation $f = mgp$ which we have seen in (1.43) gives a surjective homomorphism $p\colon X \to X/R_f$, an isomorphism $g\colon X/R_f \to \operatorname{Im} f$ and an injective homomorphism $m\colon \operatorname{Im} f \to Y$ (the inclusion of a substructure).

Exercises and complements. (a) Any binary relation R in a semigroup *generates* a congruence E, the least one containing R.

(b) A congruence of monoids in $(\mathbb{N}, +)$ is not determined by the class $[0]$.

1.5.8 Quotients of groups

Quotients of groups extend the quotients of abelian groups, dealt with in 1.3.5.

As we have seen, if G is a group and E is a congruence of semigroups on G (i.e. it satisfies the consistency condition (1.141)), then the canonical projection $p\colon G \to G/E$ is a homomorphism of groups, and determines an invariant subgroup $\operatorname{Ker} p$ of G. It is also determined by the latter, as

$$x\,E\,y \;\Leftrightarrow\; p(x) = p(y) \;\Leftrightarrow\; xy^{-1} \in \operatorname{Ker} p. \tag{1.144}$$

Conversely, if $H \lhd G$, we define the *associated congruence* as:

$$x \equiv_H y \;\Leftrightarrow\; xy^{-1} \in H. \tag{1.145}$$

This is indeed an equivalence relation

$$\begin{gathered} xx^{-1} = 1 \in H, \\ xy^{-1} \in H \;\Rightarrow\; yx^{-1} \in H, \\ (xy^{-1} \in H \text{ and } yz^{-1} \in H) \;\Rightarrow\; (xz^{-1} = (xy^{-1}).(yz^{-1}) \in H). \end{gathered} \tag{1.146}$$

Moreover, it is consistent with the operation, *because of the invariance* of H. In fact, the relations $x \equiv_H y$ and $z \equiv_H t$ imply that $xz \equiv_H yt$, as proved by the following computation

$$(xz).(t^{-1}y^{-1}) = (xy^{-1}).y(zt^{-1})y^{-1} \in H.$$

For a fixed group G, there is thus a natural bijection between congruences in G and *invariant subgroups* of G

$$\begin{aligned} E &\mapsto \operatorname{Ker}(G \to G/E) = \{x \in G \mid x\,E\,1_G\}, \\ H &\mapsto \;\equiv_H, \qquad x \equiv_H y \;\Leftrightarrow\; xy^{-1} \in H. \end{aligned} \tag{1.147}$$

The quotient of G modulo the associated congruence $\equiv_H$ is denoted as G/H. In this quotient, the equivalence class of any element x is determined in the following forms (see Exercise 1.5.5(b)):

$$[x] = xH = \{xh \mid h \in H\} = Hx = \{hx \mid h \in H\}, \tag{1.148}$$

and called a *coset* of x (with respect to the invariant subgroup H).

Again, the trivial subgroup $\{1\}$ determines the equality congruence $x = y$, and $G/\{1\}$ can be identified with G. The total subgroup G determines the coarsest congruence $x, y \in G$, and G/G is a trivial group.

For a homomorphism $f\colon G \to G'$ of groups, we already know that the equivalence relation R_f coincides with the congruence of G associated to the invariant subgroup $\operatorname{Ker} f$. The canonical factorisation of the mapping f (in 1.2.5) can thus be rewritten as for abelian groups

$$\begin{array}{ccc} G & \xrightarrow{f} & G' \\ {\scriptstyle p}\downarrow & & \uparrow{\scriptstyle m} \\ G/\operatorname{Ker} f & \xrightarrow[g]{} & \operatorname{Im} f \end{array} \qquad f = mgp, \qquad (1.149)$$

where p is a surjective homomorphism, m is an injective homomorphism and g is an isomorphism of groups.

The canonical projection $G \to G/[G, G]$ will be used in 5.4.3(d), as a universal way of 'making G commutative'.

1.5.9 Lagrange's Theorem and finite groups

For any subgroup $H \subset G$, *possibly non-invariant*, one can consider the relation

$$xRy \;\Leftrightarrow\; xy^{-1} \in H, \qquad (1.150)$$

which is an equivalence relation, as already proved in (1.146). The associated partition of the set G consists of the right cosets of H

$$[x]_R = \{y \in H \mid yx^{-1} \in H\} = Hx. \qquad (1.151)$$

(The equivalence classes of the relation $x^{-1}y \in H$ are the *left* cosets xH. These relations are transformed one into the other by the bijection $\rho\colon |G| \to |G|$ sending x to x^{-1}, see (1.124). The two equivalence relations coincide if and only if H is invariant in G.)

As in the case of abelian groups, in 1.3.9(b), the cosets Hx form a partition of the set G; each of them is equipotent to the subgroup $H = H.1_G$, through a canonical bijection

$$\varphi\colon H \rightleftarrows Hx : \psi, \qquad \varphi(h) = hx, \qquad \psi(x') = h^{-1}x'.$$

If G is finite, we deduce that $\sharp G = (\sharp H).(\sharp(G/R))$, where G/R is only a set. (It is a group when H is invariant.) Therefore, the order of any subgroup of G divides the order of G, a result known as *Lagrange's Theorem* (for groups).

The *index* of H in G is $\sharp(G/R)$, i.e. the cardinal of the set of right cosets of H, or equivalently of its left cosets (also when G is infinite).

It is easy to see that a subgroup $H \subset G$ of index 2 is necessarily normal in G. In fact, H has two right cosets in G, namely H and $G \setminus H$; similarly, it has two left cosets, which must be the same. Finally

$$\begin{aligned} xH &= H = Hx, \qquad && \text{for } x \in H, \\ xH &= G \setminus H = Hx, \qquad && \text{for } x \notin H. \end{aligned}$$

In this case, the quotient G/H is isomorphic to $\mathbb{Z}/2$: see Exercise (a), below.

Exercises and complements. (a) A finite group of prime order p is cyclic and isomorphic to $\mathbb{Z}/p$. Every group of order < 6 is commutative. Classify all these cases, up to isomorphism.

*(b) Write the multiplication table of the symmetric group $\underline{S}_3$, a non-commutative group of order 6.

1.6 Complements on groups and semigroups

We begin by studying free algebraic structures, in the domain of abelian groups, monoids and groups.

Then we investigate other topics, also important, like actions of groups and simple groups; a reader can skip this second part and come back when referred to.

1.6.1 Free abelian groups

As we have seen in 1.3.3(a), an abelian group A is generated by a subset X if every element $a \in A$ can be written as a finite linear combination $\sum_i \lambda_i x_i$ of elements of X, with integral coefficients; or equivalently as an essentially finite linear combination $\sum_{x \in X} \lambda_x x$, indexed by X itself. We say that A is *freely generated* by X if, for every $a \in A$, this expression is unique. All this can be said in a more formal way.

Starting from a set X, we construct an abelian group $\mathbb{Z}X$ as a direct sum of copies of the abelian group $\mathbb{Z}$, indexed by the elements of X

$$\mathbb{Z}X = \textstyle\bigoplus_{x \in X} \mathbb{Z} = \{\lambda\colon X \to |\mathbb{Z}| \mid \text{supp}(\lambda) \text{ is finite}\}. \qquad (1.152)$$

An element λ is thus a quasi null mapping $\lambda\colon X \to |\mathbb{Z}|$, often written as a quasi null family $(\lambda_x)_{x \in X}$ of integers, or simply as (λ_x). They are added componentwise: $(\lambda_x) + (\mu_x) = (\lambda_x + \mu_x)$.

The set X is embedded in this abelian group, by a canonical (injective) mapping that sends an element x to the family $\eta(x)$, which takes value 1 at x and 0 elsewhere

$$\eta\colon X \to \mathbb{Z}X, \qquad (\eta(x))_x = 1, \qquad (\eta(x))_y = 0 \quad (\text{for } y \neq x). \tag{1.153}$$

The element $\eta(x)$ is often written as e_x or identified with x. Each quasi null family (λ_x) of integers is an essentially finite linear combination $\sum_{x\in X} \lambda_x e_x$ in the group $\mathbb{Z}X$, and is called a *formal linear combination of elements of* X.

Now, the pair $(\mathbb{Z}X, \eta)$ has a *universal property*, saying that each mapping $f\colon X \to A$ with values in an abelian group can be uniquely extended to $\mathbb{Z}X$; more precisely, there is a unique *homomorphism* $g\colon \mathbb{Z}X \to A$ of abelian groups such that

$$\begin{array}{ccc} X & \xrightarrow{\eta} & \mathbb{Z}X \\ & {\scriptstyle f}\searrow & \downarrow{\scriptstyle g} \\ & & A \end{array} \qquad g\eta = f. \tag{1.154}$$

In fact, if such a homomorphism g exists, then it preserves linear combinations with integral coefficients, and its value on a quasi null family (λ_x) of integers is determined

$$g(\textstyle\sum_{x\in X} \lambda_x e_x) = \sum_{x\in X} \lambda_x f(x). \tag{1.155}$$

as an essentially finite linear combination in A. Conversely, the mapping g defined as above is easily seen to be a homomorphism of abelian groups.

We express this universal property saying that $\mathbb{Z}X$ is the *free abelian group* generated by the set X, and that the mapping $\eta\colon X \to \mathbb{Z}X$ is the insertion of X as the *canonical basis* $(e_x)_{x\in X}$ of $\mathbb{Z}X$.

More generally, a *free abelian group* is, by definition, any abelian group A isomorphic to some $\mathbb{Z}X$. This means that there exists a family $(e_x)_{x\in X}$ of elements of A, indexed by some set X, such that every element of A can be written, in a unique way, as an essentially finite linear combination $\sum_{x\in X} \lambda_x e_x$ (with integral coefficients). The basis $(e_x)_{x\in X}$ can equivalently be viewed as the corresponding *subset* $\{e_x \mid x \in X\}$ of A.

When studying a certain mathematical structure, with its own 'privileged mappings' (as in 1.2.7), *the free structure over a set is always defined by the appropriate universal property*, as above for abelian groups. We will see below various instances of free equational algebras, and a general approach in Section 5.4.

Exercises and complements. (a) The trivial group is free, with an empty basis. A power $\mathbb{Z}^n$ is free, with canonical basis

$$e_1 = (1, 0, ..., 0), \quad e_2 = (0, 1, 0, ..., 0), ... \quad e_n = (0, ..., 0, 1).$$

(b) Every free abelian group A is torsion free (see 1.3.9(h)).

(c) The additive group $\mathbb{Q}$ of rational numbers is torsion free, and is not a free abelian group.

*(d) Every subgroup of a free abelian group is free. The proof of this important fact is not easy, and depends on the Axiom of Choice: see [Fu], Theorem 12.1.

1.6.2 Free monoids and abelian monoids

For a set X, we interpret an n-tuple $w = (x_1, x_2, ..., x_n) \in X^n$ as an n-ary *word* in the alphabet X. Two words are multiplied by *concatenation*

$$(x_1, ..., x_p).(y_1, ..., y_q) = (x_1, ..., x_p, y_1, ..., y_q). \tag{1.156}$$

We are also considering the empty word $e = (\,) \in X^0$, of length 0, which acts as a unit for this operation: $ew = w = we$, for every word w.

The disjoint union $M(X)$ of all sets X^n (for $n \in \mathbb{N}$) is thus a monoid. It is actually the *free monoid* on the set X, as made clear by the mapping 'of unary words'

$$\eta\colon X \to M(X), \qquad \eta(x) = (x). \tag{1.157}$$

Its *universal property* is similar to that of free abelian groups (in 1.6.1): every mapping $f\colon X \to S$ with values in a monoid can be uniquely extended to $M(X)$; more precisely, there is a unique *homomorphism* $g\colon M(X) \to S$ of monoids such that

$$\begin{array}{ccc} X & \xrightarrow{\eta} & M(X) \\ & {\scriptstyle f}\searrow & \downarrow{\scriptstyle g} \\ & & S \end{array} \qquad g\eta = f. \tag{1.158}$$

As to uniqueness, g must send the empty word to 1_S, while a non-empty word w can be expressed as a product of unary words

$$w = (x_1, x_2, ..., x_n) = (x_1).(x_2). \dots .(x_n),$$

and must be sent to the product $f(x_1).f(x_2). \dots .f(x_n)$ in S. As to existence, if we define g in this way the condition $g\eta = f$ is satisfied, and we do have a homomorphism: the unit is preserved and

$$\begin{aligned} &g((x_1, ..., x_p).(y_1, ..., y_q)) = g(x_1, ..., x_p, y_1, ..., y_q) \\ &= f(x_1). \dots .f(x_p).f(y_1). \dots .f(y_q) = g(x_1, ..., x_p).g(y_1, ..., y_q). \end{aligned}$$

Also here we can identify the element $x \in X$ with the unary word $\eta(x) = (x)$, so that each element of $M(X)$ can be written in a unique way as a finite product $x_1.x_2.\dots.x_n$ of elements of the basis (including the empty product $e = \prod_{i\in\emptyset} x_i$).

Exercises and complements. (a) The *free semigroup* on the set X is similarly obtained, discarding the empty word.

(b) Construct the *free abelian monoid* on a set X.

Solutions. (b) We can adapt the construction of the free abelian group on the set X, in 1.6.1, replacing the abelian group $\mathbb{Z}$ with the abelian (additive) monoid $\mathbb{N}$. We form thus the set of quasi null families of natural numbers, indexed by X

$$\mathbb{N}X = \{\lambda\colon X \to |\mathbb{N}| \mid \text{supp}(\lambda) \text{ is finite}\}, \tag{1.159}$$

which is an abelian monoid with the obvious pointwise sum. (It is a direct sum $\bigoplus_{x\in X}\mathbb{N}$ of abelian monoids, essentially defined as in the case of abelian groups.)

Again, the set X is embedded in this abelian monoid, by a canonical (injective) mapping

$$\eta\colon X \to \mathbb{N}X, \qquad (\eta(x))_x = 1, \quad (\eta(x))_y = 0, \text{ for } y \neq x. \tag{1.160}$$

The element $\eta(x)$ can be written as e_x or identified with x.

Each quasi null family (λ_x) of natural numbers is an essentially finite linear combination $\sum_{x\in X} \lambda_x x$ of elements of X, with coefficients in $\mathbb{N}$. The universal property works as in (1.154), and proves that $\mathbb{N}X$ is indeed the free abelian monoid generated by X.

1.6.3 Free groups

The *free group* on the set X is constructed in a more complex way than the free monoid $M(X)$, because we must provide each $x \in X$ with an element x^* that will become its inverse.

We begin by forming the disjoint union of two 'copies' of X, namely the following set (where $0, 1 \in \mathbb{N}$)

$$Y = (X\times\{0\}) \cup (X\times\{1\}). \tag{1.161}$$

For the sake of simplicity, each $x \in X$ will be 'identified' with the pair $(x, 0)$, and we write $x^* = (x, 1)$. Now we consider the free monoid $M(Y)$ on the set Y, and the congruence of monoids E on $M(Y)$ generated (see Exercise 1.5.7(a)) by relating each word (x, x^*) and each word (x^*, x) with the empty word (for $x \in X$).

Finally we take the quotient monoid

$$G(X) = M(Y)/E, \tag{1.162}$$

writing as $[y_1, ..., y_n]$ the equivalence class of the word $(y_1, ..., y_n) \in M(Y)$.

Thus $[x, x^*] = [e] = [x^*, x]$, for all $x \in X$, which means that $[x]$ and $[x^*]$ are inverse to each other, and every 'unary class' $[y]$ is invertible (for

$y \in Y$). But every class $[y_1, ..., y_n]$ is a product of unary classes, whence $G(X)$ is a group.

The canonical mapping of 'classes of unary words'

$$\eta\colon X \to G(X), \qquad \eta(x) = [x], \tag{1.163}$$

satisfies the usual *universal property* of free structures: for every mapping $f\colon X \to G$ there is a unique group-homomorphism $g\colon G(X) \to G$ such that

$$\begin{array}{ccc} X & \xrightarrow{\eta} & G(X) \\ & {\scriptstyle f}\searrow & \downarrow{\scriptstyle g} \\ & & G \end{array} \qquad g\eta = f. \tag{1.164}$$

As to its existence, G is a monoid, and we have a unique homomorphism of monoids $h\colon M(Y) \to G$ determined as follows on the basis Y:

$$h(x) = f(x), \qquad h(x^*) = (f(x))^{-1} \qquad \text{(for } x \in X). \tag{1.165}$$

This homomorphism takes each word (x, x^*) and each word (x^*, x) to 1_G. It is thus invariant under the congruence E, and induces a homomorphism $g\colon M(Y)/E \to G$ of monoids; as both are groups, it is a homomorphism of groups. Moreover $g[x] = h(x) = f(x)$, for every $x \in X$.

As to uniqueness, every element $[y_1, ..., y_n]$ of $G(X)$ is a product of unary classes of type $[x]$ or $[x^*] = [x]^{-1}$ (for $x \in X$), on which the value of g is determined, as $f(x)$ or $(f(x))^{-1}$, respectively.

The 'embedding' η of the basis is indeed an injective mapping; this 'seemingly' obvious fact can be deduced from the universal property without giving an explicit construction of the congruence E.

In fact, each set X has an injective mapping $f\colon X \to G$ with values in a group: if $X \neq \emptyset$ it is sufficient to equip X itself with a group structure, as one can always do (see Exercise 1.3.6(f)). Then we can factorise $f = g\eta$, and the mapping η must be injective as well. (Alternatively, we can also embed X in the free abelian group $\mathbb{Z}X$.)

1.6.4 Actions of groups

Let G be a group. A *G-set* is a set X equipped with a *left action* of G. This means a mapping

$$\varphi\colon X \times G \to X, \qquad \varphi(x, g) = gx \qquad \text{(for } x \in X,\ g \in G), \tag{1.166}$$

that satisfies the following axioms (for all $x \in X$ and $g, h \in G$):

(*i*) (*Compatibility*) $h(gx) = (hg)x$,

(*ii*) (*Unitarity*) $1_G\, x = x$.

We also say that G *acts*, or *operates*, on the set X (on the left). The elements of G can be called *operators*, and those of X *points*. The *congruence* $x \equiv_G x'$ is defined by the existence of an operator g such that $gx = x'$; an equivalence class $Gx = \{gx \mid g \in G\}$ of this relation is called an *orbit* of the action, and the quotient set $X/G = X/\equiv_G$ is called the *orbit set*.

A *G-morphism* $F\colon X \to Y$ is a mapping between G-sets consistent with the G-actions: $F(gx) = g.F(x)$, for all $x \in X$ and $g \in G$.

Every group has a *trivial action* on each set, with $gx = x$ for all operators and points. On the empty set and any singleton this is the unique action of a given group.

The symmetric group $\mathrm{Sym}(X)$ of permutations of the set X (see 1.5.5(c)) has a *canonical action* on the latter, in the obvious way: $fx = f(x)$, for $f \in \mathrm{Sym}(X)$ and $x \in X$.

A G-action φ on the set X gives a group-homomorphism

$$\psi\colon G \to \mathrm{Sym}(X), \qquad \psi(g) = \varphi(-, g)\colon X \to X, \tag{1.167}$$

where $\psi(g)(x) = gx$. In fact, the mapping $\psi(g)$ has inverse $\psi(g^{-1})$, and the compatibility axiom says that $\psi(hg) = \psi(h).\psi(g)$. Conversely, we readily see that a homomorphism $G \to \mathrm{Sym}(X)$ gives such an action.

An action is said to be *transitive* if it has only one orbit:

- for every $x, x' \in X$ there is some $g \in G$ such that $x' = gx$.

It is said to be *free*, or *fixed-point free*, if it satisfies the following readily equivalent properties:

- for every $x, x' \in X$ there is at most one $g \in G$ such that $x' = gx$,

- for every $g, h \in G$, if $gx = hx$ for some $x \in X$, then $g = h$,

- if an operator $g \in G$ has a *fixed point* x (i.e. $gx = x$) then $g = 1_G$.

The action is transitive and free if and only if, for every $x \in X$, the mapping

$$\varphi_x = \varphi(x, -)\colon G \to X, \qquad g \mapsto gx, \tag{1.168}$$

is bijective. The canonical action of $\mathrm{Sym}(X)$ on the set X is transitive. Free actions will be revisited in 5.4.2(b).

A *right action* of G on X is written as xg, and has a 'reversed' compatibility axiom: $(xh)g = x(hg)$, where hg operates *first* by h and *then* by g. This is equivalent to a left action of the opposite group G^{op} on X, and it is sufficient to consider one case. An 'action' will mean a left action, unless differently specified.

1.6.5 Actions of monoids and semigroups

More generally, a *left action* of a monoid S on a set X is assumed to satisfy the same axioms above. We speak of an S*-set*, and use the same terminology as above, for *operators, points, orbits, transitive* and *free actions.*

The monoid $\mathrm{End}(X)$ of endomappings of the set X acts on the latter as above: $fx = f(x)$, for $f \in \mathrm{End}(X)$ and $x \in X$. An S-action on the set X amounts to a homomorphism $S \to \mathrm{End}(X)$ of monoids.

Even more generally, one can consider actions of semigroups, leaving out the unit axiom 1.6.4(ii).

1.6.6 Simple groups

A non-trivial group G is said to be *simple* if the only normal subgroups of G are the trivial and the total one.

In the variety of groups, the simple groups are structurally distinguished as the non-trivial objects G which have no quotients other than the necessary ones: G and the trivial group.

Exercises and complements. (a) Prove that an abelian group is simple if and only if it is a finite cyclic group of prime order p, isomorphic to $\mathbb{Z}/p$. (Note that, in this domain, the order determines the type of isomorphism.)

*(b) The study of finite, non-commutative, simple groups has been an important field of research, since E. Galois proved (in 1831) that, for every $n \geqslant 5$, the alternating group $\underline{A}_n$ is simple, and deduced from this fact that the general algebraic equation of this order cannot be solved 'by radicals'. ($\underline{A}_n$ is a subgroup of $\underline{S}_n$, of index 2 and order $n!/2$: see 1.6.7.)

A complete description of finite simple groups was achieved in 2004, and is the subject of ponderous books. The smallest non-commutative, simple group is the alternating group $\underline{A}_5$, with 60 elements. There exist non-isomorphic simple groups with the same finite order: the smallest such order is 20160.

1.6.7 Exercises and complements (Finite symmetric groups)

As we have seen in 1.5.5(c), the symmetric group on n objects $\underline{S}_n$ is formed of all the permutations of the finite set $\{1, ..., n\}$; from Combinatorics, we know that it has $n!$ elements.

The *alternating group* $\underline{A}_n \subset \underline{S}_n$ is defined below.

(a) The groups $\underline{S}_0$ and $\underline{S}_1$ are trivial; $\underline{S}_2$ is isomorphic to $\mathbb{Z}/2$. $\underline{S}_3$ is non-commutative, as all the higher symmetric groups.

(b) Classify the proper, non-trivial subgroups of $\underline{S}_3$, proving that one of them is invariant and three others are not.

(c) Let $n \geqslant 2$, so that $\underline{S}_n$ is not trivial. For $f \in \underline{S}_n$, we write as $n(f)$

the *number of inversions* of the permutation f: this is the number of pairs (i, j) such that $1 \leqslant i < j \leqslant n$ and $f(i) > f(j)$. Then the mapping

$$\varepsilon \colon \underline{S}_n \to \{-1, 1\}, \qquad \varepsilon(f) = (-1)^{n(f)}, \tag{1.169}$$

with values in the multiplicative group $\{-1, 1\} = \mathrm{Inv}(\mathbb{Z})$ is called the *sign homomorphism*, and is surjective (for $n \geqslant 2$).

It is indeed a homomorphism: the proof is not easy, and can be found in Chapter 7.

(d) The permutation f is said to be *even* if $n(f)$ is even, and $\varepsilon(f) = 1$; it is *odd* otherwise.

The (normal) *alternating subgroup* $\underline{A}_n = \mathrm{Ker}\,(\varepsilon) \subset \underline{S}_n$ is, by definition, the group of even permutations of n objects. It has index 2 in $\underline{S}_n$, because $\underline{S}_n/\underline{A}_n \cong \mathbb{Z}/2$.

In particular, $\underline{A}_3$ is the unique invariant subgroup of $\underline{S}_3$, after the trivial and the total subgroup.

*(e) The permutation that interchanges two indices $i \neq j$ and leaves the others unchanged is called a *transposition* and often written as (i, j). One proves that $\underline{S}_n$ is generated by its transpositions (see [Boul], I.5.7, Proposition 8), and that every transposition is an odd permutation. It follows that a permutation is even if and only if it can be obtained as a product of an even number of transpositions.

*1.6.8 *An exercise* (Subgroups of cyclic groups)

The lattice $X = \mathrm{Sub}(\mathbb{Z})$ of subgroups of the abelian group $\mathbb{Z}$ is distributive. It follows that the same holds for all cyclic groups.

Hints. We know, from Exercise 1.4.5(d), that X is anti-isomorphic to the divisibility lattice $\mathbb{N}_D$ of natural numbers. Then we can prove that the latter is distributive, a (hopefully amusing) exercise based on the prime factor decomposition of integers.

1.7 Complements on the real line and set theory

After reviewing the intervals of the real line, we explore some consequences of the completeness axiom, involving the square root of a positive real number.

Then we say something more on set theory, about the axiom of choice and cardinals.

1.7.1 Real intervals

The intervals of the real line $\mathbb{R}$ are well known. Let $a \leqslant b$. We use the usual notation for a bounded closed interval

$$[a,b] = \{x \in \mathbb{R} \mid a \leqslant x \leqslant b\}, \tag{1.170}$$

where $a = \min\,[a,b]$ and $b = \max\,[a,b]$. In particular, the singleton $[a,a] = \{a\}$ is a degenerate interval.

For a bounded open interval we will use a less usual notation, similar to Bourbaki's

$$]a,b[\, = \{x \in \mathbb{R} \mid a < x < b\}, \tag{1.171}$$

which has the advantage of distinguishing this interval from the pair (a,b).

Note. Nicolas Bourbaki is a collective pseudonym for a group of (mostly French) mathematicians, who published an influential treatise, 'Éléments de Mathématique', from the 1930's on. (Here we only refer to some of these books, on Algebra and Topology, in the English version.)

The empty subset $]a,a[\, = \emptyset$ is also a degenerate interval. If $a < b$, then

$$a = \inf\,]a,b[, \qquad\qquad b = \sup\,]a,b[,$$

and $]a,b[$ has no minimum nor maximum.

The other kinds of intervals will be denoted in a consistent way, as in the following examples (for $a < b$ in $\mathbb{R}$):

$$[a,b[\, = \{x \in \mathbb{R} \mid a \leqslant x < b\}, \qquad [a,+\infty[\, = \{x \in \mathbb{R} \mid a \leqslant x\},$$
$$]a,+\infty[\, = \{x \in \mathbb{R} \mid a < x\}, \qquad]-\infty,+\infty[\, = \mathbb{R}.$$

1.7.2 The extended line

Studying the intervals of $\mathbb{R}$, it can be convenient to work in the *extended real line*, where we add two points 'at infinity'

$$\overline{\mathbb{R}} = \{-\infty\} \cup \mathbb{R} \cup \{+\infty\}, \tag{1.172}$$

and extend the order letting $-\infty < x < +\infty$, for all $x \in \mathbb{R}$.

In $\overline{\mathbb{R}}$ every subset has some lower and some upper bound. Thus $\sup A = +\infty$, for every A upper unbounded *in* $\mathbb{R}$, while $\sup \emptyset = -\infty$, because every element of $\overline{\mathbb{R}}$ is (trivially) an upper bound of the empty subset.

Finally, *every* subset of the extended line has infimum and supremum, so that $\overline{\mathbb{R}}$ is a complete lattice (but not a field). For instance:

$$\begin{aligned} -\infty &= \inf \mathbb{R} = \inf \mathbb{Z} = \min \overline{\mathbb{R}} = \sup \emptyset, \\ +\infty &= \sup \mathbb{R} = \sup \mathbb{Z} = \max \overline{\mathbb{R}} = \inf \emptyset. \end{aligned} \tag{1.173}$$

1.7.3 Proposition (Convexity of the intervals)

(a) A subset A of $\mathbb{R}$ is an interval if and only if it satisfies the following condition (which says that A is a convex *subset of the ordered set $\mathbb{R}$):*

(i) if $a, b \in A$ and $a \leqslant x \leqslant b$ in $\mathbb{R}$, then $x \in A$.

(b) A subset A of an interval I is an interval if and only if:

(ii) if $a, b \in A$ and $a \leqslant x \leqslant b$ in I, then $x \in A$.

Proof (a) Plainly, any kind of interval of the real line satisfies this condition.

Suppose that $A \subset \overline{\mathbb{R}}$ satisfies condition (i), and let $\alpha = \inf A$, $\beta = \sup A$, in the extended real line.

Since the empty subset is an interval, we can suppose that A has some element ξ, and we deduce that $\alpha \leqslant \xi \leqslant \beta$, whence $\alpha \leqslant \beta$, and $A \subset [\alpha, \beta]$. But we can also prove that $]\alpha, \beta[\subset A$:

- if $\alpha < x$, then x cannot be a lower bound of A, and therefore $x > a$ for some $a \in A$,

- if $x < \beta$, then x cannot be an upper bound of A, and therefore $x < b$ for some $b \in A$.

Thus $\alpha < x < \beta$ implies $a < x < b$ with $a, b \in A$, and $x \in A$, by (i).

We have shown that $]\alpha, \beta[\subset A \subset [\alpha, \beta]$, so that A is one of the following four intervals of $\overline{\mathbb{R}}$

$$]\alpha, \beta[, \qquad [\alpha, \beta[, \qquad]\alpha, \beta], \qquad [\alpha, \beta].$$

If $A \subset \mathbb{R}$, we simply have to discard the intervals of $\overline{\mathbb{R}}$ that contain $-\infty$ or $+\infty$, and deduce that A is an interval of $\mathbb{R}$.

(b) An obvious consequence. □

1.7.4 Proposition and Definition (Square roots)

On the real interval $J = [0, +\infty[$, the squaring mapping:

$$f\colon J \to J, \qquad f(x) = x^2, \tag{1.174}$$

is an isomorphism of ordered sets: it preserves and reflects the order relation, and is bijective.

The inverse mapping $J \to J$ is called the (main) *square root*, and denoted as $\sqrt{-}$ or $(-)^{1/2}$.

Proof (We give here an elementary proof of the existence of square roots, without using the continuity of f and the Intermediate Value Theorem, to be dealt with in Chapters 3 and 4. See Exercise 4.1.9(c).)

First, the function f is strictly increasing, because of axiom (A.15): if $0 \leqslant x < y$, then $x^2 \leqslant xy < y^2$. Therefore it is injective; moreover, since the interval J is totally ordered, the function f reflects the ordering: from $x^2 \leqslant y^2$, it follows that $x \leqslant y$ (as $y < x$ would give $y^2 < x^2$).

We still have to prove that f is surjective; it is sufficient to prove that any real number $a \geqslant 1$ belongs to $\operatorname{Im} f$, because $1/a$ will then do the same (and the number 0 certainly does).

We introduce the set

$$A = \{x \in J \mid x^2 \leqslant a\} \subset [0, a[, \tag{1.175}$$

which contains 1 and is upper bounded; we take the real number $\xi = \sup A \in [1, a]$ and want to prove that $\xi^2 = a$. (This is not trivial: if we replace J with $\mathbb{N}$ and take $a = 2$, we find $A = \{0, 1\}$ and $\sup A = 1$.)

If $\xi^2 < a$, we let $\varepsilon = \min(a - \xi^2, 1) > 0$, and remark that $0 < \xi/a \leqslant 1$. Then we get a contradiction, because $\xi + \varepsilon/(3a)$ belongs to A:

$$(\xi + \varepsilon/(3a))^2 = \xi^2 + 2\varepsilon\xi/(3a) + \varepsilon^2/(3a)^2 \leqslant \xi^2 + 2\varepsilon/3 + \varepsilon/9 < \xi^2 + \varepsilon \leqslant a.$$

If $\xi^2 > a$, we let $\varepsilon = \xi^2 - a > 0$; we still have $0 < \xi/a \leqslant 1$ (and $-\xi/a \geqslant -1$). Then $\xi - \varepsilon/(3a)$ does not belong to A:

$$(\xi - \varepsilon/(3a))^2 = \xi^2 - 2\varepsilon\xi/(3a) + \varepsilon^2/(3a)^2 > \xi^2 - 2\varepsilon/3 > \xi^2 - \varepsilon = a.$$

Therefore all the elements of A are smaller than $\xi - \varepsilon/(3a)$ and $\sup A < \xi$, again a contradiction. We are left with $\xi^2 = a$.

□

1.7.5 Exercises and complements (Rationals and irrationals)

(a) The square root $\sqrt{2}$ is an irrational number. The same is true of the square root of every natural number which is not the square of a natural number.

Hints. Following a classical proof, going back to the Pythagorean school, one assumes that $a^2/b^2 = 2$, where a and b are coprime natural numbers, and finds a contradiction.

(b) Every non-degenerate interval of $\mathbb{R}$ contains rational numbers and irrational numbers – infinitely many of both kinds.

(c) Every real number a can be expressed as a least upper bound of some set of rational numbers, which is non-empty and upper bounded in $\mathbb{Q}$. For instance:

$$a = \sup\{x \in \mathbb{Q} \mid x \leqslant a\}. \tag{1.176}$$

1.7.6 The axiom of choice

Speaking of 'Set Theory' we generally mean *Zermelo–Fraenkel theory*, denoted as (ZF), *with the addition of the Axiom of Choice* (AC); the complex is denoted as (ZFC). (ZF) is named after Ernst Zermelo and Abraham Fraenkel.

The Axiom of Choice can be formulated in various forms, equivalent in (ZF):

(i) given a family $(A_i)_{i \in I}$ of non-empty sets, one can (globally) choose an element $x_i \in A_i$ in each of them,

(ii) the cartesian product of a family $(A_i)_{i \in I}$ of non-empty sets is non-empty,

(iii) if $g\colon Y \to X$ is a surjective mapping, there is a mapping $f\colon X \to Y$ such that $gf = \mathrm{id}\, X$.

The equivalence of (i) and (ii) is obvious. Admitting (i), we prove (iii) by considering the family of subsets $g^{-1}\{x\} \subset Y$, indexed by $x \in X$. Each of them (if any) is non-empty (because g is assumed to be surjective), and we can choose an element $y_x \in g^{-1}\{x\}$ in each of them. This defines a mapping $f\colon X \to Y$, with $f(x) = y_x$, and $g(f(x)) = x$, for each $x \in X$.

Conversely, admitting (iii) and given a family $(A_i)_{i \in I}$ of non-empty sets, we construct their disjoint union, as in Exercise 1.2.3(d)

$$Y = \textstyle\bigcup_i A_i \times \{i\}. \tag{1.177}$$

We can now define a mapping $g\colon Y \to I$ by letting $g(y, i) = i$, for every $y \in A_i$. This mapping is surjective, as we know that, for each $i \in I$, $A_i \neq \emptyset$. There is thus a mapping $f\colon I \to Y$ such that $gf = \mathrm{id}\, I$, and for $i \in I$ the element $f(i) = (x_i, i) \in A_i \times \{i\}$ gives a global choice $x_i \in A_i$.

1.7.7 Zorn's Lemma

This important result of set theory was proved by Kazimierz Kuratowski in 1922, and independently by Max Zorn in 1935. It is stated here without a proof.

We first introduce some terminology, for an ordered set X and a subset $A \subset X$. We say that A is a *chain* of X if its induced order is total: for every $x, y \in A$, we have $x \leqslant y$ or $y \leqslant x$, in X. We say that an element $a \in A$ is *maximal* in A if it has no strict upper bound in A: if $x \in A$ and $a \leqslant x$, then $x = a$. Dually, a is *minimal* in A if it has no strict lower bound in A.

Now, Zorn's Lemma, also called the Kuratowski–Zorn Lemma, says the following.

Every non-empty ordered set where every chain has an upper bound admits a maximal element.

In many domains of mathematics, this lemma is used to prove crucial facts, like:

- every vector space has a basis (in the infinitely generated case): see 2.4.4,

- in a commutative unital ring, every proper ideal is contained in a maximal ideal: see 2.2.3(f),

- every product of compact spaces is compact (Tychonoff's Theorem, for an infinite product): see 4.2.4.

Zorn's lemma is equivalent to the axiom of choice, and also to the *well-ordering theorem* (stating that every set can be given an ordering where every non-empty subset has a minimum). More precisely, any one of these three statements is sufficient to prove the other two, in Zermelo–Fraenkel set theory.

Exercises and complements. (a) If the preordered set X has a minimum $\bot$, an atom of X (see (1.92)) is the same as a minimal element of the preordered subset $X' = \{x \in X \mid x \nsim \bot\}$.

1.7.8 Cardinals

The theory of cardinals is an important part of set theory. We say now something more about it, after 1.2.1, still at an informative level and without proving the main facts.

The *equipotence relation* $X \sim Y$ between sets is defined by the existence of a bijective mapping $X \to Y$. This relation is plainly reflexive, symmetric and transitive (even though it relates terms that do not form a set). One can construct, for every set X, an equipotent set $\alpha = \sharp X$, called the *cardinal* of X, so that X and Y are equipotent if and only if they have the same cardinal.

A set is finite if and only if its cardinal belongs to $\mathbb{N}$. The cardinal of $\mathbb{N}$ is written as $\aleph_0$ (read as *aleph zero*), and is the smallest infinite cardinal. A set is said to be *countable* if its cardinal is finite or $\aleph_0$.

More generally, we say that $X \prec Y$ when there exists an injective mapping $f \colon X \to Y$. This relation is reflexive and transitive. We also note that, provided that $X \neq \emptyset$, this property is equivalent to the existence of a surjective mapping $g : Y \to X$.

Indeed, given g we apply (AC) in the form (iii) and obtain an injective mapping $f \colon X \to Y$; given f, we form g by sending each element $y \in \mathrm{Im}\, f$ to the unique $x \in X$ such that $f(x) = y$, and each $y \in Y \setminus \mathrm{Im}\, f$ to a fixed element $x_0 \in X$ (there must be some).

There are now important results.

(a) (*Schröder–Bernstein Theorem*) X and Y are equipotent if (and only if) $X \prec Y$ and $Y \prec X$.

(b) Cardinal sets are thus ordered by the restricted relation $\alpha \prec \beta$, preferably written as $\alpha \leqslant \beta$.

(c) This order is total: given two cardinal sets α and β, we have $\alpha \leqslant \beta$ or $\beta \leqslant \alpha$.

For cardinal sets α, β one defines:

$$\begin{aligned} &\alpha + \beta = \sharp((\alpha \times \{0\}) \cup (\beta \times \{1\})), \quad &\alpha.\beta = \sharp(\alpha \times \beta), \\ &\beta^\alpha = \sharp(\mathrm{Map}(\alpha, \beta)) &(2^\alpha = \sharp(\mathcal{P}\alpha)). \end{aligned} \tag{1.178}$$

(Here 0 and 1 are natural numbers.) One proves that, if one at least of these cardinals is infinite and $\alpha, \beta > 0$

$$\alpha + \beta = \alpha.\beta = \max(\alpha, \beta). \tag{1.179}$$

It is always true that $\alpha < 2^\alpha$. The cardinal of the set $\mathbb{Q}$ is $\aleph_0$ (it is easy to form an injective mapping $\mathbb{Q} \to \mathbb{Z} \times \mathbb{Z}$). The cardinal of the set $\mathbb{R}$ is $2^{\aleph_0}$, strictly bigger than $\aleph_0$.

Finally, it will be useful to know that any countable union of countable sets is countable. More generally

- if $X = \bigcup_{i \in I} X_i$, where the set I is infinite and each X_i is countable, then $\sharp X \leqslant \sharp I$.

In fact, using the previous properties, we readily have:

$$\sharp X \leqslant \sharp(\textstyle\bigcup_{i \in I}(\mathbb{N} \times \{i\})) = \sharp(\mathbb{N} \times I) = \sharp I.$$

2
Algebraic structures, II

We continue our study of the basic properties of algebraic structures, examining rings and fields, modules and vector spaces, and algebras on a ring.

We rely as little as possible on Calculus: the trigonometric functions *sine* and *cosine* will only be used in marginal points, while dealing with complex numbers and groups of invertible matrices.

2.1 Rings and fields

The theory of rings follows – initially – a pattern similar to that of groups. A quotient of a ring R is determined by a special kind of subring, a *bilateral ideal* $I \lhd R$, much in the same way as a quotient of a group is determined by a normal subgroup. Then an important role is played by the study of ideals, mainly promoted by Emmy Noether in the 1920's.

The study of commutative rings and their modules is a broad domain of research, called *Commutative Algebra*, closely related to *Algebraic Geometry*; a classical reference is the book by Oscar Zariski and Pierre Samuel [ZS], published in 1958.

2.1.1 Main definitions

Reviewing what we already saw in Section 1.1, a *ring* is a set R equipped with two operations, written as $x+y$ and xy, that satisfy the axioms (A.1–6) of Subsection 1.1.1.

In other words, R is an additive abelian group equipped with a multiplication xy which is associative and distributes over the addition, on both sides. As we have seen, $0x = 0 = x0$ (for all x); the multiplication also distributes over opposites and the subtraction operation $x - y$.

The ring R is *unital* if it satisfies (A.7), i.e. its multiplication has an identity element 1, called the *unit* of R. Then its invertible elements form a multiplicative group $\mathrm{Inv}(R)$.

The ring R is *commutative* if its multiplication is, as in axiom (A.8).

Finally a *field* K satisfies the axioms (A.1–9). In other words, it is a commutative unital ring where $1 \neq 0$ and each element $x \neq 0$ has an inverse, written as x^{-1}. Then $\mathrm{Inv}(K) = K^* = K \setminus \{0\}$.

$\mathbb{Z}$ is a commutative unital ring. $\mathbb{Q}$ and $\mathbb{R}$ are fields.

For a ring R we write as R^{op} the *opposite* ring, with the same addition and reversed multiplication

$$x * y = yx \qquad (x, y \in R). \tag{2.1}$$

It coincides with R if and only if R is commutative.

Exercise and complements. (a) In a ring R the additive multiples kx (for $x \in R$ and $k \in \mathbb{Z}$) have the following properties with respect to the ring multiplication (written here as $x.y$, to better distinguish it from multiples)

$$(kx).y = k(x.y) = x.(ky) \qquad (x, y \in R,\ k \in \mathbb{Z}), \tag{2.2}$$

$$(hx).(ky) = (hk)(x.y) \qquad (x, y \in R,\ k \in \mathbb{Z}). \tag{2.3}$$

(b) Prove that a boolean algebra X is equivalent to an idempotent commutative unital ring, where 'idempotent' means that $xx = x$, for all elements.
Hints. Define the symmetric difference, extending Exercise 1.3.0(e).
Note. The interest of this exercise lies in showing the coincidence of two structures which look far from each other. The proof is not difficult but rather long. The reader might be satisfied with writing the essential points, then looking the rest in the solution.

*(c) A *skew-field* is only assumed to satisfy the axioms (A.1–7) and (A.9): the product need not be commutative. See 2.7.2(c).

2.1.2 Homomorphisms and subrings

A homomorphism $f \colon R \to S$ of rings is a mapping that preserves the operations:

$$f(x + y) = f(x) + f(y), \qquad f(xy) = f(x).f(y) \qquad (x, y \in R). \tag{2.4}$$

As we have already seen in 1.1.5, these mappings are closed under composition. Their partial composition law is associative (when legitimate), and any identity homomorphism $\mathrm{id}\, R$ acts as a unit for every legitimate composition. A homomorphism of rings gives a homomorphism of the underlying abelian groups, and what we have seen in Section 1.3 will be applied, in an enriched form.

An *isomorphism* $f: R \to S$ of rings is a homomorphism which has an inverse f^{-1}; this happens if and only if f is a bijective mapping. In this case the rings R and S are said to be *isomorphic*, and we write $R \cong S$. The mapping $\rho: R \to R^{\mathrm{op}}$ which is the identity of the underlying sets is an *anti-isomorphism*, as it reverses the multiplication.

A *subring* of a ring R is a subset $R' \subset R$ such that:

(i) R' is a subgroup of the additive group R,

(ii) R' is a subsemigroup of the multiplicative semigroup R.

Then R' is a ring, under the restricted operations, and the inclusion $R' \to R$ is a homomorphism. There is always the null subring $\{0\}$ and the total subring R.

A homomorphism $f: R \to S$ of rings determines a subring of its domain, called the *kernel*, and a subring of its codomain, called the *image*

$$\begin{aligned} \operatorname{Ker} f &= \{x \in R \mid f(x) = 0\} \subset R, \\ \operatorname{Im} f &= \{y \in S \mid \text{there is } x \in R \text{ such that } f(x) = y\} \subset S. \end{aligned} \tag{2.5}$$

We already know that f is injective if and only if $\operatorname{Ker} f = \{0\}$, and surjective if and only if $\operatorname{Im} f = S$. ($\operatorname{Ker} f$ has a stronger property than being a subring, investigated in the next subsection.)

A *homomorphism* $f: R \to S$ *of unital rings*, or *unital homomorphism*, also has to preserve the unit. Its image is a unital subring of S.

A *homomorphism* $f: K \to K'$ *of fields* is a homomorphism of unital rings, between fields; it is always injective, as we have seen in 1.1.6(b).

Exercise and complements. (a) For a ring R, determine all the ring homomorphisms $f: \mathbb{Z} \to R$.

(b) If R is unital, determine all the unital homomorphisms $f: \mathbb{Z} \to R$.

(c) In a ring R, the *centre*

$$\mathrm{Cnt}(R) = \{x \in R \mid \text{for all } y \in R,\ xy = yx\}, \tag{2.6}$$

is a (commutative) subring. If R is unital, 1 belongs to the centre.

2.1.3 Ideals of a ring

A subset $I \subset R$ is said to be a *left ideal* of R if it is an additive subgroup and

$$RI \subset I \qquad (\text{i.e. } xh \in I, \text{ for all } x \in R \text{ and } h \in I), \tag{2.7}$$

which implies that I is a subring of R. The least left ideal is $\{0\}$, and the greatest is R; a *proper* left ideal is assumed to be different from R.

Symmetrically, a subset $I \subset R$ is said to be a *right ideal* of R if it is an additive subgroup and

$$IR \subset I \qquad \text{(i.e. } hx \in I\text{, for all } x \in R \text{ and } h \in I\text{).} \tag{2.8}$$

The canonical anti-isomorphism $R \to R^{\text{op}}$ turns left ideals into right ones, and conversely.

A *bilateral ideal* I, or *two-sided ideal*, is both a left and a right ideal; we will use the notation $I \triangleleft R$ to denote this fact, which plays a role similar to the invariance of subgroups: *the kernel* $\operatorname{Ker} f$ *of a homomorphism* $f\colon R \to S$ *is obviously a bilateral ideal of* R.

In a commutative ring, these three kinds of ideal coincide, and we only speak of 'ideals'. In a unital ring any left (or right) ideal containing an invertible element is total, so that a proper left (or right) ideal cannot be a unital subring.

In a commutative unital ring R, the set

$$(a) = Ra = \{xa \mid x \in R\}, \tag{2.9}$$

is the ideal generated by the element a, i.e. the least ideal containing this element. An ideal generated by a single element is called a *principal ideal.*

More generally we write as

$$(a_1, a_2, ..., a_n) = \{x_1a_1 + x_2a_2 + ... + x_na_n \mid x_i \in R\}, \tag{2.10}$$

the ideal generated by the elements $a_1, a_2, ..., a_n$ of R; such an ideal is said to be *finitely generated.*

Exercise and comments. (a) A non-zero commutative unital ring R is a field if and only if every ideal (or principal ideal, equivalently) is either (0) or R.

(b) The injectivity of a homomorphism $f\colon K \to K'$ of fields can now be derived from the fact that $\operatorname{Ker} f$ is an ideal of K, and a proper one as $f(1) = 1$.

2.1.4 Cartesian products

For a family of rings $(R_i)_{i \in I}$ indexed by a set I, the cartesian product $R = \prod_{i \in I} R_i$ is a ring under the *componentwise addition* and *componentwise multiplication*

$$(x_i) + (y_i) = (x_i + y_i), \quad (x_i)(y_i) = (x_iy_i) \qquad (x_i, y_i \in R_i), \tag{2.11}$$

where it is understood that each sum $x_i + y_i$ and each product x_iy_i is computed in R_i.

The cartesian product comes with a family of *cartesian projections*

$$p_i\colon R \to R_i, \qquad p_i((x_i)_{i \in I}) = x_i, \tag{2.12}$$

which are homomorphisms.

The *universal property of the product* (of rings) works as usual, and determines its solution up to canonical isomorphism. Namely, the cartesian product R can be viewed as a ring provided with a family of homomorphisms $p_i \colon R \to R_i$ (for $i \in I$), such that:

(i) for every similar pair $(S, (f_i \colon S \to R_i)_{i\in I})$ formed of a ring S and a family of homomorphisms $f_i \colon S \to R_i$, there exists precisely one homomorphism $f \colon S \to R$ such that

$$\begin{array}{ccc} S & \xrightarrow{f} & R \\ & \searrow_{f_i} & \downarrow p_i \\ & & R_i \end{array} \qquad p_i f = f_i \ \text{ (for } i \in I). \tag{2.13}$$

The cartesian product of a family of unital rings is unital, with a unit $(1_i)_{i\in I}$ whose components are the units of the factors R_i. Then the canonical projections are unital.

On the other hand, if K and K' are fields, the ring $K \times K'$ is not a field, as we have already remarked in 1.2.8. Fields are not equational algebras, and behave in a way strongly different from semigroups, monoids, groups, abelian groups, vector spaces, and so on.

2.1.5 Quotients of rings

Let R be a ring. A *congruence* (of rings) in R is an equivalence relation E consistent with both operations. Precisely, this means that:

$$\text{if } x \, E \, x' \text{ and } y \, E \, y', \text{ then } (x+y)E(x'+y') \text{ and } xy \, E \, x'y'. \tag{2.14}$$

The *quotient ring* R/E is the quotient set, with the induced operations

$$[x] + [y] = [x+y], \qquad [x].[y] = [xy], \tag{2.15}$$

and the canonical projection

$$p \colon R \to R/E, \qquad p(x) = [x], \tag{2.16}$$

is a homomorphism. The congruence E determines a bilateral ideal $\mathrm{Ker}\, p$ of R. It is also determined by the latter, as we already know (from Section 1.3) that

$$x \, E \, y \;\Leftrightarrow\; p(x) = p(y) \;\Leftrightarrow\; x - y \in \mathrm{Ker}\, p. \tag{2.17}$$

Conversely, if $I \triangleleft R$ (i.e. I is a bilateral ideal of R) we define the *associated congruence* as:

$$x \equiv_I y \;\Leftrightarrow\; x - y \in I. \tag{2.18}$$

We already know that this is a congruence of abelian groups. As to the multiplicative part, if $x - y \in I$ and $z - t \in I$, we deduce that

$$xz - yt = (xz - yz) + (yz - yt) = (x - y)z + y(z - t) \in I. \qquad (2.19)$$

For a fixed ring R, there is thus a natural bijection between congruences in R and *bilateral ideals* of R

$$\begin{aligned} &E \mapsto \mathrm{Ker}\,(R \to R/E) = \{x \in R \mid x\,E\,0\}, \\ &I \mapsto \equiv_I, \qquad x \equiv_I y \;\Leftrightarrow\; x - y \in I. \end{aligned} \qquad (2.20)$$

The quotient of R modulo the congruence $\equiv_I$ is denoted as R/I. In this quotient, the equivalence class $[x] = x + I$ of any element x is still called a *coset* of x (with respect to the ideal I).

Again, the trivial ideal (0) determines the equality congruence $x = y$, and $R/(0)$ is identified with R. The total ideal R determines the coarsest congruence $x, y \in R$, and R/R is a trivial ring.

The canonical factorisation of a homomorphism $f\colon R \to S$ of rings can be written as for abelian groups

$$\begin{array}{ccc} R & \xrightarrow{f} & S \\ {\scriptstyle p}\downarrow & & \uparrow{\scriptstyle m} \\ R/\mathrm{Ker}\, f & \xrightarrow[g]{} & \mathrm{Im}\, f \end{array} \qquad f = mgp, \qquad (2.21)$$

where p is a surjective homomorphism, m is an injective homomorphism and g is an isomorphism of rings.

If R is a unital ring, so is any quotient R/I, with identity $[1_R]$. (But note that the ideal I is not a unital subring, unless $I = R$.)

If R is commutative, so is R/I.

In a field K the only ideals are (0) and K, and the only quotient field is $K/(0) = K$.

Exercises and complements. (a) Let R be a commutative ring. We write as $\mathrm{Idl}(R)$ the ordered set of its ideals. Prove that it is a complete lattice.

(b) Prove that it is a sublattice of the lattice $\mathrm{Sub}(R,+)$ of the lattice of subgroups of the underlying abelian group $(R,+)$, with meets and joins computed as in the latter

$$I \wedge J = I \cap J, \qquad I \vee J = I + J = \{x + y \mid x \in I, y \in J\}, \qquad (2.22)$$

and therefore a modular lattice (by 1.4.7(d)).

(c) We have already seen that fields do not form a variety of algebras.

It is interesting to note that, in the variety of commutative unital rings, the fields are structurally distinguished as the non-trivial objects R which have no quotients other than the necessary ones: R and the trivial ring; see Exercise 2.1.3(a). (A characterisation corresponding to simple groups in the variety of all groups, see 1.6.6.)

2.1.6 Exercises and complements (Modular arithmetic, II)

We now show that the quotient $\mathbb{Z}/n$ is actually a quotient of rings, and modular arithmetic also works for multiplication – as we all know in the particular case of 'casting out nines'.

(a) In the ring $\mathbb{Z}$ of integers, each subgroup is of the form $n\mathbb{Z}$, with $n \geqslant 0$ (by 1.3.6(b)), and each of them is an ideal. In other words, subgroups, subrings and ideals of $\mathbb{Z}$ coincide. Every ideal is principal, and the ideal $n\mathbb{Z}$ is often written as (n), as anticipated in Section 1.3.

(b) As a consequence, each quotient ring of $\mathbb{Z}$ is of the form

$$\mathbb{Z}/n = \mathbb{Z}/(n\mathbb{Z}) \qquad (n \geqslant 0), \tag{2.23}$$

with multiplication $[h].[k] = [hk]$.

(c) The ring $\mathbb{Z}/n$ is a field if and only if n is a prime number. In particular, $\mathbb{Z}/2$ is isomorphic to the field $\mathbb{F}_2$ of 1.1.3(j).

(d) Every non-empty set can be given a structure of commutative unital ring. (This is not true of fields: see 2.4.7(e).)

Solutions. (a), (b) Any multiple of some $kn \in n\mathbb{Z}$ is a multiple of n.

(c) If p is a prime number and $1 \leqslant a < p$, we have $(a) \vee (p) = (a \wedge_D p) = (1) = \mathbb{Z}$, and we can write $1 = ha + kp$ (for some $h, k \in \mathbb{Z}$). Then $[a]$ is invertible in $\mathbb{Z}/p$, with inverse $[h]$. Conversely, if $\mathbb{Z}/n$ is a field and $n = ab$ with $1 \leqslant a, b \leqslant n$, we have $[a].[b] = 0$ in $\mathbb{Z}/n$, whence $[a]$ or $[b]$ is 0, which means that $a = n$ (and $b = 1$) or $b = n$ (and $a = 1$).

(d) For a set X with a finite number $n > 0$ of elements, it is sufficient to consider the ring $\mathbb{Z}/n$, and transport its structure to X (see 1.2.7).

If X is infinite, we can take the product ring $\prod_{x \in X} \mathbb{Z}$. *More interestingly, we can use the ring of polynomials with integral coefficients and indeterminates belonging to the set X (see Exercise 5.4.2(d)).*

2.1.7 Integral domains and pids

We are now interested in the cancellative property of multiplication, in a ring R. Of course we must let out the element 0, that annihilates every element, and consider which non-zero elements are cancellable; in a field, all of them are.

More precisely, we say that an element $a \in R$ is *left-cancellable* in R if $ax = ay$ implies $x = y$ (for all $x, y \in R$), or equivalently if $ax = 0$ implies $x = 0$ (for all $x \in R$). Symmetrically, a is *right-cancellable* in R if $xa = ya$ implies $x = y$, or equivalently if $xa = 0$ implies $x = 0$ (for all $x, y \in R$).

An *integral domain* R is defined as a non-zero unital commutative ring in which, equivalently:

(i) every non-zero element is (left- and right-) cancellable,

(ii) the product of two non-zero elements is always non-zero,

(iii) if $xy = 0$ then $x = 0$ or $y = 0$.

This property is also expressed saying that the only *zero-divisor* in R is 0 itself, or that R has *no proper zero-divisors*.

A *pid*, or *principal ideal domain*, is an integral domain where every ideal is principal, i.e. the ideal (x) generated by some element (see 2.1.1).

We know that the ring $\mathbb{Z}$ is a pid, with ideals $(n) = n\mathbb{Z}$ (Exercise 2.1.6(a)). Every field is – trivially – a principal ideal domain. We have already seen that the quotient $\mathbb{Z}/n$ is a field when n is prime; otherwise, it is not even an integral domain (for $n > 0$): any decomposition $n = hk$ with $1 < h, k < n$ gives $[h].[k] = 0$.

Every subring of a field (or an integral domain) is an integral domain. Conversely, we will see in 2.2.4 that any integral domain R can be embedded as a subring in a field, by constructing the 'field of fractions' of R. An integral domain need not be a pid: we will see an important example in Exercise 2.5.3(i), by a ring of polynomials.

2.1.8 Exercises and complements

(a) Let $p\colon R \to R/H$ be the projection of the commutative unital ring R on its quotient modulo the ideal H. Images and preimages along p can be restricted to order-preserving mappings

$$\begin{aligned} &p_*\colon \mathrm{Idl}(R) \to \mathrm{Idl}(R/H), && p^*\colon \mathrm{Idl}(R/H) \to \mathrm{Idl}(R), \\ &p^*p_*(I) = I + H, && p_*p^*(J) = J, \end{aligned} \tag{2.24}$$

where $I \lhd R$ and $J \lhd R/H$. A further restriction gives an isomorphism of lattices

$$p^*\colon \mathrm{Idl}(R/H) \to \{I \in \mathrm{Idl}(R) \mid I \supset H\}. \tag{2.25}$$

(b) A ring which is a cartesian product of two integral domains, or pids, is not an integral domain: integral domains and pids are not equational algebras.

(c) (*Divisibility in commutative rings*) Let R be a commutative unital ring. Extending Exercise 1.4.2(d), we write as R_D the underlying set $|R|$, equipped with the divisibility relation $a|b$, namely there exists some $\lambda \in R$ such that $b = \lambda a$. It is a preorder, characterised as:

$$a \mid b \;\Leftrightarrow\; b \in (a) \;\Leftrightarrow\; (b) \subset (a), \tag{2.26}$$

with minimum 1 (or any invertible element) and maximum 0.

If $a \mid b$ and $b \mid a$, or equivalently $(a) = (b)$, the elements a, b are said to be *associated* in R, written as $a \sim_D b$.

The mapping

$$\varphi\colon R_D \to \mathrm{Idl}(R), \qquad \varphi(a) = (a), \tag{2.27}$$

induces an anti-isomorphism from the ordered set $R_D/{\sim_D}$ to the set of principal ideals of R, ordered by inclusion.

(d) (*Divisibility in integral domains*) Divisibility is preferably studied in an integral domain R. Then $(a) = (b)$ if and only if $a = \lambda b$ for some $\lambda \in \mathrm{Inv}(R)$.

Studying divisibility, the invertible elements (associated to 1) are often called 'units', but here this term is reserved to the unit 1_R of the ring. Prime elements will be examined in 2.2.2.

2.2 Rings and fields, continued

We take on the study of rings and fields.

2.2.1 *Characteristic of rings and fields*

Let R be a unital ring. The unique unital homomorphism $f\colon \mathbb{Z} \to R$ is defined by $f(k) = k1_R$ (see Exercise 2.1.2(b)). Its image

$$R_0 = \{k1_R \mid k \in \mathbb{Z}\}, \tag{2.28}$$

is a unital subring of R, contained in any other; it is called the *minimal unital subring* of R.

By the canonical factorisation of f, in 2.1.5, R_0 is isomorphic to a quotient of $\mathbb{Z}$, which we already know to be of the form $\mathbb{Z}/n$, for a unique $n \in \mathbb{N}$. All natural numbers can occur in this way, because the smallest unital subring of $\mathbb{Z}/n$ is the total one.

This number $n \geqslant 0$ is called the *characteristic of the ring* R. Note that in characteristic 0 (and only there), the ring $R_0 \cong \mathbb{Z}$ is infinite. (A ring of characteristic zero is also called a 'ring of infinite characteristic', with reference to the cardinal of R_0.)

Exercises and complements. (a) (*Characteristic of fields*) If K is a field, its characteristic is either 0 or a prime number p.

K has a smallest subfield K_0 (called the *minimal subfield*) which is either isomorphic to the rational field $\mathbb{Q}$ (in characteristic 0) or coincides with the minimal subring $R_0 \cong \mathbb{Z}/p$ (in characteristic p). Speaking of a *field of characteristic* p we always refer to the second case.

An (injective) homomorphism of fields $f\colon K \to K'$ restricts to an isomorphism between the minimal fields of K and K', which have thus the same characteristic.

The theory of vector spaces will show that a finite field, necessarily of characteristic p, has cardinal p^n, for a (plainly unique) positive integer n (see Exercise 2.4.7(e)). There is thus no field of 6 elements. Moreover, all cardinals p^n can be realised (see Exercise 2.4.7(f)).

(b) Find a proper subfield of $\mathbb{R}$ that properly contains $\mathbb{Q}$. *Hints:* use $\sqrt{2}$.

2.2.2 Prime and maximal ideals

Let R be a commutative unital ring and I an ideal of R. We say that I is a *prime ideal* (resp. a *maximal ideal*) if it is proper and the quotient ring R/I is an integral domain (resp. a field).

Every maximal ideal is prime. The ideal (0) is prime in R if and only if R is an integral domain, and is maximal if and only if R is a field.

Let R be an integral domain. A *prime* element is any non-zero, non-invertible element $a \in R$ such that, if $a \mid xy$, then $a \mid x$ or $a \mid y$ (for all $x, y \in R$). An *irreducible* element is any non-zero, non-invertible element a which has no proper divisors: if $a = bc$ in R, one of b and c is invertible; or, equivalently, one of them is associated to a.

A prime element a is always irreducible: if $a = bc$, then $a \mid b$ or $a \mid c$, which means that a is associated to b or c.

The converse need not be true. But prime and irreducible elements coincide in $\mathbb{Z}$ and any pid (see Exercise 2.2.3(e)), which 'explains' why prime integers are usually defined by the property of being irreducible (as in 1.2.6(c)).

2.2.3 Exercises and complements

Let R be a commutative unital ring, let I be an ideal of R and $a \in R$.

(a) I is a prime ideal if and only if it is proper and:

- for all $x, y \in R$, the condition $xy \in I$ implies: $x \in I$ or $y \in I$.

(b) I is a maximal ideal if and only if it is a maximal element of the set of proper ideals of R, ordered by inclusion.

(c) If R is an integral domain, the following properties are equivalent:

- the element a is prime,
- $a \neq 0$ and the ideal (a) is prime,
- $a \neq 0$ and $R/(a)$ is an integral domain.

(We recall that a prime element is always irreducible.)

(d) If R is an integral domain, the following properties are equivalent:

- the element $a \in R$ is irreducible,
- $a \neq 0$ and (a) is maximal in the ordered set of proper principal ideals of R,
- $a \neq 0$ is an atom of the preordered set R_D, defined in 2.1.8(c).

(Note that, if R is a field, 0 is an atom of R_D.)

(e) If R is a pid, a non-zero element a is prime if and only if it is irrreducible, if and only if the ideal (a) is prime, if and only if (a) is maximal. In particular this holds in the ring $\mathbb{Z}$ of integers.

(f) Any proper ideal I of R is contained in a maximal ideal. *Hints*: use Zorn's Lemma.

2.2.4 Theorem and Definition (The field of fractions)

Let R be an integral domain. There is a field $Q(R)$ and an injective homomorphism of unital rings

$$\eta\colon R \to Q(R), \tag{2.29}$$

which is universal: *for every injective unital homomorphism $f\colon R \to K$ with values in a field, there is a unique homomorphism $g\colon Q(R) \to K$ of fields such that $g\eta = f$.*

$Q(R)$ is called the field of fractions *of R.*

Proof This argument is a natural extension of the construction of the rational numbers as fractions h/k of integers, with $k \neq 0$. The proof is an interesting exercise: the reader is invited to write it down, either autonomously or after reading the first lines.

(a) We start from the set $A = R \times R^*$ of all pairs (a, b) of elements of the ring, with $b \neq 0$. It is equipped with a relation

$$(a, b) \sim (c, d) \;\Leftrightarrow\; ad = cb, \tag{2.30}$$

which is easily seen to be an equivalence relation. An equivalence class is written as $a/b = [(a, b)]$, and called a *fraction* of R; their set is written as $Q(R) = (R \times R^*)/\!\sim$.

The set A is equipped with two operations (inspired again by the construction of $\mathbb{Q}$):

$$\begin{gathered}(a, b) + (a', b') = (ab' + a'b, bb'),\\ (a, b).(a', b') = (aa', bb').\end{gathered} \tag{2.31}$$

At this stage, the algebraic structure we obtain is not really good, as it will appear below. But the structure becomes good in the quotient $Q(R)$.

(b) The operations of A are consistent with the equivalence relation: if $(a,b) \sim (c,d)$ and $(a',b') \sim (c',d')$, straightforward computations show that:

$$(ab' + a'b, bb') \sim (cd' + c'd, dd'), \qquad (aa', bb') \sim (cc', dd').$$

The induced operations on $Q(R)$ are thus:

$$a/b + a'/b' = (ab' + a'b)/(bb'), \qquad (a/b).(a'/b') = (aa')/(bb'). \qquad (2.32)$$

(c) The set $Q(R)$ is an abelian group with respect to the sum. In fact $(A,+)$ is already an abelian monoid, with identity $(0,1)$: associativity is easily verified and the rest is obvious. But $(A,+)$ lacks opposites (generally), which exist in the quotient $Q(R)$

$$a/b + (-a)/b = (ab - ba)/b^2 = 0/1 = \underline{0}.$$

(d) It is even more easy to see that $(A,.)$ is a commutative monoid, with identity $(1,1)$. This also holds in the quotient $Q(R)$, with unit $1/1$. Distributivity holds in the quotient

$$\begin{aligned}(a/b + a'/b').(c/d) &= ((ab' + a'b)/(bb')).(c/d) \\ &= (ab'c + a'bc)/(bb'd), \\ (a/b).(c/d) + (a'/b').(c/d) &= ac/bd + a'c/b'd \\ &= (acb'd + bda'c)/(bdb'd) = (acb' + ba'c)/(bb'd).\end{aligned}$$

(e) $Q(R)$ is thus a commutative unital ring; it is actually a field, because $a/b \neq \underline{0} = 0/1$ means that $a \neq 0$, and then $(a/b)^{-1} = b/a$. The mapping

$$\eta\colon R \to Q(R), \qquad \eta(a) = a/1, \qquad (2.33)$$

is (plainly) an injective homomorphism of unital rings. We will identify every $a \in R$ with $a/1$.

Let us note that, if $b \neq 0$ in R, then $1/b = (b/1)^{-1} = b^{-1}$ in $Q(R)$, so that $a/b = ab^{-1}$.

(f) Finally, let $f\colon R \to K$ be an injective homomorphism of unital rings with values in a field. There is at most one extension g which is a field homomorphism

$$g\colon Q(R) \to K, \qquad g(a/b) = f(a).(f(b))^{-1}. \qquad (2.34)$$

On the other hand, if we define g in this way (legitimate, because f is injective), we do get a field homomorphism. In fact, g preserves the sum

$$g(a/b + a'/b') = g((ab' + a'b)/(bb')) = (f(ab') + f(a'b)).f(bb')^{-1}$$
$$= f(a).f(b)^{-1} + f(a').f(b')^{-1} = g(a/b) + g(a'/b'),$$

and the rest is obvious.

If R is already a field, $Q(R)$ is canonically isomorphic to R.

□

2.2.5 Comments and complements

(a) In the variety of commutative unital rings, the integral domains are characterised as the non-trivial subrings of fields; the latter are characterised as in 2.1.5(c).

(b) On the ring of integers we get $Q(\mathbb{Z}) = \mathbb{Q}$, the rational field.

(c) In a different perspective, one *could* develop the theory of algebraic structures up to the present point, ignoring $\mathbb{Q}$ (as well as $\mathbb{R}$) and only build now, the rational field, as the field of fractions of $\mathbb{Z}$.

Looking further back, one *could* even ignore the ring $\mathbb{Z}$ up to the present point, and build it now (before $\mathbb{Q}$) from the 'semiring' $\mathbb{N}$ (which simply comes out of the theory of finite sets), following the outline sketched below, in 2.2.7. Didactically, such an organisation is ineffective, as we would have to study a long, abstract theory without the support of the main examples which are at its basis. Yet, from a foundational point of view, all this makes sense, as we will reconsider in 4.6.7.

2.2.6 Ordered algebraic structures

From the axioms (A.10–15) of Section 1.1, one can easily guess the corrected notion of the main structures combining operations and ordering, like: ordered abelian group, ordered ring, ordered field, ordered vector space on the real field.

In each of these cases, the ordering is determined by the 'positive cone' of the elements $\geqslant 0$. The adequate transformations are the monotone homomorphisms.

A reader can like to develop the basic part of this discipline, either autonomously or beginning from the following layout.

Exercises and complements. (a) An *ordered abelian group* is an abelian group A

equipped with an order relation $x \leqslant y$ which is consistent with the addition, as in (A.14):

(i) for every $x, y, z \in A$, if $x \leqslant y$ then $x + z \leqslant y + z$.

Equivalently, one can equip the abelian group A with a *positive cone*, namely a submonoid A^+ (stable in A under addition and containing 0) such that

$$A^+ \cap (-A^+) = \{0\} \qquad \textit{(anti-symmetry)}, \qquad (2.35)$$

where $-A^+ = \{-x \mid x \in A^+\}$. Given the ordering, one lets $A^+ = \{x \in A \mid x \geqslant 0\}$; given A^+, one defines $x \leqslant y$ by the relation $y - x \in A^+$.

The order is total if and only if $A = A^+ \cup (-A^+)$.

A homomorphism $f \colon A \to B$ between ordered abelian groups is monotone if and only if it takes the positive cone of A to the positive cone of B.

(b) An *ordered ring* is a ring R equipped with an order relation consistent with addition and multiplication: after (i), it also has to satisfy:

(ii) for every $x, y, z \in R$, if $x \leqslant y$ and $z \geqslant 0$, then $xz \leqslant yz$ and $zx \leqslant zy$.

The positive cone is now required to be also stable under multiplication. One proves as in 1.1.9(c) that $x^2 \geqslant 0$ (for every $x \in R$). Thus, in an ordered unital ring, $0 \leqslant 1$, and $0 < 1$ in a non-trivial one.

(c) An *ordered vector space* on the real field is a vector space X equipped with an order relation consistent with addition and scalar multiplication: after (i), it has also to satisfy:

(iii) for every $x, y \in X$ and $\lambda \geqslant 0$ in $\mathbb{R}$, if $x \leqslant y$ then $\lambda x \leqslant \lambda y$.

The positive cone is now required to be stable under addition and scalar multiplication by scalars $\lambda \geqslant 0$.

One can similarly define an ordered vector space on any ordered field.

(d) Every abelian group has a trivial order, the discrete one. The natural order makes $\mathbb{Z}$ into a totally ordered abelian group. An ordered abelian group A whose order is not discrete is necessarily infinite.

Hints. There is some element $a > 0$ in A, and we can construct an injective monotone homomorphism $\mathbb{Z} \to A$.

(e) After examining cartesian products of the present ordered structures, we see that the abelian group $\mathbb{Z}^n$ has a natural (consistent) order, non-total for $n \geqslant 2$. Similarly, the real vector space $\mathbb{R}^n$ has a natural order.

(f) Direct sums can also be easily described, for ordered abelian groups and ordered vector spaces.

(g) Ordered abelian groups can be easily extended to *preordered abelian groups*, that in many respects give a better framework: we are dropping the anti-symmetry condition (2.35).

If A and B are preordered abelian groups, the abelian group $H = \mathrm{Hom}(A, B)$ of all homomorphisms $A \to B$ (in Exercise 1.3.2(e)) is preordered, with positive cone H^+ formed of the monotone homomorphisms. Note that, if A has the discrete order, $\mathrm{Hom}(A, B)$ has the *indiscrete preorder* (even if B is ordered). We also note that $f \prec g$ in H means that $g - f \in H^+$, that is $f(x) \prec g(x)$ for all $x \in A^+$: *this does not agree with the pointwise preorder of mappings* $A \to B$.

*Category theory can solve this disagreement. The important *forgetful functor* $U \colon \mathsf{pAb} \to \mathsf{Set}$ of the category pAb of preordered abelian groups and monotone homomorphisms (see 5.1.3) sends the object A to the set A^+, *not* to the whole underlying set.

For a reader interested in categories, let us add that the importance of U comes out of a symmetric monoidal closed structure of pAb, with the internal hom described above. The identity of this structure is the totally ordered group $\mathbb{Z}$; the associated forgetful functor sends A to the set $\mathsf{pAb}(\mathbb{Z}, A)$, which can be identified with A^+. (See [G1], Section 2.1.1)*

2.2.7 *Semirings

Loosely speaking, a *semiring* is a 'ring without opposites (possibly)'. More precisely, it is an additive abelian monoid A with an associative multiplication that distributes over the sum and satisfies the relation $0x = 0 = x0$, for all $x \in A$. (The latter is not a consequence, generally; it is when A satisfies the additive cancellation law.)

Let A be a semiring that satisfies the additive cancellation law. Then there is a universal embedding $\eta\colon A \to D(A)$ into a *ring of formal differences.*

The construction is similar to the field of fractions: here we want to create additive inverses (for all elements) instead of the multiplicative ones (for non-zero elements). We start from the set $B = A \times A$, equipped with the equivalence relation

$$(a, b) \sim (c, d) \;\Leftrightarrow\; a + d = c + b, \tag{2.36}$$

and let $D(A) = (A \times A)/{\sim}$. An equivalence class is written as $a - b = [(a, b)]$. (Note that, if A is already a ring, the relation (2.36) amounts to $a - b = c - d$, and $D(A)$ can be identified with A.)

The set $A \times A$ is equipped with two operations:

$$\begin{aligned} (a, b) + (a', b') &= (a + a', b + b'), \\ (a, b).(a', b') &= (aa' + bb', ab' + ba'). \end{aligned} \tag{2.37}$$

Straightforward verifications, as in 2.2.4, prove that we have two induced operations on the quotient, which make $D(A)$ into a ring: the opposite of $a - b$ is $b - a$.

The embedding $\eta\colon A \to D(A)$ is defined by $\eta(a) = a - 0$. The universal property says that every homomorphism $f\colon A \to R$ with values in a ring has a unique extension $g\colon D(A) \to R$, where $g(a - b) = f(a) - f(b)$.

Applying this procedure to the commutative, unital semiring $\mathbb{N}$, we get the ring of integers. (Using some further properties of finite sets, including the complement of a subset, we readily see that a difference $a - b$ either belongs to $\mathbb{N}$ or is the opposite of an element of $\mathbb{N}^*$.)

2.2.8 *The uniqueness of the real field

The axioms (A.1–16) of Section 1.1 determine the real field up to isomorphism. More precisely, if K and K' are two complete totally ordered fields, there exists a unique isomorphism $f\colon K \to K'$ of totally ordered fields (that preserves addition, multiplication and ordering).

The interested reader can write the proof, or a sketch of it, solving the following list of exercises. Most of the missing points can be found in the solutions of Chapter 7.

Exercises and complements. (a) A non-trivial totally ordered ring R has characteristic 0.

(b) For a totally ordered field K, there is a unique embedding of the rational field $\mathbb{Q}$ (preserving addition, multiplication and ordering)

$$f\colon \mathbb{Q} \to K, \qquad f(h/k) = (h1_K).(k1_K)^{-1}, \tag{2.38}$$

where $h, k \in \mathbb{Z}$ and $k \neq 0$.

(c) If K and K' are two complete totally ordered fields, with minimal subfields Q and Q', we have a unique isomorphism $f\colon Q \to Q'$ of ordered fields. This can be extended to a unique isomorphism $K \to K'$ of ordered fields.

2.3 Modules and vector spaces

Vector spaces on a field K have been briefly introduced in Section 1.1. When the scalars belong to a ring R we have a more general structure, called an R-module. For the ring $\mathbb{Z}$ of integers, a $\mathbb{Z}$-module is 'the same' as an abelian group.

The theory of vector spaces, formally introduced by G. Peano in 1888 and closely related to Mathematical Physics, predates that of modules and has peculiar properties, which still justify the use of a different name.

For the sake of simplicity, we assume that R is a commutative unital ring; its elements are called scalars. In the non-commutative case one has to distinguish left modules from right modules; this case is only marginally mentioned.

This section is a part of Linear Algebra [Bou1, La1, La2]. However, as we are mostly referring to commutative rings, it is also a part of Commutative Algebra [ZS].

2.3.1 Main definitions

A *module* M on the commutative unital ring R, or *R-module*, is an abelian group equipped with a *scalar multiplication*

$$R{\times}M \to M, \qquad (\lambda, x) \mapsto \lambda x, \tag{2.39}$$

satisfying the axioms (VS.5–8) of 1.1.7. This means that, for every $x, y \in M$ and $\lambda, \mu \in R$:

(i) (*Distributivity, I*) $\lambda(x+y) = \lambda x + \lambda y,$

(ii) (*Distributivity, II*) $(\lambda + \mu).x = \lambda x + \mu x,$

(iii) (*Compatibility*) $(\lambda\mu).x = \lambda.(\mu x),$

(iv) (*Unitarity*) $1_R.x = x.$

(The last two axioms say that the multiplicative monoid $(R, .)$ acts on the set underlying M: see 1.6.5.)

The ring R is a module on itself, with the 'internal' multiplication of the ring. More generally, for every $n \in \mathbb{N}$, the abelian group R^n has a canonical structure of R-module, with pointwise scalar multiplication

$$\lambda(x_1, ..., x_n) = (\lambda x_1, ..., \lambda x_n) \qquad (\lambda, x_1, ..., x_n \in R). \tag{2.40}$$

(Verifying the axioms is straightforward.) In particular, we have the null module, or trivial module, $R^0 = \{0\}$.

*If the unital ring R is not assumed to be commutative, a *left R-module* M is defined as above. For a *right R-module*, the scalar multiplication is written as $x\lambda$; the axioms are rewritten accordingly, and the compatibility axiom:

(iii′) $x.(\lambda\mu) = (x\lambda).\mu,$

says a different thing: the scalar $\lambda\mu$ acts multiplying by λ, and *then* by μ. A right R-module is thus the same as a left module on the opposite ring R^{op}.*

A module on a field is generally called a *vector space*. We have already seen, in 1.2.9(a), how the structure of vector space on the real field can be presented as an equational algebra. One proceeds in a similar way for modules on a given ring.

Exercises and complements. Let R be a commutative unital ring.

(a) In an R-module M, we have:

$$0_R.x = 0, \quad \lambda(-x) = (-\lambda)x = -(\lambda x) \qquad (\lambda \in R,\ x \in M). \tag{2.41}$$

On the trivial ring, all modules are null.

(b) In an R-module M, the multiples kx (for $k \in \mathbb{Z}$) are consistent with the multiples $k\lambda$ in R, in the sense that

$$k(\lambda x) = (k\lambda).x \qquad (kx = (k1_R).x). \tag{2.42}$$

(c) An abelian group M has a unique structure of $\mathbb{Z}$-module: the scalar multiplication is defined by the multiples kx (for $k \in \mathbb{Z}$ and $x \in M$), in 1.3.3.

The algebraic structures of abelian group and $\mathbb{Z}$-module are equivalent.

(d) More generally, for a fixed natural number n, a $\mathbb{Z}/n$-module is 'the same' as an abelian group satisfying the equational axiom $nx = 0$ (for all elements x). These modules are vector spaces when n is a prime number.

(e) If I is an ideal of the ring R, a module on the quotient ring R/I is 'the same' as an R-module M such that $\lambda x = 0$, for all $\lambda \in I$ and $x \in M$. This condition is written as $IM = \{0\}$.

(f) Prove that an abelian group M can be given at most one structure of vector space on the rational field $\mathbb{Q}$, and characterise the abelian groups that admit such a structure.

Hints. As a part of the question, an abelian group M is said to be *divisible* if

- for every $x \in M$ and $n \in \mathbb{N}^*$ there is some $y \in M$ such that $ny = x$.

2.3.2 Homomorphisms and submodules

A *homomorphism* $f\colon M \to N$ *of* R*-modules*, or R*-homomorphism*, or R*-linear mapping*, is a homomorphism of abelian groups that preserves the scalar multiplication

$$f(\lambda x) = \lambda f(x) \qquad (\lambda \in R,\ x \in M). \tag{2.43}$$

Equivalently, we can ask that f preserves all linear combinations

$$f(\lambda x + \mu y) = \lambda f(x) + \mu f(y) \qquad (\lambda, \mu \in R,\ x, y \in M), \tag{2.44}$$

or also all finite linear combinations $\sum_i \lambda_i x_i$, or also all essentially finite linear combinations $\sum_i \lambda_i x_i$ (for quasi null families of scalars).

As for any algebraic structure, consecutive homomorphisms compose, in an associative way, and any identity homomorphism $\mathrm{id}\, M\colon M \to M$ acts as a unit for legitimate compositions. A bijective homomorphism $f\colon M \to N$ has an inverse homomorphism, written as f^{-1}, and is called an *isomorphism* $f\colon M \to N$ of R-modules. Then the modules M and N are said to be *isomorphic*, and we write $M \cong N$.

A *submodule* of a module M is a subset $H \subset M$ that contains 0_M and is stable in M under sum and scalar multiplication

$$x, y \in H \ \Rightarrow\ x + y \in H, \qquad x \in H, \lambda \in R \ \Rightarrow\ \lambda x \in H, \tag{2.45}$$

or, equivalently, a non-empty subset stable under linear combinations. (Note that an opposite $-x$ can be obtained as $(-1_R).x$.)

Restricting to H the operations of sum and scalar multiplication, we get a structure of R-module. It is the only structure on H that makes the inclusion $H \to M$ into a linear mapping.

An R-homomorphism $f\colon M \to N$ determines a submodule of its domain, called the *kernel*, and a submodule of its codomain, called the *image*

$$\begin{gathered} \mathrm{Ker}\, f = \{x \in M \mid f(x) = 0\} \subset M, \\ \mathrm{Im}\, f = \{y \in N \mid f(x) = y \text{ for some } x \in M\} \subset N. \end{gathered} \tag{2.46}$$

When R is a field, a submodule is also called a *vector subspace*, or a *linear subspace.*

2.3.3 Exercises and complements

R is always a commutative unital ring. The following points are important and should be kept in mind; their proof is easy, and left to the reader.

(a) When $R = \mathbb{Z}$, a homomorphism of abelian groups is automatically $\mathbb{Z}$-linear (see 1.3.3(d)), and a subgroup is automatically a submodule.

(b) (*Ideals as submodules*) An ideal of the ring R is the same as a submodule of R, as a module on itself.

(c) For a module M, the submodule $\langle S\rangle$ generated by a subset $S \subset M$ is formed by all essentially finite linear combinations $\sum_i \lambda_i x_i$ of the elements of S. (Note that, if S is empty, we still get 0_M as the empty linear combination.)

(d) Consider the vector space $\mathbb{R}^3$ on the real field $\mathbb{R}$. Each of the following subsets is a linear subspace:

- the trivial subspace $\{0\}$, any line and any plane through the origin, the total space.

It is less obvious that there are no others. The theory of linear dependence in vector spaces, developed below, gives an evident proof: each vector subspace of $\mathbb{R}^3$ has dimension $n \leqslant 3$, and a basis consisting of n linearly independent vectors. The cases $n = 0, 1, 2, 3$ amount to the previous ones.

(e) The ordered set $\mathrm{Sub}(M)$ of submodules of the R-module M is a complete modular lattice, with the same operations described in 1.4.5(c) for the lattice of all subgroups of M

$$\wedge H_i = \bigcap H_i, \qquad \vee H_i = \textstyle\sum H_i, \tag{2.47}$$

where $\sum H_i$ denotes the set of essentially finite sums $\sum_i h_i$, with $h_i \in H_i$ (for every index i). $\mathrm{Sub}(M)$ is thus a complete sublattice of $\mathrm{Sub}_{\mathbb{Z}}(M)$, the lattice of subgroups of M.

This shows again that the ideals of the ring R form a complete modular lattice $\mathrm{Idl}(R)$, as we have already seen in 2.1.5.

(f) (*Change of rings*) Given a homomorphism $\varphi\colon R \to S$ of commutative unital rings, every S-module M can be converted to an R-module $\varphi^*(M)$ having the same underlying abelian group, and R-multiplication

$$\lambda x = \varphi(\lambda).x \qquad (\lambda \in R,\ x \in M). \tag{2.48}$$

We have already used this procedure in 2.3.1(e), for the projection homomorphism $p\colon R \to R/I$. Using it for the inclusion $R \to S$ of a unital subring, one simply restricts the scalars to R.

Coming back to the general procedure, any S-homomorphism $f\colon M \to N$ becomes an R-homomorphism. *(All this forms a covariant functor φ^* from the category of S-modules to that of R-modules.)*

(g) (*The module of homomorphisms*) The abelian group $\mathrm{Hom}(A, B)$ of Exercise 1.3.2(e) can be easily extended to R-modules. For two R-modules M, N, the set $\mathrm{Hom}_R(M, N)$ of all homomorphisms from M to N is an R-module, when equipped with the pointwise operations

$$(\lambda f + \mu g)(x) = \lambda f(x) + \mu g(x) \qquad (x \in M), \tag{2.49}$$

where $f, g \in \mathrm{Hom}(M, N)$ and $\lambda, \mu \in R$.

For any R-module M, the module $\mathrm{Hom}_R(R, M)$ is canonically isomorphic to M.

2.3.4 Cartesian products and direct sums

For a family of R-modules $(M_i)_{i\in I}$ indexed by a set I, the *cartesian product* $M = \prod_{i\in I} M_i$ is an R-module under the obvious, componentwise operations

$$(x_i)_{i\in I} + (y_i)_{i\in I} = (x_i + y_i)_{i\in I}, \qquad \lambda(x_i)_{i\in I} = (\lambda x_i)_{i\in I}. \tag{2.50}$$

The cartesian projections are linear

$$p_i\colon M \to M_i, \qquad p_i((x_i)_{i\in I}) = x_i, \tag{2.51}$$

and satisfy the usual universal property.

In particular we have the *cartesian power* of a module

$$M^I = \textstyle\prod_{i\in I} M, \tag{2.52}$$

which is the same as the module $\mathrm{Map}(I, M)$ of all mappings $I \to M$, with pointwise addition and scalar multiplication (already considered in 1.1.7, in a particular case).

As for abelian groups, in (1.3.8), we are also interested in the *direct sum*

$$N = \textstyle\bigoplus_{i\in I} M_i = \{(x_i) \in \prod_i M_i \mid \mathrm{supp}(x_i) \text{ is finite}\}. \tag{2.53}$$

Again, this submodule of $\prod_i M_i$ comes equipped with a family of *canonical injections*

$$u_i\colon M_i \to N, \tag{2.54}$$

that satisfy the universal property of the sum, in the domain of R-modules and R-homomorphisms (as in (1.87)).

Given a module M and two submodules A, B, there is a canonical homomorphism

$$f\colon A \oplus B \to M, \qquad f(a,b) = a + b, \tag{2.55}$$

determined by the inclusions $A \to M$ and $B \to M$. We say that M is the (internal) *direct sum of the submodules* A, B if this homomorphism is an isomorphism. The reader will easily verify that this is equivalent to the conjunction of two conditions in the lattice $\mathrm{Sub}(M)$

$$A + B = M, \qquad A \cap B = \{0\}, \tag{2.56}$$

and will see that this is stronger than simply requiring that M be isomorphic to $A \oplus B$ (in Exercise (e), below); this is why we are insisting on the 'internal character' of the current direct sum, as concerned with *sub*modules of M.

We also say that M *splits*, or *decomposes, as an internal direct sum* $A \oplus B$. This decomposition is *trivial* if one submodule is trivial (and the other is total).

We say that the module M is *decomposable* if has a non-trivial decomposition; that it is *indecomposable* if this is not the case.

This terminology also applies to abelian groups, as $\mathbb{Z}$-modules.

Exercises and complements. (a) Any direct sum $M = A \oplus B$ of R-modules is the internal direct sum of two submodules A' and B', isomorphic to A and B, respectively.

(b) Any R-module R^n with $n \geqslant 2$ is decomposable.

(c) The abelian group $\mathbb{Z}$ is indecomposable (as a $\mathbb{Z}$-module).

(d) For a prime number p, a cyclic group $\mathbb{Z}/p$ is indecomposable. $\mathbb{Z}/4$ is also, but $\mathbb{Z}/6$ is decomposable.

(e) Note that the subgroup $2\mathbb{Z}$ is isomorphic to $\mathbb{Z}$, and therefore $\mathbb{Z} \cong 2\mathbb{Z} \oplus \{0\}$. But of course $\mathbb{Z}$ is not the *internal* direct sum of the subgroups $2\mathbb{Z}$ and $\{0\}$, whose join is $2\mathbb{Z}$.

(f) If the module M decomposes as two internal direct sums $M = A \oplus B = A \oplus C$, there is a canonical isomorphism $B \to C$.

An isomorphism $f\colon A \oplus B \to A \oplus C$ of R-modules is not sufficient to imply $B \cong C$; one should add convenient hypotheses on the relationship of f and A.

(g) Given a module M and a family $(M_i)_{i \in I}$ of submodules, there is a canonical homomorphism

$$f\colon \textstyle\bigoplus_{i \in I} M_i \to M, \qquad f((x_i)_{i \in I}) = \sum x_i, \tag{2.57}$$

determined by the family of embeddings $f_i\colon M_i \to M$. We say that M is the (internal) *direct sum* of the *sub*modules M_i (for $i \in I$) if this homomorphism f is an isomorphism. This amounts to:

$$\textstyle\sum_i M_i = M, \qquad (\sum_{i \neq j} M_i) \cap M_j = \{0\}, \text{ for all } j \in I. \tag{2.58}$$

2.3.5 Quotients of modules

Quotients of R-modules are an obvious enrichment of quotients of abelian groups, in 1.3.5. The reader can easily 'upgrade' what we have seen there, adding the scalar multiplication, or read the following.

Let E be an equivalence relation in the R-module M. We say that E is a *congruence* (of R-modules) if it is consistent with the operations in M, which means that (for all $\lambda \in R$ and $x, x', y, y' \in M$)

$$\begin{gathered} \text{if } \; x\,E\,x' \text{ and } y\,E\,y', \text{ then } (x+y)\,E\,(x'+y'), \\ \text{if } \; x\,E\,x' \text{ then } \lambda x\,E\,\lambda x'. \end{gathered} \tag{2.59}$$

The *quotient module* M/E is the quotient set, with the induced operations

$$[x] + [y] = [x+y], \qquad \lambda[x] = [\lambda x]. \tag{2.60}$$

This is the only structure on the quotient set that makes the canonical projection R-linear

$$p \colon M \to M/E, \qquad p(x) = [x]. \tag{2.61}$$

The congruence E determines a submodule $\operatorname{Ker} p$ of M, and it is determined by the latter, as we already know from the theory of abelian groups (in Section 1.3):

$$x\,E\,y \;\Leftrightarrow\; p(x) = p(y) \;\Leftrightarrow\; x - y \in \operatorname{Ker} p. \tag{2.62}$$

For a fixed module M, there is a natural bijection between congruences E and submodules H

$$\begin{gathered} E \mapsto \{x \in M \mid x\,E\,0_M\} = \operatorname{Ker} p, \\ H \mapsto \equiv_H, \qquad x \equiv_H x' \;\Leftrightarrow\; x - x' \in H. \end{gathered} \tag{2.63}$$

The quotient of M modulo the associated congruence $\equiv_H$ is denoted as M/H. In this quotient, the equivalence class of any element x is a coset $[x] = x + H$ of H. The quotients of modules are usually presented in the form M/H.

For an R-homomorphism $f \colon M \to N$, the equivalence relation E_f coincides with the congruence of M associated to the submodule $\operatorname{Ker} f$

$$f(x) = f(y) \;\Leftrightarrow\; (x-y) \in \operatorname{Ker} f \qquad (x, y \in M). \tag{2.64}$$

The canonical factorisation of f works as in 1.3.7.

2.3.6 Free modules

The extension of what we have already seen for free abelian groups, in 1.6.1, is straightforward: one simply replaces the ring $\mathbb{Z}$ with a general commutative unital ring R. We rewrite the whole argument, because of its importance.

We suppose that the ring R is non-trivial, or the current topic becomes trivial as well (see the Remarks below).

The *free R-module* on a set X is a direct sum of copies of R (viewed as an R-module):

$$RX = \bigoplus_{x \in X} R = \{\lambda\colon X \to |R| \mid \mathrm{supp}(\lambda) \text{ is finite}\}. \tag{2.65}$$

An element λ is thus a quasi null mapping $\lambda\colon X \to |R|$; it can be written as a quasi null family $(\lambda_x)_{x \in X}$ of scalars, or simply as (λ_x). They are added componentwise, and multiplied componentwise by a scalar. The set X has a canonical injective mapping, called the insertion of X as the *canonical basis* of RX

$$\eta\colon X \to RX, \qquad (\eta(x))_x = 1, \quad (\eta(x))_y = 0 \ \text{ for } x \neq y. \tag{2.66}$$

The element $\eta(x)$ is often written as e_x, and can be identified with x. Each quasi null family (λ_x) of scalars is an essentially finite linear combination $\Sigma_{x \in X} \lambda_x e_x$.

The universal property of the pair (RX, η) says that each mapping $f\colon X \to N$ with values in an R-module can be uniquely extended to an R-homomorphism $g\colon RX \to N$, so that $g\eta = f$. In fact the homomorphism g is the R-linear extension of f, computed as

$$g(\Sigma_{x \in X} \lambda_x e_x) = \Sigma_{x \in X} \lambda_x f(x), \tag{2.67}$$

by essentially finite linear combinations in the module N.

More generally, a *free R-module* is, by definition, any module M isomorphic to some RX. This means that there exists a family $(e_x)_{x \in X}$ of elements of M, indexed by some set X, such that every element of M can be written as a linear combination $\Sigma_{x \in X} \lambda_x e_x$, by a unique quasi null family (λ_x) of scalars in R. The basis $(e_x)_{x \in X}$ can equivalently be viewed as the corresponding subset $B = \{e_x \mid x \in X\}$ of M. M is said to be *freely generated* over R, by the family (e_x), or the set B.

The trivial module is free, with an empty basis. A power R^n is free, with a canonical basis of n elements

$$e_1 = (1, 0, ..., 0), \quad e_2 = (0, 1, 0, ..., 0), ..., \quad e_n = (0, ..., 0, 1). \qquad (2.68)$$

We have already seen that a $\mathbb{Z}$-module need not be free. We prove below that *all vector spaces are free*: this crucial fact is the source of the 'simplicity' of the theory of vector spaces.

In particular, the vector space $\mathbb{R}^n$ on the real field has a canonical basis as above: the n unit points of its canonical cartesian axes.

Remarks and complements. (a) If R is the trivial ring $\{0\}$, all R-modules are trivial (i.e. singletons), and the free object RX only contains the mapping $X \to \{0\}$: in other words, the trivial R-module is *free on each set.*

Everything written above holds true, *except* one point: the mapping $\eta\colon X \to RX$ *need not be injective*: all e_x are the same, and we cannot identify e_x with x. Of course this case is of little interest, but marginal situations should not be ignored: modules on the trivial ring are still a (poor!) equational algebraic structure, and every such structure has a free object on each set.

(b) (*Multiplicative notation and physical dimensions*) An R-module M can also be written in multiplicative notation: a linear combination $\sum_i \lambda_i x_i$ is then rewritten as a product of powers, with quasi null exponents in R:

$$\Pi_i \, (x_i)^{\lambda_i}. \qquad (2.69)$$

Such a notation may come out in a natural way: for instance, dimensional analysis, in Physics, uses $\mathbb{Q}$-vector spaces in multiplicative notation. Thus the vector space $\mathcal{M}$ of dimensional analysis in Mechanics is freely generated by 3 elements: L (the dimension of *length*), M (*mass*) and T (*time*). The dimension of a *force* is LMT^{-2} (writing it in additive notation, as $\mathsf{L} + \mathsf{M} - 2\mathsf{T}$ would be confusing).

The theory of linear dependence on vector spaces will show that any *basis* of $\mathcal{M}$ has 3 elements.

2.4 Linear dependence and bases

We come back to a *non-trivial* commutative unital ring R, so that a basis of a free R-module can be seen as a subset.

The theory of *linear dependence* over R leads to important results, which can be outlined as follows:

- if R is a field, every vector space M is free (a consequence of the axiom of choice), and all its bases have the same cardinal, called the dimension of M (Theorems 2.4.4 and 2.4.5),

- if $R = \mathbb{Z}$, we already know that an R-module (i.e. an abelian group) need not be free,

- if R is a commutative unital ring, all the bases of a free module have the same cardinal: this is quickly proved in the infinite dimensional case (see Lemma 2.4.3), but requires otherwise higher tools (see [Boul], II.7.2, Corollary of Proposition 3),

- if R is not commutative, there can be counterexamples (see *loc. cit.*).

Zariski–Samuel's book has a formal theory of dependence, that also applies to other cases in Commutative Algebra ([ZS], Section I.21).

2.4.1 Main definitions

Let M be an R-module.

We say that a subset $L \subset M$ is *linearly independent* (over R), or a *free subset*, if a linear combination $\sum_{x \in L} \lambda_x x$ only annihilates when all its coefficients λ_x are zero. Equivalently, this means that any linear combination $\sum_{x \in L} \lambda_x x$ determines its coefficients. If this is not the case, our subset is said to be *linearly dependent.*

We say that a subset S *generates* M, or is a *set of generators* (over R), if $\langle S \rangle = M$, which means that every element of M is an (essentially finite) linear combination $\sum_{x \in S} \lambda_x x$ of elements of S.

Thus, the subset B is a basis of M, as defined in 2.3.6, if and only if it is linearly independent and a set of generators. The module M is said to be *finitely generated* if it has a finite set of generators.

An injective R-homomorphism preserves the free subsets, while a surjective R-homomorphism preserves the subsets of generators.

Exercises and complements. (a) In any R-module, if a set of generators S is contained in a free subset L, then $S = L$ is a basis.

(b) In any R-module, a basis B is a minimal set of generators and a maximal linearly independent subset. (See 2.4.2 for a partial converse.)

*(c) If R is a pid, every submodule of a free R-module is free (extending the similar result for abelian groups). See [Nr], Section 9.1, Theorem 3.

*(d) Conversely, a non-trivial commutative unital ring R with this property is necessarily a pid. *Hints:* R is a free module on itself.

*(e) If, in the previous case, every R-module is free, then R is a field.

2.4.2 Theorem (Linear dependence in vector spaces)

The following points hold for a K-vector space M and a subset $X \subset M$. (All of them fail for $\mathbb{Z}$-modules.)

(a) If X is linearly independent and $y \notin X$, then $y \in \langle X \rangle$ if and only if $X \cup \{y\}$ is linearly dependent.

(b) A subset of M is a basis if and only if it is a minimal set of generators, if and only if it is a maximal linearly independent subset.

(c) (Exchange property) *If* $z \in \langle X \cup \{y\}\rangle \setminus \langle X\rangle$ *then* $\langle X \cup \{y\}\rangle = \langle X \cup \{z\}\rangle$.

Proof Let $X' = X \cup \{y\}$.

(a) If $y = \sum_{x \in X} \lambda_x x$ then X' is obviously linearly dependent. Conversely, if X' is linearly dependent, there is a non-trivial relation $\sum_{x \in X'} \lambda_x x = 0$, where $\lambda_y \neq 0$, because X is linearly independent. Then λ_y has an inverse λ in the *field* K, and $y = -\sum_{x \in X} \lambda\lambda_x x \in \langle X\rangle$.

(In the $\mathbb{Z}$-module $\mathbb{Z}$, the subset $X = \{2\}$ is linearly independent and does not generate 1, but the set $X' = \{1, 2\}$ is linearly dependent.)

(b) Half of the statement holds for general modules, by Exercise 2.4.1(b).

For the second half, suppose first that B is a minimal set of generators. Then B is linearly independent, because a relation $\sum_{x \in B} \lambda_x x = 0$ with $\lambda_y \neq 0$ for some $y \in B$ would give a proper subset $X = B \setminus \{y\}$ which generates y, and therefore the whole vector space.

Suppose now that B is a maximal linearly independent subset; then any vector $y \notin B$ gives a linearly dependent subset $B \cup \{y\}$, and $y \in \langle B\rangle$.

(In the $\mathbb{Z}$-module $\mathbb{Z}$, the subset $\{2, 3\}$ is a set of generators, because $\langle 2\rangle \vee \langle 3\rangle = \langle 1\rangle$, and a minimal one, but it is linearly dependent. The subset $\{2\}$ is linearly independent, and a maximal one, because any $k \in \mathbb{Z}$ has $k.2 - 2.k = 0$, but it does not generate $\mathbb{Z}$.)

(c) We obviously have $\langle X \cup \{z\}\rangle \subset \langle X'\rangle$. For the other inclusion, we can write $z = \sum_{x \in X'} \lambda_x x = \sum_{x \in X} \lambda_x + \lambda_y y$, with $\lambda_y \neq 0$ (because $z \notin \langle X\rangle$). Letting $\lambda = (\lambda_y)^{-1}$ we have $y = \lambda z - \sum_{x \in X} \lambda\lambda_x x \in \langle X \cup \{z\}\rangle$, which proves that $\langle X'\rangle = \langle X \cup \{y\}\rangle \subset \langle X \cup \{z\}\rangle$.

(In the $\mathbb{Z}$-module $\mathbb{Z}$, the subset $X = \emptyset$ does not generate 2, but $2 \in \langle 1\rangle$ and $1 \notin \langle 2\rangle$.)

□

2.4.3 Lemma

Suppose that the R-module M has an infinite basis B. Then each set of generators S has $\sharp S \geqslant \sharp B$, and all bases of M are equipotent.

A finitely generated module can only have finite bases.

Proof The argument is based on a few points on cardinals, reviewed in 1.7.8.

Each element $x \in M$ can be uniquely written as $\sum_{x \in B} \lambda_x x$. For each $x \in S$ we denote as $B_x \subset B$ the finite support of its family of coefficients. Then $B = \bigcup_{x \in S} B_x$, or we would have a proper subset of B that generates M.

Now S is infinite as well (or B would be finite), and $\sharp B \leqslant \sharp S$ (by (1.7.8)).

If S is also a basis, the opposite inequality holds, and set theory says that $\sharp B = \sharp S$ (see 1.7.8(a)).

The last claim is an obvious consequence.

□

2.4.4 Theorem (All vector spaces are free)

Let M be a vector space on the field K. Let S' be a linearly independent subset of M contained in a set S'' of generators. Then there exists a basis B of M such that $S' \subset B \subset S''$.

Notes. (a) One can always take $S' = \emptyset$ and $S'' = |M|$, so that every vector space has a basis.

(b) The proof uses Zorn's Lemma (see 1.7.7), and thus the Axiom of Choice. However, if the complement $S'' \setminus S'$ is a finite set, the statement can be simply proved by induction: the existence of a basis of a finitely generated vector space does not depend on (AC).

Proof Consider the set $\mathcal{S}$ of all subsets $S \subset S''$, which are linearly independent and contain S', ordered by inclusion. It is not empty, as $S' \in \mathcal{S}$.

Any chain $\mathcal{X}$ in $\mathcal{S}$ has an upper bound, namely its union $X = \bigcup \mathcal{X}$, because a condition $\Sigma_{x \in X} \lambda_x x = 0$ (for a quasi null family of scalars (λ_x)) involves a finite number of elements of X (the support of the family); each of them belongs to some element of the chain $\mathcal{X}$ and all of them belong to the biggest, where they are linearly independent and force all scalars λ_x to be null.

Now we take a maximal element $B \in \mathcal{S}$. It is a linearly independent set of M, with $S' \subset B \subset S''$. Moreover it generates all the elements of S'': if $x \in S'' \setminus B$, the subset $B' = B \cup \{x\}$ is linearly dependent (or B would not be maximal), and $x \in \langle B \rangle$ by Theorem 2.4.2(a). But S'' generates M, whence B is also a set of generators.

If the complement $S'' \setminus S'$ is a finite set, the argument can be reorganised without using Zorn's Lemma, by induction on $n = \sharp(S'' \setminus S')$. In fact, if $n = 0$, then $S' = S''$ and we already have a basis. Now we assume that the claim holds for $n - 1 \geqslant 0$ and prove it for $n > 0$.

There is some $x \in S'' \setminus S'$. If $S' \cup \{x\}$ is linearly independent, we replace S' with the latter, and by the induction hypothesis there is a basis B with $S' \cup \{x\} \subset B \subset S''$. If $S' \cup \{x\}$ is linearly dependent, we prove as above that x is generated by S', and we replace S'' with $S'' \setminus \{x\}$, which is still a set of generators; by the inductive assumption there is now a basis B with $S' \subset B \subset S'' \setminus \{x\}$. □

2.4.5 Theorem and Definition (Dimension Theorem of vector spaces)

(a) Let M be a vector space on the field K. Any two bases of M have the same cardinal, that is called the dimension *of M (over K), and written as* $\dim M$, *or* $\dim_K M$.

(b) Two vector spaces are isomorphic if and only if they have the same dimension. A vector space is indecomposable if and only if its dimension is 1.

(c) In a vector space of finite dimension n, every set of n linearly independent vectors is a basis, and every set of n generators is a basis.

Proof (a). We already know that, if M has an infinite basis, all of them have the same cardinal (by Lemma 2.4.3). Therefore, we assume that there are two finite bases B, B' and prove that $\sharp B' = \sharp B$, by induction on $n = \sharp B$.

If $n = 0$ the vector space is trivial and B' is empty as well. Suppose that our property holds for some $n \geqslant 0$, and let $\sharp B = n + 1$. Taking $y \in B$, the set $L = B \setminus \{y\}$ does not generate M (by Theorem 2.4.2(b)), and cannot generate B'. There is thus some $z \in B'$ such that $z \in \langle B' \rangle \setminus \langle L \rangle$, and the Exchange property (Theorem 2.4.2(c)) says that $M = \langle L \cup \{y\} \rangle = \langle L \cup \{z\} \rangle$.

Now $B'' = L \cup \{z\}$ is a set of generators, and is linearly independent (or z would be generated by L, by Theorem 2.4.2(a)), so it is a basis.

Decomposing the bases B'' and B' as disjoint unions (of linearly independent subsets)

$$B'' = L \cup \{z\}, \qquad B' = L' \cup \{z\},$$

the vector space M decomposes in two internal direct sums (see 2.3.4)

$$M = \langle L \rangle \oplus \langle z \rangle = \langle L' \rangle \oplus \langle z \rangle.$$

Applying Exercise 2.3.4(f), $\langle L \rangle \cong \langle L' \rangle$. By the inductive assumption their bases L, L' have both n elements, and $B' = L' \cup \{z\}$ has $n + 1$.

Now (b) and (c) are obvious consequences, using Theorem 2.4.4 for the second point. □

2.4.6 Algebras on a ring

We are still considering a commutative unital ring R.

An *R-algebra* A is an R-module which is also equipped with an (internal) product $x.y$, so that:

(i) the underlying abelian group $(A, +)$ becomes a ring, with this product (i.e. the product is associative and distributes over the sum, on both sides),

(ii) the scalar multiplication and the product are consistent, in the sense that

$$(\lambda x).y = \lambda(x.y) = x.(\lambda y) \qquad (\lambda \in R,\ x, y \in A). \tag{2.70}$$

We reserve the name of *multiplication* to the external operation, defined on $R \times A$, and the name of *product* to the internal one, defined on $A \times A$. An R-algebra is said to be unital, or commutative, if it is as a ring.

In the unital case, (2.70) can be equivalently replaced with

$$(\lambda 1_A).x = \lambda x = x.(\lambda 1_A), \tag{2.71}$$

because from this we deduce:

$$(\lambda x).y = (\lambda 1_A)x.y = \lambda(x.y), \quad x.(\lambda y) = x(\lambda 1_A)y = (\lambda x)y = \lambda(x.y).$$

Any commutative ring is an algebra on itself, with multiplication given by the product. A ring is 'the same' as a $\mathbb{Z}$-algebra, with multiplication given by multiples.

A *homomorphism* of R-algebras $f\colon A \to B$ has to preserve all operations: it is both a homomorphism of R-modules and of rings. A *subalgebra* of A is a subset that is both a submodule and a subring. An *R-ideal* of A is a sub-R-module of A which is an ideal for the ring structure.

2.4.7 ***Exercises and complements*** (Algebras on rings)

(a) Study the structure of R-algebras, examining subalgebras, the kernel of a homomorphism, quotient algebras modulo bilateral R-ideals, the canonical factorisation of a homomorphism, and the cartesian product $\prod_{i \in I} A_i$ of a family of R-algebras.

(b) Let X be a unital ring. Enriching its structure to an R-algebra is equivalent to assigning a homomorphism $f\colon R \to \mathrm{Cnt}(X)$ of unital rings, with values in the centre of X (Exercise 2.1.2(c)). In particular, if R is a unital subring of $\mathrm{Cnt}(X)$, then X is canonically an R-algebra.

(c) A K-algebra A over a field has a dimension $\dim_K A$ as a vector space over K, also written as $[A : K]$ and called the *degree* of A over K. If K is finite of order k and $\dim_K A = n$, then A is finite of order k^n. If A is finite and non-trivial, K also must be finite.

In particular, an *extension* of K is a field K' containing K, with the natural structure of K-algebra given by the inclusion of K. All this will become particularly important for algebraic extensions, in Section 2.5.

(d) A boolean algebra X has a natural structure of (idempotent) algebra on the two-element field $\mathbb{F}_2 = \mathbb{Z}/2$. If X is finite, then it has a finite dimension n on the field $\mathbb{F}_2$, and order 2^n; the last point is already known, from Exercise 1.4.7(g).

(e) (*Finite fields, I*). A finite field is also called a *Galois field*, after E. Galois, who studied them in relation to algebraic equations, in the first half of the 19th century. Prove that the order of a finite field F is a power p^n of a prime number, with $n \geqslant 1$. Then F is an algebra over the field $\mathbb{Z}/p$.

*(f) (*Finite fields, II*). For every power $q = p^n$ of a prime number, with $n \geqslant 1$, there is a Galois field of order q, and only one, up to isomorphism (see [ZS], Section II.8); it is denoted as $\mathbb{F}_q$ or $GF(q)$. This gives a complete classification of finite fields, and shows that two finite fields with the same number of elements are always isomorphic.

The construction of $\mathbb{F}_q$ is non-elementary: it can be obtained as a quotient of the polynomial ring $\mathbb{F}_p[X]$ in one indeterminate ([ZS], loc. cit.).

2.5 Algebras of polynomials

For a commutative unital ring R, we introduce here the R-algebra $R[X]$ of polynomials in one indeterminate. Polynomials in several indeterminates will be briefly considered later, taking full advantage of universal properties (in Section 5.4).

The study of polynomial rings and algebraic equations is a basic part of Commutative Algebra [ZS] and Algebraic Geometry.

2.5.1 Introducing polynomials

(a) The reader is probably familiar with *polynomials in real variables*. In one variable, a polynomial function can be expressed as

$$f \colon \mathbb{R} \to \mathbb{R}, \qquad f(x) = a_0 + a_1x + a_2x^2 + ... + a_nx^n. \tag{2.72}$$

The numbers $a_i \in \mathbb{R}$ are called the *coefficients* of the polynomial and a_0 is the *constant term*. If $a_n \neq 0$, then it is called the *leading coefficient*, and we say that the *degree* of f is n, written as $\deg f = n \in \mathbb{N}$. The zero polynomial, where all coefficients are zero, is given $\deg(0) = -\infty$.

As a crucial fact, let us note that two polynomial functions, as above, coincide if and only if they have the same coefficient in each degree.

This will be proved later, in more general situations; but a reader familiar with elementary Calculus can readily verify this fact, noting that the coefficients

a_i are determined by the derivatives of the function f at the origin of $\mathbb{R}$

$$\begin{aligned} f(0) = a_0, \qquad & f'(0) = a_1, \qquad f''(0) = 2a_2, \\ ..., \qquad & f^{(n)}(0) = n!\, a_n, \quad f^{(n+1)}(0) = 0, ... \end{aligned} \tag{2.73}$$

The set of these polynomials is denoted as $\mathbb{R}[x]$, where x is the name of the variable we are using. It has obvious operations, of sum, product and multiplication by scalars $\lambda \in \mathbb{R}$, which we examine now in a wider structure.

For every set S, the set $F_S = F(S, \mathbb{R}) = \mathrm{Map}(S, \mathbb{R})$ of functions $S \to \mathbb{R}$ has a canonical structure of $\mathbb{R}$-algebra, with pointwise operations computed in $\mathbb{R}$

$$\begin{aligned} (f+g)(x) = f(x) + g(x), \qquad & (\lambda f)(x) = \lambda.f(x), \\ (fg)(x) = f(x).g(x). & \end{aligned} \tag{2.74}$$

already considered in 1.1.7 as a real vector space.

The axioms are obviously satisfied, since they hold in $\mathbb{R}$ as an algebra on itself; or because F_S is a cartesian power $\mathbb{R}^S = \prod_{x \in S} \mathbb{R}$ of $\mathbb{R}$-algebras. (Note that, here, fg does not stand for composition of functions.)

Taking $S = \mathbb{R}$ (or, more precisely, the set $|\mathbb{R}|$ of real numbers), we have the algebra $F_{\mathbb{R}}$ of real functions in one variable, and our set $\mathbb{R}[x]$ is the subalgebra of $F_{\mathbb{R}}$ generated by all the power-functions $f_n(x) = x^n$ $(n \in \mathbb{N})$.

Indeed we can write the polynomial functions as linear combinations of the latter, with quasi null coefficients in $\mathbb{R}$

$$f(x) = \Sigma_{i \in \mathbb{N}}\, a_i x^i, \qquad g(x) = \Sigma_{i \in \mathbb{N}}\, b_i x^i, \tag{2.75}$$

a form that allows an effective way of computing the previous operations:

$$\begin{aligned} (f+g)(x) &= \Sigma_{i \in \mathbb{N}}\, (a_i + b_i) x^i, \\ (fg)(x) &= (a_0 b_0) + (a_1 b_0 + a_0 b_1) x + ... \\ &= \Sigma_{n \in \mathbb{N}}\, (\Sigma_{i+j=n}\, a_i b_j)\, x^n, \\ (\lambda f)(x) &= \Sigma_{i \in \mathbb{N}}\, (\lambda a_i) x^i. \end{aligned} \tag{2.76}$$

(Form (2.72) is less effective in expressing the sum of polynomials of different degree.)

Extending the sum of natural numbers to $\mathbb{N} \cup \{-\infty\}$, in the obvious way (making $-\infty$ an absorbing element), we have

$$\begin{aligned} & \deg(f+g) \leqslant \max(\deg f, \deg g), \\ & \deg(fg) = \deg f + \deg g, \qquad \deg(\lambda f) = \deg f \ \ (\text{for } \lambda \neq 0), \end{aligned} \tag{2.77}$$

and $f + g$ can have any degree less than or equal to $\max(\deg f, \deg g)$.

A *root* of a polynomial $f \neq 0$ is a number $a \in \mathbb{R}$ such that $f(a) = 0$; this is also expressed saying that a is a *solution*, or a *root*, of the algebraic equation $f(x) = 0$. For instance, the polynomial $(x+3)(x-2)^2$ has two roots: -3 (of *multiplicity* 1) and 2 (of *multiplicity 2*); the polynomial x^2+1 has none, as $x^2+1>0$ for all $x \in \mathbb{R}$.

The reader likely knows that a polynomial of degree n has at most n roots in the real field, and may know that – if we extend our range to the complex field $\mathbb{C}$ – there are always n roots 'counted with their multiplicity' (as will be made clear, below).

(b) All this is better studied in a more general situation, replacing the real field $\mathbb{R}$ by an arbitrary commutative unital ring R, viewed again as an algebra on itself.

We have now the R-algebra $F_R = F(|R|, R)$, and the subalgebra $A \subset F_R$ of *polynomial functions*, generated by the sequence of functions

$$f_n \colon R \to R, \qquad f_n(x) = x^n \quad (n \in \mathbb{N}).$$

A polynomial function can still be written as a linear combination $f(x) = \sum_{i \in \mathbb{N}} a_i x^i$, with quasi null coefficients in R, but now the function f need not determine the coefficients, as will be seen in Exercises 2.5.3(a), (b).

In this general case, we distinguish a polynomial function from a *polynomial*: the latter is a formal expression $f = \sum_{i \in \mathbb{N}} a_i X^i$, equivalent to assigning a quasi null sequence $(a_i)_{i \in \mathbb{N}}$ of scalars. The symbol X is called an *indeterminate*.

(c) A polynomial over R in two indeterminates X and Y is again a formal expression

$$\begin{aligned} f = a_{00} &+ (a_{10}X + a_{01}Y) + (a_{20}X^2 + a_{11}XY + a_{02}Y^2) + \dots \\ &+ (a_{n0}X^n + a_{n-1,1}X^{n-1}Y + \dots + a_{0n}Y^n), \end{aligned} \tag{2.78}$$

corresponding to the family of its coefficients (a_{ij}). There is now a polynomial function $|R|^2 \to R$ taking a pair (x, y) to the obvious value $f(x, y) \in R$.

More generally, we can consider several indeterminates $X_1, \dots, X_n$ or even an arbitrary set $\mathcal{X}$ of indeterminates. The polynomial ring $R[\mathcal{X}]$ will be considered in 5.4.2(d), taking advantage of universal properties. In this chapter, we only study polynomials in one indeterminate.

2.5.2 Polynomials in one indeterminate

Let R be a commutative unital ring; we want to define the polynomials in one indeterminate X, with coefficients in R. Their structure is constructed in two steps.

First we take the free commutative monoid $\mathbb{N}$ on the singleton and we rewrite it in multiplicative notation, as the free multiplicative monoid M generated by the singleton $\{X\}$: each element of M is a formal *monomial* X^n, with product $X^m.X^n = X^{m+n}$.

Secondly, we let $R[X]$ be the free R-module on the countable set M of these monomials. An element, called a *polynomial*, is a formal linear combination

$$f = \Sigma_{i\in\mathbb{N}}\, a_i X^i, \tag{2.79}$$

with quasi null coefficients $a_i \in R$. The R-module structure (isomorphic to the direct sum $\bigoplus_{i\in\mathbb{N}} R$) is described as in (2.76): taking a second polynomial $g = \Sigma_{i\in\mathbb{N}}\, b_i X^i$ and scalars $\lambda, \mu \in R$, we have

$$\lambda f + \mu g = \Sigma_{i\in\mathbb{N}}\, (\lambda a_i + \mu b_i)\, X^i. \tag{2.80}$$

This R-module is made into a (commutative, unital) R-algebra, by an R-linear extension of the product of monomials

$$\begin{aligned} fg &= (a_0 b_0) + (a_1 b_0 + a_0 b_1)X + ... \\ &= \Sigma_{n\in\mathbb{N}}\, (\Sigma_{i+j=n}\, a_i b_j)\, X^n. \end{aligned} \tag{2.81}$$

The axioms 2.4.6(i), (ii) are easily verified. $R[X]$ is called the R-algebra of *polynomials in one indeterminate* X, *over the ring* R, or *with coefficients in* R. Its universal property will be the subject of Exercise 2.5.3(g).

If $f \neq 0$, the highest coefficient $a_n \neq 0$ is called the *leading coefficient* of f, written as $c(f)$, and we say that $\deg(f) = n$; the polynomial f is said to be *monic* if $c(f) = 1$. Again, we let $\deg(0) = -\infty$ and extend the sum of $\mathbb{N}$ to $\mathbb{N} \cup \{-\infty\}$, obtaining a monoid with an absorbing element.

Evaluation at some $x \in R$ of each polynomial $f = \Sigma_i a_i X^i$ is a surjective homomorphism of R-algebras

$$\mathrm{ev}_x \colon R[X] \to R, \qquad \mathrm{ev}_x(f) = f(x) = \Sigma_{i\in\mathbb{N}}\, a_i x^i. \tag{2.82}$$

Globally, letting x vary in R, we have an *evaluation* homomorphism with values in the algebra $F_R = F(|R|, R)$

$$\mathrm{ev} \colon R[X] \to F_R, \qquad \mathrm{ev}(f) \colon x \mapsto f(x) = \Sigma_{i\in\mathbb{N}}\, a_i x^i. \tag{2.83}$$

Its image is the algebra A of *polynomial functions in one variable*, of the ring R. The evaluation gives an isomorphism $R[X] \to A$ if and only if $\mathrm{Ker}\,(\mathrm{ev}) = \{0\}$, which means that a polynomial annihilates at each $x \in R$ (if and) only if all its coefficients are zero.

When this is the case, as in 2.5.1, we can identify polynomials and polynomial functions. Examples to the contrary can be found in the exercises below.

If $f = \sum_{i \in \mathbb{N}} a_i X^i$, the coefficient $a_0 = f(0)$ is called the *constant term* of f. The ring R will be viewed as a unital subring of $R[X]$, by the obvious embedding $R \to R[X]$ that identifies a scalar λ with the *constant polynomial* formed by this constant term.

If the ring R is trivial, the R-algebra $R[X]$ is trivial as well and $X = 1X = 0$. One may prefer to leave out this case, and *assume that R is not trivial*, for the rest of this section.

2.5.3 *Exercises and complements*

R is always a commutative unital ring.

(a) If the ring R is *finite*, non-trivial, one can find a non-zero polynomial that vanishes at each $x \in R$, and therefore belongs to the kernel of the evaluation homomorphism (2.83). (A partial converse can be found in Theorem 2.5.5(c).)

(b) Find an infinite commutative unital ring R such that the polynomial $X - X^2$ vanishes at any $x \in R$.

(c) (*Degrees, I*) Prove the following properties of degrees, for $f, g \in R[X]$ and $\lambda \in R$

$$\begin{gathered} \deg(f+g) \leqslant \max(\deg f, \deg g), \qquad \deg(\lambda f) \leqslant \deg f, \\ \deg(fg) \leqslant \deg f + \deg g. \end{gathered} \tag{2.84}$$

(d) (*Degrees, II*) Let f and g be non-zero polynomials. If R is an integral domain:

$$fg \neq 0, \qquad c(fg) = c(f).c(g), \qquad \deg(fg) = \deg f + \deg g. \tag{2.85}$$

For any R, the same holds if one, at least, of the leading coefficients $c(f)$ and $c(g)$ is not a zero-divisor (see 2.1.7).

(e) The ring $R[X]$ cannot be a field. It is an integral domain if and only if R is, and then $\mathrm{Inv}(R[X]) = \mathrm{Inv}(R)$.

(f) Find a polynomial f of degree 1, over the ring $\mathbb{Z}/4$, such that $f^2 = 1$ (and $\deg(f^2) = 0$).

(g) Write and prove the universal property of $R[X]$ as the free commutative unital R-algebra on the singleton set $\{X\}$. (The general case will be treated in 5.4.2(d).)

(h) Deduce that the ring $\mathbb{Z}[X]$ of polynomials in one indeterminate, over the ring of integers, is the free commutative unital ring on the singleton.

(i) Prove that the ring $\mathbb{Z}[X]$ is not a pid. *Hints*: start from the principal ideals (2) and (X).

2.5.4 Roots of polynomials

Roots of polynomials work well when the ring of coefficients is an integral domain, but the first steps can be performed in the general case.

Let $f \neq 0$ be a polynomial in $R[X]$ of degree n, and $a \in R$. We say that a is a *root* of f if $f(a) = 0$ (an *algebraic equation*). We prove below that this is equivalent to saying that f is divisible by $X - a$, i.e. there is some polynomial g such that $f = (X - a)g$.

But a finer analysis is needed. Let us note that the polynomial $(X - a)^m$ has degree m, and is monic: this is obvious, but also follows from Exercise 2.5.3(d). By the same reason, a polynomial divisible by $(X - a)^m$ must have degree $\geqslant m$.

We can now define the *root-multiplicity* $m_f(a)$ of a, with respect to f, as the greatest $m \in \mathbb{N}$ such that f is divisible by $(X - a)^m$. Thus $m_f(a) = 0$ means that a is not a root, and for every $a \in R$ we have $m_f(a) \leqslant \deg(f)$.

A field K is said to be *algebraically closed* if every polynomial of $K[X]$ of positive degree has a root.

Exercises and complements. R is always a commutative unital ring, $f \in R[X]$ and $a \in R$.

(a) The element a is a root of f (i.e. $f(a) = 0$) if and only if f is divisible by $X - a$. Thus $\operatorname{Ker} \operatorname{ev}_a$ is the ideal $(X - a)$, and $R[X]/(X - a) \cong R$.

(b) If R is an integral domain, the polynomial $X - a$ is irreducible. An irreducible polynomial f of degree > 1 has no roots in R.

(c) If R is an integral domain and $f = (X - a)^m.g$, then $m_f(a) = m$ if and only if $g(a) \neq 0$.

(d) If R is a field, every polynomial of degree 1 is irreducible and has a root in R.

(e) Let R be an *idempotent* commutative unital ring, as in Exercise 2.5.3(b), and $f = X + X^2 = X - X^2$ (recalling that $2.1_R = 0$). We have already remarked that every $a \in R$ is a root of f (and there can be infinitely many of them). Compute $m_f(a)$, for every $a \in R$.

(f) The real field is not algebraically closed. It has irreducible polynomials of degree > 1, and reducible polynomials without any roots. (Theorem 2.5.5 will show that this cannot happen in an algebraically closed field.)

*(g) As an important result of Commutative Algebra, every field has an *algebraic closure*, i.e. a 'least' algebraically closed field containing it. The *fundamental theorem of algebra* says that the complex field $\mathbb{C}$ is algebraically closed, and actually the algebraic closure of $\mathbb{R}$.

The usual proofs use some elementary part of Algebraic Topology or Complex Analysis.

2.5.5 Theorem (Factorising polynomials)

Suppose that R is an integral domain.

(a) A polynomial $f \neq 0$ in $R[X]$, of degree n, can be factorised as an essentially finite product

$$f = f_0.\Pi_{a\in R}\,(X-a)^{m_f(a)}$$
$$= f_0.(X-a_1)^{m_1}.(X-a_2)^{m_2}.\dots.(X-a_r)^{m_r}, \qquad (2.86)$$
$$\Sigma_{a\in R}\, m_f(a) \leqslant \deg f_0 + \Sigma_{a\in R}\, m_f(a) = \deg f = n,$$

where the polynomial f_0 has no roots, all a_i are distinct and $m_i = m_f(a_i)$.

This shows that the polynomial f has a finite set of roots $\{a_1, \dots, a_r\}$ (possibly empty). There are at most n roots, even when counted with multiplicity *by the sum $\Sigma_{a\in R}\, m_f(a)$. The first factorisation of f, in* (2.86)*, is unique; the second is unique up to the order of the roots.*

(b) If R is an algebraically closed *field, formula* (2.86) *holds with $f_0 = c(f)$, the leading coefficient of f, and $\deg f$ coincides with $\Sigma_{a\in R}\, m_f(a)$, the* number of roots counted with multiplicity. *A polynomial is irreducible if and only if it has degree 1.*

(c) If the integral domain R is infinite, the evaluation homomorphism $\mathrm{ev}\colon R[X] \to F(|R|, R)$ *of* (2.83) *is injective, and polynomials can also be viewed as polynomial functions.*

Proof (a) Let us recall that $R[X]$ is also an integral domain, which allows us to cancel non-zero polynomials (in products) and to compute the degree of a product as a sum of degrees (Exercises 2.5.3(d), (e)).

If f has no roots, we take $f = f_0$ and we are done. If there is a root a_1 with positive multiplicity m_1, we can write

$$f = (X-a_1)^{m_1}.g, \qquad g(a_1) \neq 0.$$

If there are no other roots of f, our factorisation is achieved.

Otherwise, there is some other root $a_2 \neq a_1$, with positive multiplicity $m_f(a_2) = m_2$, and $g(a_2) = 0$. Now $m = m_g(a_2) \leqslant m_2$ must coincide with m_2, because we can factorise

$$g = (X-a_2)^m.h, \qquad h(a_2) \neq 0,$$
$$f = (X-a_1)^{m_1}.(X-a_2)^m.h = (X-a_2)^{m_2}.k,$$

where the polynomial $(X-a_1)^{m_1}.h = (X-a_2)^{m_2-m}.k$ cannot vanish at a_2, and $m_2 - m = 0$.

We have thus obtained $f = (X-a_1)^{m_1}.(X-a_2)^{m_2}.h$, with $h(a_i) \neq 0$ for $i = 1, 2$. Proceeding this way, we cover all the roots and prove the required formula.

(b) In an algebraically closed field, the polynomial $f_0 \neq 0$ of factorisation (2.86), having no roots, is a non-zero constant.

Consider now our polynomial $f \neq 0$: if it has degree 1, it is irreducible; otherwise, either it has degree 0 and is invertible, or it has degree > 1, and a non-trivial factorisation (2.86).

(c) By (a), every polynomial (on an integral domain) has a finite number of roots.

□

2.5.6 *Transcendental and algebraic elements*

Let R be a unital subring of a commutative unital ring R'. Then a polynomial $f = \Sigma_{i \in \mathbb{N}} a_i X^i$ over R can be computed on any $x \in R'$

$$f(x) = \Sigma_{i \in \mathbb{N}} a_i x^i \in R', \tag{2.87}$$

giving the evaluation homomorphism

$$\mathrm{ev}_x \colon R[X] \to R', \qquad \mathrm{ev}_x(f) = f(x). \tag{2.88}$$

The image of the latter is a unital subring of R', denoted as $R[x]$. We have thus an induced isomorphism

$$R[X] \,/\, \mathrm{Ker}\,(\mathrm{ev}_x) \to R[x]. \tag{2.89}$$

The element x is said to be *transcendental* over R if ev_x is injective, and induces an isomorphism from $R[X]$ to $R[x]$.

Otherwise, $\mathrm{Ker}\,(\mathrm{ev}_x) \neq \{0\}$: there exists a polynomial $f \neq 0$ over R which annihilates at $x \in R'$; we say that x is *algebraic* over R, and a *root* of f in R'.

In particular, the indeterminate $X \in R[X]$ is obviously transcendental over R.

2.5.7 *Theorem and Definition* (Euclidean division)

We have an integral domain R and two polynomials $f, g \in R[X]$; the leading coefficient of g is 1 (or, more generally, an invertible element).

Then there is a unique expression, in $R[X]$

$$f = qg + r, \qquad \deg\,(r) < \deg\,(g). \tag{2.90}$$

The polynomial q is called the quotient *and r the* remainder, *of the* euclidean division *of f by g.*

Proof We begin by writing $f = 0g + f$, which does what we want if $\deg(f) < \deg(g)$.

Otherwise, $\deg(f) = m \geqslant \deg(g) = n$ and

$$f = a_m X^m + h, \quad g = X^n + k \qquad (\deg h < m,\ \deg k < n).$$

We take $q_1 = a_m X^{m-n}$ and $r_1 = (f - q_1 g)$, so that

$$\begin{aligned} f &= q_1 g + r_1 = q_1 g + a_m X^m + h - a_m X^m - a_m X^{m-n} k \\ &= q_1 g + h - a_m X^{m-n} k, \end{aligned}$$

and $\deg(r_1) \leqslant m - 1$.

If $m_1 = \deg(r_1) < n$ we are done, otherwise we proceed as above taking $q_2 = a_m X^m + bX^{m_1}$ where b is the leading coefficient of r_1. After a finite number of steps we reach our goal. □

2.5.8 *Exercises and complements* (Polynomials on a field)

Let K be a field, and a subring of an integral domain R. In the ring $K[X]$ of polynomials on K, the euclidean division by a non-zero polynomial always works.

The following points are important; the (easy) proofs are below.

(a) $K[X]$ is a principal ideal domain: every non-trivial ideal I is generated by a non-zero polynomial of least degree in I; there is precisely one monic generator.

Therefore, if f is a non-zero polynomial on K: f is prime if and only if it is irreducible, if and only if the ideal (f) is prime, if and only if the latter is maximal (by Exercise 2.2.3(e)).

(b) Let $x \in R^*$ be algebraic on K. The *minimal polynomial* of x on K is the monic generator of the ideal $\mathrm{Ker}(\mathrm{ev}_x)$ of the polynomials that annihilate at x.

It is also defined as the monic polynomial f of least degree, having a root at x. Evaluation at x induces a canonical isomorphism $K[X]/(f) \to K[x]$.

(c) In these hypothesis, the minimal polynomial f is irreducible and its degree $n = \deg(f) \geqslant 1$ coincides with $\dim_K K[x]$. The ring $K[x]$ is a field, generally written as $K(x)$, and called an *algebraic extension of the field* K, *of degree* n. In particular, degree 1 means that $a \in K$.

(d) (*An important remark*) The algebraic extension $K(x)$ is determined, up to isomorphism, by the field K and an irreducible polynomial f (monic, if we want): we can forget about the super-ring R, and speak of *the algebraic extension* $K(x)$ *determined by the polynomial* f, irreducible in $K[X]$.

Thus, we can construct the complex field $\mathbb{C}$ as the algebraic extension of the real field determined by the polynomial X^2+1, irreducible in $\mathbb{R}[X]$, as will be done in 2.7.7.

(e) If x is transcendental on K, the dimension of the K-algebra $K[x]$ on the field K is infinite countable.

(f) (*The field of rational functions*) The field of fractions on the integral domain $K[X]$ (see Theorem 2.2.4)

$$K(X) = Q(K[X]), \tag{2.91}$$

is called the *field of rational functions* on K.

Its elements are formal fractions f/g of polynomials (with $g \neq 0$), under the equivalence relation and operations defined in 2.2.4.

Solutions. (a) We already know that $K[X]$ is an integral domain. Let I be a non-null ideal, and let g be a non-zero polynomial of least degree in I.

If $f \in I$, let $f = qg + r$ with $\deg(r) < \deg(g)$. Then $r = f - qg \in I$ must be 0, and $f \in (g)$.

(b) An obvious consequence of (a) and (2.89).

(c) The ring $K[x] \subset R$ is an integral domain, whence the ideal (f) is prime, and maximal: f is irreducible and $K[X]/(f)$ is a field.

The n elements $1, x, x^2, ..., x^{n-1}$ form a linear basis of $K[X]/(f)$ on K. They are linearly independent because every polynomial of degree $< n$ which annihilates at x is zero; moreover, every polynomial g can be written as $qf + r$, with $\deg r < n$, and $[g] = [r]$ is a linear combination of our n elements.

(e) If x is transcendental on K, $K[x]$ is isomorphic to the polynomial algebra $K[X]$, with basis $(X^n)_{n \geqslant 0}$ as a vector space on K.

2.5.9 An exercise (Modules on $R[X]$)

Let R be a (non-trivial) commutative unital ring. An $R[X]$-module M can be described as an R-module suitably enriched.

Hints. Use the scalar multiplication by X in M.

2.6 Matrices, linear and affine spaces

The study of the ring of matrices and its application to linear systems is a basic part of Linear Algebra [Ko1, La1, La2, Bou1]. Here we only sketch the main points of this topic.

Affine spaces and linear groups are briefly sketched in 2.6.6 and 2.6.7. Finally, in 2.6.8, we show how a linear differential equation with constant coefficients can be interpreted as a linear equation: the latter lives in a module of C^∞-functions over a polynomial ring $\mathbb{R}[D]$ in one indeterminate, the 'operator of derivation'.

2.6.1 Linear systems and matrices

We begin, again, by working in the real field $\mathbb{R}$, and fix two positive integers m, n. The reader probably knows that a *linear system* is a family of m linear equations in n variables $x_1, x_2, ..., x_n$, or *unknowns*

$$\begin{cases} a_{11}x_1 + a_{12}x_2 + \ ... \ + a_{1n}x_n \ = b_1 \\ a_{21}x_1 + a_{22}x_2 + \ ... \ + a_{2n}x_n \ = b_2 \\ \quad ... \\ a_{m1}x_1 + a_{m2}x_2 + \ ... \ + a_{mn}x_n = b_m \end{cases} \tag{2.92}$$

The numbers $a_{ij} \in \mathbb{R}$ are given (for $i = 1, ..., m$ and $j = 1, ..., n$), and called the *coefficients* of the system; they are indexed by pairs $(i, j) \in I = \{1, ..., m\} \times \{1, ..., n\}$, written as ij (or i, j) in subscripts. The numbers $b_1, ..., b_n$ are also given, and called the *constant terms* of the system.

Solving the system means to find all the n-tuples $(x_1, x_2, ..., x_n)$ of real numbers that satisfy all the equations above, if any.

A single linear equation is also a system, like

$$3x_1 = 2, \qquad 0x_1 = 2, \qquad 3x_1 + 0x_2 + x_3 = 2. \tag{2.93}$$

These three examples show cases where, respectively:

- there is precisely one solution, namely $x_1 = 2/3$,
- there is no solution,
- there are infinitely many solutions, namely all the triples $(x_1, x_2, 2 - 3x_1)$, for arbitrary $x_1, x_2 \in \mathbb{R}$.

It is useful to organise the family $A = (a_{ij})_{(i,j)\in I}$ of coefficients in a table with m rows and n columns, called a *matrix of type* $m \times n$, as below at the left, briefly written as $A = (a_{ij})$

$$\begin{pmatrix} a_{11} & a_{12} & ... & a_{1n} \\ a_{21} & a_{22} & ... & a_{2n} \\ ... & ... & ... & ... \\ a_{m1} & a_{m2} & ... & a_{mn} \end{pmatrix} \qquad \begin{pmatrix} x_1 \\ x_2 \\ ... \\ x_n \end{pmatrix} \qquad \begin{pmatrix} b_1 \\ b_2 \\ ... \\ b_m \end{pmatrix} \tag{2.94}$$

In particular, the n-tuple $x = (x_1, x_2, ..., x_n)$ of variables will be viewed as a matrix of type $n \times 1$ (also called a *column matrix*), as above in the middle. It will still be written as an n-tuple $(x_1, x_2, ..., x_n)$ when convenient, but should not be confused with the corresponding *row matrix* $(x_1 \ x_2 \ ... \ x_n)$, of type $1 \times n$.

In the same way the m-tuple $b = (b_1, b_2, ..., b_m)$ of constant terms is viewed as a column matrix of type $m \times 1$, as above at the write.

More generally, we consider matrices with entries in a commutative unital ring R, which allows us to study linear systems in $\mathbb{Z}$, $\mathbb{Q}$, $\mathbb{R}$ and in other rings (including rings of real functions, or the complex field).

The set of matrices of type $m \times n$, with entries in R, will be written as $\mathrm{Mat}_{mn}(R)$. In particular, we write as $M_n(R) = \mathrm{Mat}_{nn}(R)$ the set of *square matrices of order* n, with n rows and n columns.

We show below that each set $\mathrm{Mat}_{mn}(R)$ has a natural addition.

Less obviously, we also construct a product AB of matrices, which is defined *when* A is of type $m \times n$ and B is of type $n \times p$ (note the repetition of n); then the result AB is of type $m \times p$.

With this operation, the linear system (2.92) can be rewritten as the equation $A.x = b$ in the variable x, and the algebraic structure of matrices helps us to find its solutions, if any.

2.6.2 Operations on matrices

R is always a commutative unital ring. The set $\mathrm{Mat}_{mn}(R)$ has a natural structure of R-module

$$(a_{ij}) + (b_{ij}) = (a_{ij} + b_{ij}), \qquad \lambda(a_{ij}) = (\lambda a_{ij}), \tag{2.95}$$

which simply is the cartesian power R^I of R as a module on itself, indexed by the set $I = \{1, ..., m\} \times \{1, ..., n\}$ (see (2.52)).

The zero-element 0_{mn} is the *null matrix* of type $m \times n$ (all its entries are $0 \in R$), and the opposite is $-(a_{ij}) = (-a_{ij})$. The null matrix 0_{mn} can be simply written as 0, or $\underline{0}$.

The *product* AB of matrices, called *rows by columns*, is defined when $A = (a_{ij}) \in \mathrm{Mat}_{mn}(R)$ and $B = (b_{jk}) \in \mathrm{Mat}_{np}(R)$. The result $C = (c_{ik}) \in \mathrm{Mat}_{mp}(R)$ has the following general entry

$$c_{ik} = \textstyle\sum_{j=1,\dots,n} a_{ij}.b_{jk} \qquad (i = 1, ..., m;\ k = 1, ..., p), \tag{2.96}$$

obtained as a sum of entries of the i-th row of A multiplied by entries of the k-th column of b.

We already noted that the module $\mathrm{Mat}_{mn}(R)$ is isomorphic to the cartesian power R^{mn}. Yet a 2×3 matrix should not be confused with a matrix of type 3×2, nor with a column matrix of type 6×1: the product 'rows by columns' only makes sense when the number of rows and columns are determined, as specified above.

Exercises and complements. (a) The *unit matrix* $I_n \in M_n(R)$, of order n, has general entry δ_{ij}, the 'Kronecker delta'

$$\delta_{ij} = 1_R, \text{ if } i = j, \qquad \delta_{ij} = 0_R, \text{ otherwise.}$$

The product of matrices A, B, C with entries in the unital ring R satisfies the following properties, whenever the operations are legitimate (λ is a scalar)

(i) (*Associativity*) $A(BC) = (AB)C$,

(ii) (*Distributivity*) $A(B+C) = AB + AC$, $(A+B)C = AC + BC$,

(iii) (*Units*) $AI_n = A, \quad I_nB = B$,

(iv) (*Scalar multiplication*) $A(\lambda I_n) = \lambda A, \quad (\lambda I_n)B = \lambda B$.

The reader will note that the sum and product of matrices are *partial operations*, not defined on the whole set of matrices with entries in R.

*The product can be viewed as a 'composition', forming a category $\mathsf{Mat}(R)$ whose objects are the natural numbers and whose arrows $A\colon n \to m$ are the matrices of type $m \times n$; a matrix $B\colon p \to n$ gives thus a composite $AB\colon p \to m$.

The sum of matrices makes this category into an additive one. If K is a field, representing linear mappings of finite dimensional K-vector spaces by matrices, as in 2.6.4, gives an *equivalence* between the category $\mathsf{Mat}(K)$ and the category of finite dimensional K-vector spaces (see Section 5.3).*

(b) (*Rings of square matrices*) In the set $M_n(R)$ of square matrices of order n, sum, scalar multiplication and product are always defined and give a result of the same type. One gets a unital R-algebra, that 'coincides' with the ring R for $n = 1$. For $n > 1$, the ring $M_n(R)$ is only commutative when R is the trivial ring.

(c) (*Transpose matrix*) Exchanging rows and columns, a matrix $A = (a_{ij}) \in \mathrm{Mat}_{mn}(R)$ gives a *transpose* matrix $A^{\mathrm{tr}} = (a_{ji})_{ij} \in \mathrm{Mat}_{nm}(R)$, with general term $b_{ij} = a_{ji}$ (for $i = 1, ..., n$ and $j = 1, ..., m$).

In particular, a column vector $x \in \mathrm{Mat}_{n1}(R)$ is turned into a row vector $x^{\mathrm{tr}} \in \mathrm{Mat}_{1n}(R)$, and conversely.

Transposition acts in the following way (provided the operations are legitimate):

$$(A+B)^{\mathrm{tr}} = A^{\mathrm{tr}} + B^{\mathrm{tr}}, \qquad (AB)^{\mathrm{tr}} = B^{\mathrm{tr}}.A^{\mathrm{tr}},$$
$$(I_n)^{\mathrm{tr}} = I_n, \qquad (A^{\mathrm{tr}})^{\mathrm{tr}} = A. \tag{2.97}$$

(d) (*Linear systems*) The system (2.92) is equivalent to the equation $Ax = b$ in the variable $x = (x_1, x_2, ..., x_n)$. The latter belongs to the set $R^n = \mathrm{Mat}_{n1}(R)$ of n-tuples of elements of the ring R, written as column matrices, while $b \in R^m = \mathrm{Mat}_{m1}(R)$.

(Using row matrices $x \in \mathrm{Mat}_{1n}(R)$ and $b \in \mathrm{Mat}_{1m}(R)$, we should write $Ax^{\mathrm{tr}} = b^{\mathrm{tr}}$.)

2.6.3 ***Complements** (Hints at determinants and inverses)

We have seen that a linear system in n equations and n unknowns can be expressed in the form $Ax = b$, where A is a square matrix of order n and x, b are column matrices with n components.

If A is invertible in the ring $M_n(R)$, let $B = A^{-1}$, so that $AB = I_n = BA$. Then our linear system has one and only one solution: $x = Bb$. It is thus important to know which square matrices are invertible, and to compute their inverse.

Let $A = (a_{ij}) \in M_n(R)$ be a square matrix.

If $n \geqslant 2$, the *minor matrix* A_{ij} (of A) is defined as the square matrix of order $n-1$ obtained by removing the i-th row and the j-th column, for $i, j = 1, ..., n$; there are n^2 of them.

The *determinant* of A, written as $\det(A)$ or $|A|$, is an element of R recursively defined as follows:

(i) if $n = 1$, then $A = (a_{11})$ has one entry, and $\det(A) = a_{11}$,

(ii) if $n > 1$, then $\det(A) = \Sigma_j\, (-1)^{1+j}\, a_{1j}.\det(A_{1j})$.

Thus, for a matrix $A = \left(\begin{smallmatrix} a & b \\ c & d \end{smallmatrix}\right)$ we get the well-known formula $\det(A) = ad - bc$. In order 3 we get Sarrus' rule, expressing $\det(A)$ as an algebraic sum of six products of coefficients. And so on. The scalar $\det(A_{ij})$ is also called a *minor* of the matrix A.

We summarise the important properties of the determinant, without proofs (see [Ko1, La1], for scalars in a field, and [La2, Bou1] for the general case).

(iii) $|I_n| = 1, \qquad |AB| = |A|.|B| = |BA|, \qquad |A^{\text{tr}}| = |A|$,

(iv) *Laplace expansion*, for $i, k = 1, ..., n$ (an extension of (ii)):

$$\Sigma_j\, (-1)^{j+k}\, a_{ij}.|A_{kj}| = |A|.\delta_{ik} = \Sigma_j\, (-1)^{i+j}\, |A_{ji}|.a_{jk}. \tag{2.98}$$

From (iii) it follows that, if A is invertible in $M_n(R)$ with inverse B, then $\det(A)$ is invertible in R, with inverse $\det(B)$.

The converse can be deduced from (iv): if $\det(A)$ is invertible in R, the matrix A is invertible in $M_n(R)$. In fact, introducing the *adjugate matrix* $\text{adj}(A)$, with general term

$$\alpha_{ij} = (-1)^{i+j}.\det(A_{ji}), \tag{2.99}$$

formula (2.98) can be rewritten as:

$$A.\text{adj}(A) = (\det A).I_n = \text{adj}(A).A. \tag{2.100}$$

Therefore, if $\det A$ is invertible in R, the matrix $B = (\det A)^{-1}.\text{adj}(A)$ is inverse to A; its general term is:

$$b_{ij} = (-1)^{i+j}\det(A_{ji}).(\det A)^{-1}. \tag{2.101}$$

2.6.4 Matrices and linear mappings

A matrix $A \in \text{Mat}_{mn}(R)$ defines a homomorphism of R-modules

$$f_A \colon R^n \to R^m, \qquad f_A(x) = Ax, \tag{2.102}$$

where, as in Exercise 2.6.2(d), the elements of the free R-module $R^k = \mathrm{Mat}_{k1}(R)$ are viewed as column matrices of type $k \times 1$. The homomorphism f_A is said to be *represented* by the matrix A.

We have thus a mapping defined on the R-module $\mathrm{Mat}_{mn}(R) \cong R^{mn}$

$$\rho\colon \mathrm{Mat}_{mn}(R) \to \mathrm{Hom}_R(R^n, R^m), \qquad \rho(A) = f_A, \tag{2.103}$$

with values in the module of homomorphisms of Exercise 2.3.3(g). It is plainly an R-homomorphism: $(\lambda A + \mu B)x = \lambda Ax + \mu Bx$. Actually it is an isomorphism, by Exercise (a) below.

Thus every linear mapping $f\colon R^n \to R^m$ is uniquely represented by a matrix $A \in \mathrm{Mat}_{mn}(R)$, whose coefficients are determined below.

It is now convenient to allow the numbers m, n to be arbitrary natural numbers (instead of positive ones). If $m = 0$ or $n = 0$ (possibly both), $\mathrm{Hom}_R(R^n, R^m)$ is the trivial module; we introduce *for each such pair* a formal matrix $0_{mn} \in \mathrm{Mat}_{mn}(R)$ of type $m \times n$, and extend the isomorphism (2.103) to the new cases, where $m = 0$ or $n = 0$. (The matrix operations are extended in the obvious way.)

The reader will note that 0_{mn} can hardly be written as a table, which should have no rows and some column, or some row and no columns, or nothing at all – all these cases being distinguished!

This is why we are only introducing now these formal matrices, with the purpose of representing all homomorphisms between finite-dimensional free R-modules (see Exercise (c), below).

Exercises and complements. (a) The homomorphism (2.103) is an isomorphism. The matrix $A = (a_{ij})_{ij}$ is sent to the homomorphism which takes the element e_j of the canonical basis of R^n to the j-th column $(a_{1j}, a_{2j}, ..., a_{mj})$ of the matrix (for $j = 1, ..., n$).

(b) For two 'composable' matrices $A \in \mathrm{Mat}_{mn}(R)$ and $B \in \mathrm{Mat}_{np}(R)$, prove that the composed homomorphism $f_A.f_B\colon R^p \to R^m$ is represented by the product $AB \in \mathrm{Mat}_{mp}(R)$.

(c) Let N, M be free R-modules with a basis $(x_1, ..., x_n)$ and $(y_1, ..., y_m)$, respectively. The previous representation can be extended to an isomorphism $\mathrm{Mat}_{mn}(R) \to \mathrm{Hom}_R(N, M)$, depending on the given basis.

We have thus characterised $\mathrm{Hom}_R(N, M)$ for all finite-dimensional free R-modules. If R is a field, this covers all finitely generated vector spaces.

2.6.5 Linear systems and spaces of solutions

Let us come back to the linear system of the beginning

$$Ax = b, \tag{2.104}$$

where the matrix of coefficients A belongs to $\mathrm{Mat}_{mn}(R)$, the variable $x \in R^n$ is a column matrix of n rows, and the constant term $b \in R^m$ is a column matrix of m rows. Equivalently, we are considering the 'system' $f_A(x) = b$, where $f_A \colon R^n \to R^m$ is the homomorphism represented by the matrix A.

By definition, the *homogeneous system* associated to (2.104) is

$$Ax = 0, \tag{2.105}$$

and its solution is the submodule $\mathrm{Ker}\, f_A$ of R^m (not necessarily free).

Now, if there exists a solution x_0 of the original system (2.104), which means that $b \in \mathrm{Im}\, f_A$, all of them form the coset

$$f_A^{-1}(\{b\}) = x_0 + \mathrm{Ker}\, f_A, \tag{2.106}$$

described as: 'a particular solution of the original system added to the general solution of the associated homogeneous system'.

As a consequence:

- the system is *determined* (has a unique solution) if and only if $b \in \mathrm{Im}\, f_A$ and $\mathrm{Ker}\, f_A = \{0\}$,

- the system is *indeterminate* (has several solutions) if and only if $b \in \mathrm{Im}\, f_A$ and $\mathrm{Ker}\, f_A \neq \{0\}$,

- the system is *impossible* (has no solution) if and only if $b \notin \mathrm{Im}\, f_A$.

The theory of linear systems is often dealt with for vector spaces on a field K. Then the solutions of the associated homogeneous system form a vector subspace $V = \mathrm{Ker}\, f_A$ of K^m (of dimension $k \leqslant m$), and the solutions of the given system form a subset X of K^m which is either a coset $x_0 + V$ or $\emptyset$.

Such a subset is called an *affine subspace* of K^m; its dimension is defined to be $\dim_K V$, or $-\infty$. The structure of affine spaces on K is sketched below.

Remarks. (a) In the real vector space $\mathbb{R}^3$, a vector subspace V of dimension 0, or 1, or 2, or 3, is $\{0\}$, or a line through the origin, or a plane through the origin, or the total space (Exercise 2.3.3(d)). By translating these subspaces, an affine subspace of the form $x_0 + V$ is thus any singleton, or any line, or any plane, or the total space.

To all these we must add the empty subspace, which we need as the space of solutions of an impossible linear system. Or – structurally – as the intersection of disjoint affine subspaces.

(b) The terminology is somewhat inconsistent: a linear system should rather be called an 'affine system' (as its solutions form an affine subset), while the term linear should be reserved for the homogeneous case (where the solutions form a linear subspace).

Yet the historical terminology is as described above, and changing it would be more confusing than useful.

2.6.6 Affine spaces

An *affine space* on the field K is usually presented as a pair (X, V) where V is a K-vector space and X is a set equipped with a transitive, free right action of the additive group V (see 1.6.4).

An element $x \in X$ is also called a *point*, and an element $v \in V$ is also called a *vector*, or an *operator*; the action is written as $x + v$. If $X = \emptyset$, we have no use of operators: all pairs $(\emptyset, V)$ will be 'identified', and denoted as $\emptyset$.

We recall that the action satisfies the following axioms, for all $x \in X$ and $v, v' \in V$:

(i) (*Compatibility*) $\qquad (x + v) + v' = x + (v + v'),$

(ii) (*Unitarity*) $\qquad x + 0 = x.$

Saying that the action is transitive and free means that, for every $x \in X$ (if any), the mapping

$$\varphi_x \colon V \to X, \qquad v \mapsto x + v, \tag{2.107}$$

is bijective. If $X \neq \emptyset$, it is sufficient to know that this holds for some $x \in X$.

(If this is the case, any point $y \in X$ gives a bijection φ_y, because letting $y = x + v_0$, we have $\varphi_y(v) = x + v_0 + v = \varphi_x(v_0 + v)$. Therefore, φ_y is also surjective; finally $\varphi_y(v) = \varphi_y(v')$ gives $v_0 + v = v_0 + v'$ and $v = v'$.)

If $x, x' \in X$, we can thus denote as $\langle x' - x \rangle$ the unique vector $v \in V$ such that $x + v = x'$. Note that $\langle x' - x \rangle + \langle x'' - x' \rangle = \langle x'' - x \rangle$.

The *dimension* of the affine space (X, V) on the field K is defined as the dimension of V, if $X \neq \emptyset$; otherwise, we let $\dim_K(\emptyset) = -\infty$.

An *affine mapping* $f \colon (X, V) \to (Y, W)$ is a mapping $X \to Y$ such that, if $X \neq \emptyset$, the associated mapping

$$f' \colon V \to W, \qquad \langle x' - x \rangle \mapsto \langle fx' - fx \rangle \quad (x, x' \in X), \tag{2.108}$$

is K-linear. If $X = \emptyset$ there is no condition: the empty mapping $\emptyset \to (Y, W)$ is affine.

Affine mappings are plainly stable under composition, and have identities $\mathrm{id}(X, V)$, represented by $\mathrm{id}\, X$ (including $\mathrm{id}\, \emptyset$).

Exercises and complements. (a) Let $X \neq \emptyset$ and fix a point $x_0 \in X$. Then an affine mapping $f \colon (X, V) \to (Y, W)$ can be equivalently given by a pair $(y_0, f') \in Y \times \mathrm{Hom}_K(V, W)$.

(b) It is interesting to note that, given two points x, y of the affine space (X, V) and two scalars $\lambda, \mu \in K$ such that $\lambda + \mu = 1$, the linear combination $\lambda x + \mu y$ makes sense in X. In fact we have the vector $v = \langle y - x \rangle$ such that $y = x + v$, and we define $\lambda x + \mu y = x + \mu v$.

(The reader should think of the euclidean examples, in 2.6.5(a). If $x \neq y$ in $\mathbb{R}^3$, the point $x_\mu = x + \mu(y - x)$ describes the line joining the points x, y, when the parameter μ varies in $\mathbb{R}$; it describes the line segment from x to y when $0 \leqslant \mu \leqslant 1$.)

More generally, given a finite family $x_1, x_2, ..., x_n$ in X and scalars $\lambda_1, \lambda_2, ..., \lambda_n$ with $\sum \lambda_i = 1$ (the *weights*), the linear combination $\sum \lambda_i x_i$ makes sense in X. It is called the *affine combination* of the points x_i with coefficients λ_i, or also the *barycentre* of the points x_i with weights λ_i. (Note that, for $K = \mathbb{R}$, the weights can also be negative, as in the previous example.)

*(c) A reader can find interesting and engaging to develop the basic theory of affine spaces, about: subspaces, quotients, products, etc. The procedure is somewhat unusual, because our structure is based on the auxiliary structure of vector spaces.

2.6.7 *Linear groups

(a) The *general linear group* $\mathrm{GL}(n, \mathbb{R})$ $(n > 0)$ is the group of invertible elements of the unital ring $M_n(\mathbb{R})$, i.e. the multiplicative group of all square matrices of real numbers, of order n, whose determinant is non-zero. This is isomorphic to the group $\mathrm{Aut}(\mathbb{R}^n)$ of automorphisms of the real vector space $\mathbb{R}^n$.

In particular $\mathrm{GL}(1, \mathbb{R}) = \mathbb{R}^*$, and $\mathrm{GL}(2, \mathbb{R})$ is already non-commutative.

Inside $\mathrm{GL}(n, \mathbb{R})$ we have the *orthogonal group*, of orthonormal matrices of order n

$$\mathrm{O}(n, \mathbb{R}) = \{A \in M_n(\mathbb{R}) \mid AA^{\mathrm{tr}} = I_n\} \subset \mathrm{GL}(n, \mathbb{R}), \tag{2.109}$$

so that A and A^{tr} are inverse to each other; since $\det A = \det A^{\mathrm{tr}}$, all of them have determinant ± 1.

The *special orthogonal group* is a subgroup of the former, of index 2

$$\mathrm{SO}(n, \mathbb{R}) = \{A \in \mathrm{O}(n, \mathbb{R}) \mid \det(A) = 1\} \lhd \mathrm{O}(n, \mathbb{R}). \tag{2.110}$$

One proves that the linear transformation associated to the matrix A

$$f \colon \mathbb{R}^n \to \mathbb{R}^n, \qquad f(x) = Ax, \tag{2.111}$$

is an isometry (i.e. preserves the euclidean distance) if and only if $A \in \mathrm{O}(n, \mathbb{R})$; in this case, f 'preserves the orientation' (and is a rotation about the origin) if and only if $A \in \mathrm{SO}(n, \mathbb{R})$.

(b) (*Rotations in the plane*) For a real number φ, the matrix

$$r(\varphi) = \begin{pmatrix} \cos\varphi & -\sin\varphi \\ \sin\varphi & \cos\varphi \end{pmatrix} \in \mathrm{SO}(2, \mathbb{R}), \tag{2.112}$$

gives a rotation of φ radians in the real plane $\mathbb{R}^2$

$$f(x,y) = (x.\cos\varphi - y.\sin\varphi,\ x.\sin\varphi + y.\cos\varphi). \tag{2.113}$$

Therefore $r(\varphi+\vartheta)$ is the composed rotation $r(\varphi)r(\vartheta) = r(\vartheta)r(\varphi)$, and

$$r\colon \mathbb{R} \to \mathrm{SO}(2,\mathbb{R}), \tag{2.114}$$

is a surjective homomorphism, from the additive group of the real numbers to the multiplicative group $\mathrm{SO}(2,\mathbb{R})$, of rotations about the origin, in the plane. Its kernel is the subgroup $2\pi\mathbb{Z} = \{2k\pi \mid k \in \mathbb{Z}\} \subset \mathbb{R}$.

The sum formulas of the trigonometric functions just express the fact that r is a homomorphism (via the product of 2×2-matrices).

(c) (*Matrices of functions*) Interpreting φ as a real variable, in formula (2.112), we can also view r as a matrix of functions: an element of the ring $M_2(\mathcal{R})$ of matrices on the commutative unital ring $\mathcal{R} = F(|\mathbb{R}|,\mathbb{R})$.

Then $\det(r)\colon |\mathbb{R}| \to \mathbb{R}$ is the function constant at 1, which is invertible with respect to the product of the ring $\mathcal{R}$ (not with respect to composition!), as it never annihilates. Indeed, the matrix of functions r is invertible in the ring $M_2(\mathcal{R})$, with $r^{-1}(\varphi) = r(-\varphi)$.

2.6.8 *Linear differential equations with constant coefficients

Let $C^\infty(\mathbb{R})$ be the real vector space of C^∞-functions $\mathbb{R} \to \mathbb{R}$ (continuous, with all their derivatives). We write the n-th derivative of a function $\varphi \in C^\infty(\mathbb{R})$ as $D^n(\varphi)$, or $\varphi^{(n)}$, including $D(\varphi) = \varphi'$ and the original function $D^0(\varphi) = \varphi$.

A linear differential equation of order $n > 0$, with constant coefficients $a_i \in \mathbb{R}$ and known term $f \in C^\infty(\mathbb{R})$, is written as:

$$y^{(n)} + a_{n-1}y^{(n-1)} + \ldots + a_1 y' + a_0 y \;=\; f. \tag{2.115}$$

A *solution* is any C^∞-function $y\colon \mathbb{R} \to \mathbb{R}$ that satisfies it. For instance, the sine and cosine functions are solutions of the equation $y'' + y = 0$. The Cauchy Theorem on differential equations says that

- for every point $x_0 \in \mathbb{R}$ and every vector $(y_0, y_1, \ldots, y_{n-1}) \in \mathbb{R}^n$ there is precisely one solution y of (2.115) such that

$$y(x_0) = y_0,\quad y'(x_0) = y_1, \ldots,\quad y^{(n-1)}(x_0) = y_{n-1}.$$

All this can be viewed within the previous theory of linear equations on a commutative unital ring.

We equip the $\mathbb{R}$-vector space $C^\infty(\mathbb{R})$ with the linear endomorphism $D\colon C^\infty(\mathbb{R}) \to C^\infty(\mathbb{R})$ of derivation. As we have seen in Exercise 2.5.9, this makes $C^\infty(\mathbb{R})$ into a module on the integral ring $R = \mathbb{R}[D]$ of polynomials in the 'operator' D.

The differential equation (2.115) can thus be rewritten as a linear equation in the variable y, in the R-module $C^\infty(\mathbb{R})$

$$P(D).y = f, \qquad P(D) = D^n + a_{n-1}D^{n-1} + ... + a_1 D + a_0. \qquad (2.116)$$

The associated homogeneous equation is

$$P(D).y = 0. \qquad (2.117)$$

Now, the solutions of (2.117) form a sub-R-module $V \subset C^\infty(\mathbb{R})$, i.e. an $\mathbb{R}$-linear subspace stable under derivation. The Cauchy Theorem says that, for each $x_0 \in \mathbb{R}$, the $\mathbb{R}$-linear mapping

$$V \to \mathbb{R}^n, \qquad y \mapsto (y(x_0), y'(x_0), ..., y^{(n-1)}(x_0)), \qquad (2.118)$$

is an $\mathbb{R}$-isomorphism, and proves that V is a vector space of dimension n on the real field.

(A basis of V over $\mathbb{R}$ can be easily found, if all the complex roots of the *algebraic equation* $P(z) = 0$ are known, with their multiplicity.)

Now, if $\overline{y}$ is a particular solution of equation (2.116), all of them form the coset $\overline{y} + V$, an affine subspace of $C^\infty(\mathbb{R})$.

2.7 Constructing the complex field

We construct the complex field $\mathbb{C}$, starting from the real field $\mathbb{R}$.

We begin by an elementary construction, that only uses the basic notions reviewed in Section 1.1. The sine and cosine functions are used, to express a complex number in polar form.

Then we deal with more elaborate constructions of $\mathbb{C}$, as an algebraic extension of $\mathbb{R}$, or a set of real 2×2-matrices.

2.7.1 The original idea

In the Italian Renaissance, explicit formulas were found for the roots of the algebraic equations of degree 3 and 4. These formulas introduced an *imaginary* root $\underline{i}$ of the polynomial $X^2 + 1$, and *complex numbers* of the form $a + \underline{i}b$ (with $a, b \in \mathbb{R}$).

The roots are expressed in terms of these new 'numbers'. Computing them in a particular case, the imaginary terms can cancel (or not) and some roots can turn out to be real numbers. Later, this extension became a fundamental innovation, in Algebra and Functional Analysis.

The operations in the extended field $\mathbb{C}$ have an obvious form, deduced from the usual properties (of commutative rings) and the relation $\underline{i}^2 = -1$

$$\begin{aligned}(a+\underline{i}b)+(c+\underline{i}d) &= (a+c)+\underline{i}(b+d),\\ (a+\underline{i}b).(c+\underline{i}d) &= ac+\underline{i}bc+\underline{i}ad+\underline{i}^2bd \\ &= (ac-bd)+\underline{i}(bc+ad).\end{aligned} \tag{2.119}$$

One could define $\mathbb{C}$ in this way. Yet this approach is unclear, because addition and multiplication play a triple role: in the expression $a+\underline{i}b$ of an element of the extension, as operations on the new elements, and as operations on the real numbers.

It is thus preferable to replace the expression $a+\underline{i}b$ with the pair $(a,b) \in \mathbb{R}^2$.

2.7.2 The elementary construction

We start from the abelian group $\mathbb{R}^2$, of pairs of real numbers:

$$\begin{gathered}(a,b)+(c,d) = (a+c, b+d),\\ 0 = (0,0), \qquad -(a,b) = (-a,-b).\end{gathered} \tag{2.120}$$

We equip it with a product (suggested by the second formula in (2.119))

$$(a,b).(c,d) = (ac-bd, bc+ad), \tag{2.121}$$

and write as $\mathbb{C}$ the resulting structure.

With these operations the set $\mathbb{C}$ is a field (see Exercise (a)), with the following unit and inverses (for $(a,b) \neq 0$)

$$1 = (1,0), \qquad (a,b)^{-1} = (a/(a^2+b^2), -b/(a^2+b^2)). \tag{2.122}$$

We embed $\mathbb{R}$ in $\mathbb{C}$, by the injective homomorphism

$$\mathbb{R} \to \mathbb{C}, \qquad a \mapsto (a,0), \tag{2.123}$$

and we identify $a = (a,0)$. We also write $\underline{i} = (0,1)$, so that the complex numbers can be rewritten in the form

$$(a,b) = (a,0)+(0,1).(b,0) = a+\underline{i}b, \tag{2.124}$$

with obvious variations, as convenient, like $3-2\underline{i} = 3+\underline{i}(-2)$. The operations of $\mathbb{C}$ can now be rewritten as in (2.119).

The real numbers a, b are respectively called the *real part* and the *imaginary coefficient* of $a+\underline{i}b$

$$\underline{\mathrm{Re}}(a+\underline{i}b) = a, \qquad \underline{\mathrm{Im}}(a+\underline{i}b) = b. \tag{2.125}$$

The complex number $\underline{i}$ is a root of the polynomial $X^2 + 1$

$$\underline{i}^2 = (0,1).(0,1) = (-1,0) = -1. \tag{2.126}$$

The number $-\underline{i}$ is also a root of this polynomial. There are no others, by Theorem 2.5.5(a).

Exercises and complements. (a) $\mathbb{C}$ satisfies the axioms of fields, with unit and inverses as in (2.122).

(b) One cannot make $\mathbb{C}$ into a totally ordered field (defined in 1.1.8).

*(c) Complex numbers can be extended to *quaternions*

$$a + \underline{i}b + \underline{j}c + \underline{k}d \qquad (a, b, c, d \in \mathbb{R}),$$

introduced by William R. Hamilton, in 1843. They form a *non-commutative* $\mathbb{R}$-algebra $\mathbb{H}$ of dimension 4, where $\underline{i}.\underline{j} = \underline{k} = -\underline{j}.\underline{i}$ and $\underline{i}^2 = \underline{j}^2 = \underline{k}^2 = -1$.

$\mathbb{H}$ is a *division ring*, or *skew-field*: it satisfies all the axioms of fields except the commutativity of the product (see [Ko2]).

2.7.3 The complex plane

Since $\mathbb{C} = \mathbb{R} \times \mathbb{R}$ (as a set), we can represent the complex number $z = x + \underline{i}y$ as the point (x, y) of a cartesian plane (i.e. a plane with a system of cartesian coordinates, assumed to be orthogonal monometric)

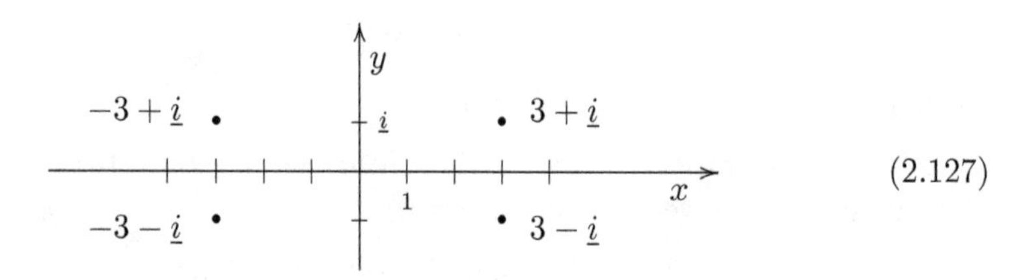

(2.127)

The real numbers $a = a + \underline{i}0$ are thus arranged on the x-axis, and the *imaginary* numbers $b\underline{i} = 0 + \underline{i}b$ on the y-axis. The plane is now called the *complex plane*, or the *Gauss plane*, or the *Argand plane*.

The *modulus* $|z|$ of the complex number $z = x + \underline{i}y$ is the euclidean distance of the point z from the origin 0, in the complex plane

$$|z| = \sqrt{x^2 + y^2} \geqslant 0 \qquad (\text{for } z = x + \underline{i}y), \tag{2.128}$$

and $z = 0$ if and only if $|z| = 0$. (This extends the modulus of a real number.)

The *conjugate* $\overline{z}$ of a complex number z has the same real part and opposite imaginary coefficient

$$\overline{z} = x - \underline{i}y \qquad (\text{for } z = x + \underline{i}y). \tag{2.129}$$

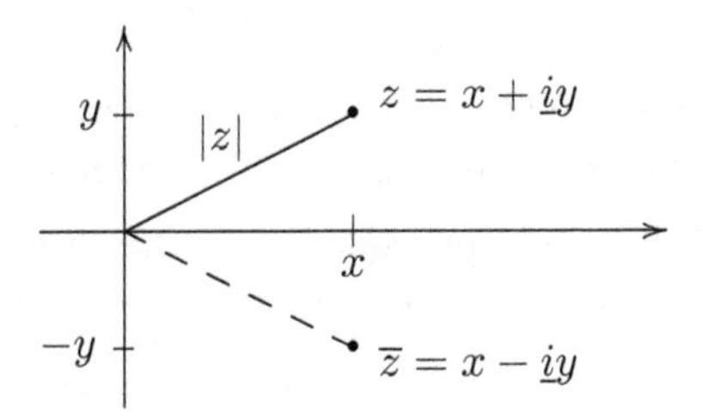

In the complex plane, z and $\overline{z}$ are symmetric with respect to the real axis, and $|z| = |\overline{z}|$. The conjugate has an important relationship with modulus and inverse:

$$\begin{aligned} z\overline{z} &= (x + \underline{i}y)(x - \underline{i}y) = x^2 - (\underline{i}y)^2 = x^2 + y^2 = |z|^2, \\ z^{-1} &= \overline{z}/|z|^2 = (x - \underline{i}y)/(x^2 + y^2) \qquad (\text{for } z \neq 0). \end{aligned} \tag{2.130}$$

2.7.4 Polar coordinates

In the complex plane, the point z can be represented in polar coordinates, with *radial coordinate* $\rho = |z| \geqslant 0$ and *angular coordinate* $\varphi \in \mathbb{R}$ (expressed in radians)

$$\begin{gathered} z = x + \underline{i}y = \rho(\cos\varphi + \underline{i}\sin\varphi), \\ \rho = |z| = \sqrt{x^2 + y^2}, \\ \cos\varphi = \underline{\mathrm{Re}}(z)/|z|, \qquad \sin\varphi = \underline{\mathrm{Im}}(z)/|z| \qquad (z \neq 0) \end{gathered} \tag{2.131}$$

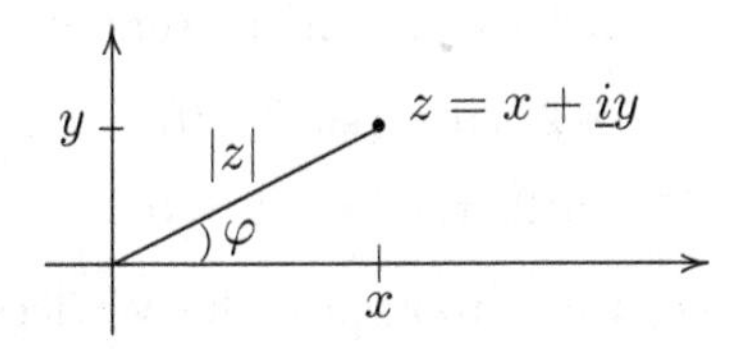

There is the usual 'ambiguity' of polar coordinates: the radial coordinate $\rho = |z| \geqslant 0$ is determined by z, but the angular coordinate φ is only determined up to an additional term $2k\pi \in 2\pi\mathbb{Z}$ (as a coset in the additive group $\mathbb{R}/2\pi\mathbb{Z}$) when $z \neq 0$, and is even arbitrary for $z = 0$.

Restricting the angular coordinate to the interval $[0, 2\pi[$ would reduce this ambiguity, *but is ineffective*, because computations get easily out of these bounds – as the next formula shows.

In polar coordinates, the product of $z = \rho(\cos\varphi + \underline{i}\sin\varphi)$ and $w = \rho'(\cos\vartheta + \underline{i}\sin\vartheta)$ is expressed as:

$$z.w = \rho\rho'(\cos(\varphi + \vartheta) + \underline{i}\,\sin(\varphi + \vartheta)), \tag{2.132}$$

where:

- the modulus is the product of the moduli of the factors,
- the angular coordinate is the sum of (any) angular coordinates of z, w.

This is a straightforward consequence of the product formula (2.121), together with the sum formulas of the trigonometric functions:

$$\begin{aligned}\cos(\varphi + \vartheta) &= \cos\varphi.\cos\vartheta - \sin\varphi.\sin\vartheta,\\ \sin(\varphi + \vartheta) &= \sin\varphi.\cos\vartheta + \cos\varphi.\sin\vartheta.\end{aligned} \tag{2.133}$$

(The other way round, formula (2.132) is easy to remember and allows us to reconstruct the sum formulas, from the product of complex numbers.)

2.7.5 *Complex roots*

From the product formula (2.132) we get the following expression, for the n-th power of a complex number $z = \rho(\cos\varphi + \underline{i}\sin\varphi)$

$$z^n = \rho^n(\cos(n\varphi) + \underline{i}\sin(n\varphi)). \tag{2.134}$$

For $n > 0$, the complex number z is an n-th root of 1, that is

$$z^n = 1 = 1(\cos 0 + \underline{i}\sin 0),$$

if and only if its polar coordinates satisfy the conditions:

$$\rho = 1, \qquad \varphi = 2k\pi/n \qquad (\text{for some } k \in \mathbb{Z}). \tag{2.135}$$

Two solutions corresponding to $k, k' \in \mathbb{Z}$ coincide if and only if $k - k' \in n\mathbb{Z}$. There are thus n distinct roots – the powers of a *generating root* w

$$\begin{aligned} w &= \cos 2\pi/n + \underline{i}\sin 2\pi/n,\\ w^k &= \cos 2k\pi/n + \underline{i}\sin 2k\pi/n \qquad (k = 0, 1, ..., n-1).\end{aligned} \tag{2.136}$$

In the complex plane, these roots are the vertices of a regular n-gon inscribed in the standard circle, with one vertex at 1. For instance, the six complex roots of the equation $z^6 = 1$ form the vertices of a regular hexagon, the powers of $w = \cos 2\pi/6 + \underline{i}\sin 2\pi/6 = 1/2 + \underline{i}\sqrt{3}/2$

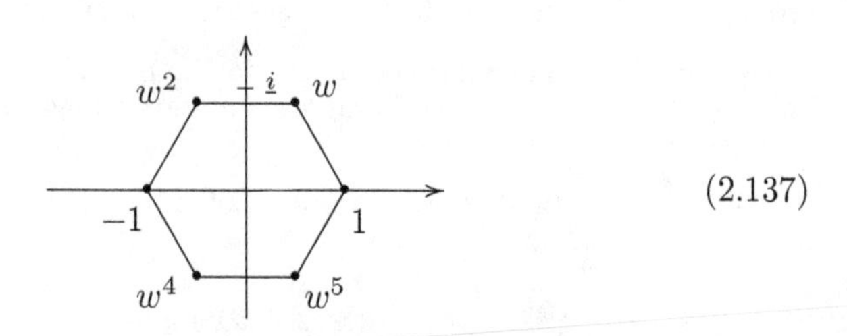

(2.137)

*More generally, the n-th roots of a complex number $z = \rho(\cos\varphi + \underline{i}\sin\varphi) \neq 0$ have modulus $|z|^{1/n}$ (a real n-th root of a positive number) and angular coordinate $(\varphi + 2k\pi)/n$; we get all of them (once), taking $k = 0, 1, ..., n-1$.

We are now relying on another elementary fact about continuous real functions: the invertibility of the n-th power function on real positive numbers, which is well known in Calculus (and can be found in Exercise 4.1.9(c)).*

2.7.6 Some multiplicative groups

All this is obvious, but becomes clearer using what we know of abelian groups, from Section 1.3. In the multiplicative group $\mathbb{C}^*$ of non-zero complex numbers, the standard circle of the complex plane

$$\mathbb{S}^1 = \{z \in \mathbb{C} \mid |z| = 1\}, \tag{2.138}$$

forms a subgroup, the *circle group* of the unimodular complex numbers.

The n-th power

$$\mathbb{S}^1 \to \mathbb{S}^1, \qquad z \mapsto z^n \qquad (n \geqslant 1), \tag{2.139}$$

is a homomorphism; its kernel is a subgroup, formed by the n-th roots of 1 in $\mathbb{C}$ computed above

$$\begin{aligned} U_n &= \{z \in \mathbb{C} \mid z^n = 1\} \\ &= \{\cos 2k\pi/n + \underline{i}\sin 2k\pi/n \mid k = 0, 1, ..., n-1\}. \end{aligned} \tag{2.140}$$

U_n is generated by the 'primitive' n-th root of 1

$$w_n = \cos 2\pi/n + \underline{i}\sin 2\pi/n,$$

and therefore a cyclic subgroup of $\mathbb{S}^1$, of order n: U_n is canonically isomorphic to the (additive) cyclic group $\mathbb{Z}/n$, sending $(w_n)^k$ to $[k]$.

It is also interesting to note that there is an isomorphism of commutative groups

$$\mathbb{R}^*_+ \times \mathbb{S}^1 \to \mathbb{C}^*, \qquad (\rho, z) \mapsto \rho z, \tag{2.141}$$

where $\mathbb{R}^*_+ =]0, +\infty[$ is the multiplicative group of positive real numbers.

2.7.7 The complex field as an algebraic extension

Using the theory of algebraic extensions of a field, developed in Section 2.5, one can define $\mathbb{C}$ as the quotient

$$\mathbb{C} = \mathbb{R}[X] \,/\, (X^2 + 1) \tag{2.142}$$

of the polynomial ring $\mathbb{R}[X]$ modulo the ideal generated by the irreducible polynomial $f = X^2 + 1$.

The polynomial f has no roots in $\mathbb{R}$, because $x^2 + 1 \geqslant 1$ for all $x \in \mathbb{R}$. Therefore f cannot have a factor of degree 1 (which would have a root) and is irreducible: in a factorisation $f = gh$, one of the factors has degree 0 (and is invertible) and the other has degree 2. (All this was already used, in Exercise 2.5.4(f).)

But $\mathbb{R}[X]$ is a principal ideal domain, whence (X^2+1) is a maximal ideal and the quotient (2.142) is a field. We define $\underline{i}$ as the equivalence class of the indeterminate X in the quotient, so that $\underline{i}^2 + 1 = 0$.

The embedding of $\mathbb{R}$ in $\mathbb{R}[X]$ induces a field-homomorphism $\mathbb{R} \to \mathbb{C}$, necessarily injective, which allows us to embed $\mathbb{R}$ into $\mathbb{C}$.

Finally, $\mathbb{C} = \mathbb{R}[\underline{i}]$ is an algebraic extension of $\mathbb{R}$, with minimal polynomial $f = X^2 + 1$ in $\mathbb{R}[X]$.

2.7.8 *Complex numbers as square matrices*

The injective mapping

$$\mathbb{C} \to M_2(\mathbb{R}), \qquad x + \underline{i}y \mapsto A(x,y) = \begin{pmatrix} x & -y \\ y & x \end{pmatrix}, \tag{2.143}$$

is easily seen to be a homomorphism of unital rings. The preservation of the product comes from:

$$\begin{pmatrix} a & -b \\ b & a \end{pmatrix} \cdot \begin{pmatrix} c & -d \\ d & c \end{pmatrix} = \begin{pmatrix} ac - bd & -ad - bc \\ bc + ad & -bd + ac \end{pmatrix}. \tag{2.144}$$

Therefore the field $\mathbb{C}$ is isomorphic to the subring $C \subset M_2(\mathbb{R})$ formed by all the real 2×2-matrices (a_{ij}) of the form (2.143), which means that $a_{11} = a_{22}$ and $a_{12} = -a_{21}$.

One can also define $\mathbb{C}$ in this way: C is a unital subring of $M_2(\mathbb{R})$, and a field in its own right, because $\det(A(x,y)) = x^2 + y^2$ is zero if and only if $A(x,y) = 0$.

The idea of representing the complex numbers in this way comes out of the product formula in polar coordinates, that we have seen in (2.132)

$$z.w = \rho\rho'(\cos(\varphi + \vartheta) + \underline{i}\,\sin(\varphi + \vartheta)). \tag{2.145}$$

The multiplication $w \mapsto zw$ by z amounts thus to the composite of two transformations of the complex plane (that commute with each other)

- a scaling by a factor $\rho = |z| \geqslant 0$ (a homothety, when $z \neq 0$),

- a rotation about the origin, of φ radians.

As we have seen in (2.112), this rotation is expressed by the matrix

$$r(\varphi) = \begin{pmatrix} \cos\varphi & -\sin\varphi \\ \sin\varphi & \cos\varphi \end{pmatrix} \in \mathrm{SO}(2,\mathbb{R}). \tag{2.146}$$

Multiplying this matrix by the scaling factor $\rho \geqslant 0$ we get an arbitrary matrix of C.

2.7.9 *The complex exponential

The reader may know that the exponential of the complex number $z = x + \underline{i}y$ is defined as

$$e^z = e^x(\cos y + \underline{i}\sin y), \tag{2.147}$$

so that the exponential mapping is still a homomorphism of an additive group into a multiplicative one (as for real numbers, in Exercise 1.5.2(f))

$$\mathbb{C} \to \mathbb{C}^*, \qquad z \mapsto e^z. \tag{2.148}$$

In fact, for $w = x' + \underline{i}y'$

$$\begin{aligned} e^z.e^w &= e^x e^y(\cos y + \underline{i}\sin y).(\cos y' + \underline{i}\sin y') \\ &= e^{x+y}(\cos(y+y') + \underline{i}\sin(y+y')) = e^{z+w}. \end{aligned}$$

This extension of the real exponential e^x is no longer injective: the kernel of the homomorphism (2.148) is not trivial

$$e^{x+\underline{i}y} = 1 \quad \Leftrightarrow \quad (x = 0 \text{ and } y \in 2\pi\mathbb{Z}).$$

Restricting (2.148), we get a homomorphism

$$f\colon \mathbb{R} \to \mathbb{C}^*, \qquad f(t) = \cos t + \underline{i}\sin t = e^{\underline{i}t}, \tag{2.149}$$

of the additive group $(\mathbb{R}, +)$ into the multiplicative group $\mathbb{C}^*$.

Its kernel $f^{-1}\{1\} = 2\pi\mathbb{Z}$ is the subgroup of $(\mathbb{R}, +)$ generated by 2π. The image is the circle subgroup $\mathbb{S}^1 = \{z \in \mathbb{C} \mid |z| = 1\}$ of $\mathbb{C}^*$, and f induces an isomorphism of groups (from additive to multiplicative notation, again)

$$g\colon \mathbb{R}/(2\pi\mathbb{Z}) \to \mathbb{S}^1, \qquad g([t]) = \cos t + \underline{i}\sin t = e^{\underline{i}t}. \tag{2.150}$$

The Taylor series $e^x = \sum_n x^n/n!$ of the real exponential extends to a Taylor series $e^z = \sum_n z^n/n!$ in a complex variable.

3

Topological structures, I

Topology is also a main discipline of mathematics: it explores the idea of continuous transformations, 'without jumps', from a space to another. It also tells us when two spaces have 'the same topological shape', meaning that we can transform one into the other by a bijective mapping, which is continuous as well as its inverse.

In order to give a preliminary idea of the theory, we begin by studying 'spaces' that are contained in the euclidean $\mathbb{R}^n$, where approaching a point and continuous transformations have an intuitive meanining. From this relatively concrete situation we abstract the features which detect continuity: neighbourhoods, open sets, closed sets, etc.

Then we give the basic definitions, about topological spaces and continuous mappings, in Sections 3.2 and 3.3. Subspaces and quotients are dealt with in Section 3.4, topological products and sums in Section 3.5, limits and the Hausdorff separation property in Section 3.6.

Other fundamental topics of Topology are deferred to Chapter 4.

Hystorically, the notion of continuity is an old one, but a formal definition for real functions was given relatively late, by Bernard Bolzano in 1817. General topological spaces were introduced by Felix Hausdorff, in 1914 [Hd], although in a form stronger than the current one: see Section 3.6. Then the discipline underwent an impressive growth, and its core was already complete in the late 1930's.

Literature. To further explore this domain, there are many books, like Munkres [Mu], Kelley [Ke] and Bourbaki [Bou2]. The last is particularly advisable for filters, uniform continuity and uniform spaces.

Algebraic Topology studies topological problems by translating them into simpler algebraic problems. For this fascinating field, rich in research, there are elementary textbooks like Vick [Vi] and Massey [Mas2], and more advanced ones, like Hilton–Wylie [HiW], Spanier [Sp] and Hatcher [Hat]; the last is freely downloadable.

3.1 Introducing continuity

We want to study 'continuous' mappings $f\colon X \to Y$ between 'spaces'; a space should be a set equipped with a 'topological structure', or 'topology', sufficient to make sense of this, and sufficiently general to produce a good theory.

This section investigates spaces of a 'concrete type', highlighting the aspects that will lead to a general notion of topological space, and of continuous mapping between such spaces. When some intuition is developed, the reader may prefer to jump to the next section, to find a more formal way of going on.

3.1.1 Euclidean spaces

We look again for inspiration at the set $\mathbb{R}$ of real numbers, and more generally at the sets $\mathbb{R}^n$ (for $n \geqslant 1$), without focusing on their linear structure – which in this chapter is only used as a tool. In particular, $\mathbb{R}^3$ can be viewed as representing the physical space (in a classical sense), where we have fixed a system of cartesian coordinates, so that each point corresponds, bijectively, to a triple (x_1, x_2, x_3) of real numbers.

More generally, an arbitrary subset $X \subset \mathbb{R}^n$ is considered here as a *euclidean space*, and an element $x = (x_1, ..., x_n) \in X$ is also called a *point* of X.

An important role will be played by the *euclidean distance* $d(x, y)$, or *euclidean metric*, between two points $x = (x_1, ..., x_n)$, $y = (y_1, ..., y_n)$

$$d(x,y) = \sqrt{\Sigma_i\,(x_i - y_i)^2} \qquad (x, y \in \mathbb{R}^n), \tag{3.1}$$

which reduces to $d(x, y) = |x - y|$ for $n = 1$ (see 1.1.9(d)).

This is a real number $\geqslant 0$, and $d(x, y) = d(y, x)$. The *triangle inequality* says that

$$d(x,y) + d(y,z) \geqslant d(x,z) \qquad (x, y, z \in \mathbb{R}^n), \tag{3.2}$$

a well known property of elementary geometry: for $n \geqslant 2$ it comes out of the triangle of vertices x, y, z, in a plane of $\mathbb{R}^n$. (See Exercise (b) for an analytic proof.)

In particular, the *euclidean norm* $||x||$ of a point x is its distance from the origin $0 = (0, ..., 0)$ of $\mathbb{R}^n$

$$||x|| = d(0, x) = (\,\Sigma_i\, x_i^2\,)^{1/2} \qquad (x \in \mathbb{R}^n), \tag{3.3}$$

and $d(x, y) = ||x - y||$. For $n = 1$, $||x|| = |x|$.

Intervals of the real line are always denoted by square brackets, as specified in 1.7.2. When we write $x \leqslant y$, or $x < y$, we always mean the natural order between real numbers, unless some other ordered set is explicitly mentioned.

The complex plane $\mathbb{C}$, studied in Section 2.7, can be identified with the euclidean plane $\mathbb{R}^2$. The complex modulus $|z|$ amounts to the euclidean norm of $\mathbb{R}^2$, and the distance in $\mathbb{C}$ can be expressed as

$$d(z, w) = |z - w| \qquad (z, w \in \mathbb{C}). \tag{3.4}$$

Exercises and complements. (a) Extending a property of the modulus in $\mathbb{R}$ (see 1.1.9(d)), we have

$$||x + y|| \leqslant ||x|| + ||y|| \qquad (x, y \in \mathbb{R}^n). \tag{3.5}$$

The proof is not easy and the reader is not expected to find it, but can be interested in reading it, in Chapter 7.

(b) (*Triangle inequality*) The inequality (3.2) follows easily from (3.5).

(c) The spaces $\mathbb{R}^n$ can be viewed as nested, identifying each $(x_1, ..., x_n) \in \mathbb{R}^n$ with $(x_1, ..., x_n, 0) \in \mathbb{R}^{n+1}$. Distance and norm are invariant with respect to this embedding.

3.1.2 The shape of a euclidean space

The continuity of a mapping $f \colon X \to Y$, from a subset $X \subset \mathbb{R}^n$ to $Y \subset \mathbb{R}^m$, is easy to define using the euclidean distance in X and Y. Likely, it is already known to the reader.

We say that the mapping f is *continuous at the point* $a \in X$ if:

(*) for every $\varepsilon > 0$ there is some $\delta > 0$ such that, if $x \in X$ and $d(x, a) < \delta$, then $d(f(x), f(a)) < \varepsilon$.

This formalises the idea that, as the point x *tends to* a in X, the point $f(x)$ *tends to* $f(a)$ in Y. The mapping f is *continuous* (on its domain X) if it is at each $a \in X$. (Let us note that, if we replace Y with the whole space $\mathbb{R}^m$, we get an equivalent condition.)

Continuous mappings between euclidean spaces are stable under composition, and include all the identity mappings of these spaces. A *homeomorphism* of euclidean spaces is a continuous mapping $f \colon X \to Y$ that has a continuous inverse $g \colon Y \to X$.

According to general terminology, this is an 'isomorphism' of the structure we are considering, but the historical term 'homeomorphism' is still preferred, and quite useful when different structures – e.g. topological and algebraic structures – are present on the same sets. (For instance, the translation $f\colon \mathbb{R} \to \mathbb{R}$, $f(x) = x + 1$ is a homeomorphism, but is not an isomorphism of fields or abelian groups.)

Two euclidean spaces X, Y are *homeomorphic* if there exists a homeomorphism $X \to Y$. We also say that X, Y *have the same (topological) shape.*

To prove that two given spaces have the same shape can require long computations, in order to build a homeomorphism between them, but is generally a 'confined' problem.

To prove that they are *not* homeomorphic can be difficult, and enticing, even when we clearly 'see' that they have a different shape: we are to prove that *there cannot exist* a homeomorphism between them. The proof is generally based on a 'topological property' (invariant up to homeomorphism) which holds in one of them but not in the other. This can require advanced parts of the theory, or prompt a researcher to establish such a discipline.

3.1.3 Examples and complements

Nevertheless, we can already have some idea of these problems, and of some solution, at an informal level; precise definitions and theorems will be developed below, or are referred to.

(a) The real line is homeomorphic to the open interval $]0, +\infty[$ as shown by the exponential function and natural logarithm, which are inverse continuous functions. $\mathbb{R}$ is also homeomorphic to the open interval $I =]-\pi/2, \pi/2[$, as shown by the function $\arctan\colon I \to \mathbb{R}$ and the restriction of the tangent function to the interval I.

(All this can also be proved using rational functions: see Exercise 3.4.4(d).)

The reader can already see that the topological shape we are investigating forgets relevant parts of the information contained in the euclidean distance: for instance, the previous interval I is 'metrically bounded', as $d(x, y) < \pi$, for all $x, y \in I$, while the real line is not. More importantly, we will see that $\mathbb{R}$ is 'metrically complete', while I is not (in Section 4.5). The topological shape is concerned with more basic aspects.

(b) (*Diameter*) Every subset $A \subset \mathbb{R}^n$ has a *diameter* in $\overline{\mathbb{R}}$

$$\mathrm{diam}(A) = \sup\{d(x, y) \mid x, y \in A\} \in \overline{\mathbb{R}}, \qquad (3.6)$$

and is said to be *bounded* if its diameter is not $+\infty$, which means that the real number $d(x, y)$ is upper bounded, when $x, y \in A$. For $n = 1$ we find again the bounded subsets of the line, as defined in 1.1.8.

The diameter of a circle or a disc is its geometric diameter; the diameter of a rectangle is the length of its diagonal.

(c) The *standard circle* $\mathbb{S}^1$ is homeomorphic to the space A drawn below (the union of the four edges of a square)

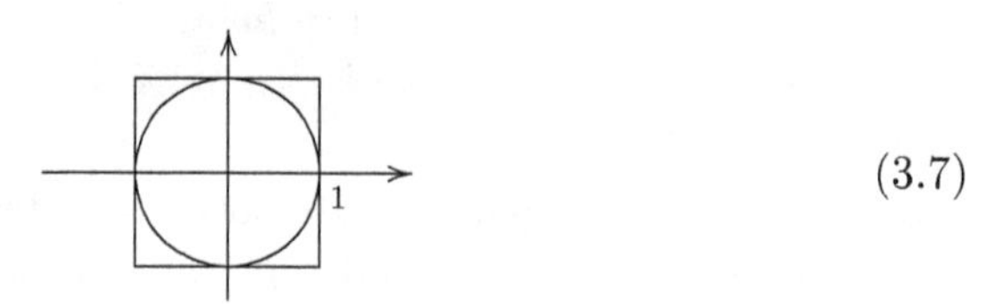

(3.7)

$$\mathbb{S}^1 = \{(x_1, x_2) \in \mathbb{R}^2 \mid x_1^2 + x_2^2 = 1\} \subset \mathbb{R}^2,$$
$$A = \{(x_1, x_2) \in \mathbb{R}^2 \mid \max(|x_1|, |x_2|) = 1\} \subset \mathbb{R}^2,$$

showing once more that topological shape ignores characters – here the 'corners' – that are important in other contexts, like Differential Geometry.

In fact, it is easy to give a continuous bijective mapping from A to the circle, by a geometric 'projection' called *normalisation*

$$p\colon A \to \mathbb{S}^1, \qquad p(x) = x/||x||, \tag{3.8}$$

using the linear structure of $\mathbb{R}^2$ and the euclidean norm. The continuity of the inverse mapping is also geometrically obvious; to prove it analytically would require some work, that we can spare: both our spaces are 'compact Hausdorff spaces', and the theory will show that in this context all continuous bijective mappings are homeomorphisms (Theorem 4.2.7).

(d) The sphere (the surface of a ball) and the 'torus' (the surface of a life buoy, examined in 4.4.6) have a different shape

Sphere and Torus (3.9)

This looks quite obvious to our intuition, but is not easy to prove.

Loosely speaking, the best way to approach this fact is to think of a 'loop' (a closed path) on these surfaces. On the sphere, every loop can be continuously deformed to a point (a constant path), but on the torus we can easily guess that a tight rope around the life buoy cannot be further contracted (staying in the space we are considering).

All this can be made precise with elementary results of Algebraic Topology: the first homology group (or the fundamental group) of the sphere is trivial, which is not true of the torus [Vi, Mas2, HiW, Sp, Hat].

(Drawing the sphere in (3.9), we have also drawn a great circle, to suggest the curvature of the surface.)

(e) One can also use Algebraic Topology to prove the Theorem of Topological Dimension: if two euclidean spaces $\mathbb{R}^m$ and $\mathbb{R}^n$ are homeomorphic, then $m = n$. The reader will note that this is a crucial fact, without which we could not speak of 'dimension' in a topological sense. (Something more on this point will be said in 5.5.7(d).)

The invariance of *linear* dimension of the vector spaces $\mathbb{R}^n$ is more easily proved (in Theorem 2.4.5), but is a different thing.

3.1.4 Nearness and continuity for euclidean spaces

We come back to a safer, less adventurous analysis of euclidean spaces $X \subset \mathbb{R}^n$. We will restrict our study of continuity to real-valued functions $X \to \mathbb{R}$, because the continuity of a mapping $f: X \to \mathbb{R}^m$ can be reduced to the continuity of its components $f_i: X \to \mathbb{R}$ (by the universal property of a topological product).

In the euclidean space $X \subset \mathbb{R}^n$, the *open ball* of centre $a \in X$ and radius $r > 0$ is defined as the subset

$$B_X(a, r) = \{x \in X \mid d(x, a) < r\}. \tag{3.10}$$

We also write $B_n(a, r)$ when $X = \mathbb{R}^n$. In particular, an open ball of the real line is an open interval of $\mathbb{R}$

$$B_1(a, r) = \{x \in \mathbb{R} \mid |x - a| < r\} =]a - r, a + r[, \tag{3.11}$$

while an open ball of $\mathbb{R}^2$ (or $\mathbb{C}$) is a disc without the boundary circle.

(The reader is warned that the term 'open ball' is geometrically adequate when $X = \mathbb{R}^n$, so that $B_n(a, r)$ is indeed an n-dimensional ball deprived of the binding sphere; but an 'open ball' of a proper subset X of $\mathbb{R}^n$ can be quite far from that: see Exercise 3.1.7(c). Nevertheless, this is the accepted name.)

We write as $\mathcal{B}_X(a) \subset \mathcal{P}X$ the collection of all the subsets $B_X(a, r)$, for $r > 0$. This collection is used as a 'measure of nearness' to the point a (in X): if $N \in \mathcal{B}_X(a)$, any point $x \in N$ is said to be *near to a of order N*.

Now the continuity of the function $f: X \to \mathbb{R}$ at the point a can be expressed in the following equivalent forms (which begin to make abstraction of the distance):

(i) for every $\varepsilon > 0$ there exists $\delta > 0$ such that $f(B_X(a, \delta)) \subset B_1(f(a), \varepsilon)$,

(ii) for every $N \in \mathcal{B}_1(f(a))$ there exists $N' \in \mathcal{B}_X(a)$ such that $f(N') \subset N$.

(In fact, the euclidean distance gives us a stronger, symmetric information: how near two points $x, y \in X$ are *to each other*. But this leads to a stronger notion of 'uniform continuity', that will be examined later, in Section 4.5.)

It is well known, from elementary Calculus, that any polynomial function

$f: \mathbb{R}^n \to \mathbb{R}$ in n variables is continuous. We will review this fact in 3.3.6. Here we begin by noting that any cartesian projection

$$p_i: \mathbb{R}^n \to \mathbb{R}, \qquad p_i(x_1, ..., x_n) = x_i \tag{3.12}$$

is continuous: in fact, for every $i = 1, ..., n$, we have: $d((x_i), (a_i)) \geqslant |x_i - a_i|$, so that, for every $\varepsilon > 0$, $p_i(B_n(a, \varepsilon)) \subset B_1(f(a), \varepsilon)$. In other words, property (i) is always satisfied taking $\delta = \varepsilon$ (or any smaller positive number).

3.1.5 Neighbourhoods, open and closed subsets

We are still examining an euclidean space $X \subset \mathbb{R}^n$. Nearness in X is based on the sets $\mathcal{B}_X(a)$ of open balls $B_X(a, r)$, at each point $a \in X$.

A *neighbourhood* of a point a in X is any subset N that contains an open ball $B_X(a, r)$, for some $r > 0$; intuitively, it contains all points of X 'sufficiently near' to a. We write as $\mathcal{N}_X(a)$, or just $\mathcal{N}(a)$, the set of all neighbourhoods of a point a in X. (The term 'neighbourhood' can be shortened to *nbd.*)

A subset $U \subset X$ is said to be an *open subset* (of X), or *open* in X, if it is a neighbourhood (in X) of any point $a \in U$, or equivalently:

- for every $a \in U$ there is some $r > 0$ such that $B_X(a, r) \subset U$.

A subset $C \subset X$ is said to be a *closed subset* (of X), or *closed* in X, if its complement $X \setminus C$ is open in X, or equivalently:

- if $a \in X \setminus C$, there is some $r > 0$ such that $B_X(a, r) \cap C = \emptyset$.

We write as $\mathcal{O}(X)$ the set of all open subsets of X, as $\mathcal{C}l(X)$ the set of all closed subsets.

Remarks, exercises and complements. (a) Being open (or closed) is a relative property, which only makes sense in a specified space; moreover, 'open' and 'not closed' are different properties, and none of them implies the other – generally.

Thus, there are (many!) subsets of $\mathbb{R}^n$ which are neither open nor closed, like the interval $[0, 1[$ in $\mathbb{R}$. On the other hand, for every euclidean space X, the empty subset and the total subset are both open and closed *in* X.

Taking $X = \mathbb{R}^n$, we will see in Section 4.1 that the only subsets which are both open and closed (in it!) are the empty and the total one – a non-obvious fact and actually the definition of a 'connected space'. On the other hand, in the euclidean space $X = \mathbb{R}^*$, the interval $]-\infty, 0[$ is also open and closed, as well as $]0, +\infty[$.

(b) Prove that an open ball $B_X(a, r)$ is indeed an open subset of X. Moreover, a subset $U \subset X$ is open if and only if it is a union of open balls of X.

(c) The subsets

$$\begin{aligned} D_X(a, r) &= \{x \in X \mid d(x, a) \leqslant r\}, \\ S_X(a, r) &= \{x \in X \mid d(a, x) = r\}, \end{aligned} \tag{3.13}$$

are respectively called the *disc*, or *closed ball*, and the *sphere* of centre a and radius $r > 0$. They are both closed subsets of X. When $X = \mathbb{R}^n$ they will also be written as $D_n(a, r)$ and $S_n(a, r)$.

(d) The sets of neighbourhoods in X satisfy the following fundamental properties (for every $a \in X$):

(Nb.1) a belongs to every element of $\mathcal{N}(a)$,

(Nb.2) every subset of X that contains an element of $\mathcal{N}(a)$ belongs to $\mathcal{N}(a)$,

(Nb.3) $\mathcal{N}(a)$ is stable in $\mathcal{P}X$ with respect to finite intersections,

(Nb.4) if $N \in \mathcal{N}(a)$, there is some subset $U \subset N$ such that $a \in U$ and, for every $b \in U$, $U \in \mathcal{N}(b)$.

(e) Open sets in X satisfy the following fundamental properties:

(Op.1) $\mathcal{O}(X)$ is stable in $\mathcal{P}X$ with respect to arbitrary unions,

(Op.2) $\mathcal{O}(X)$ is stable in $\mathcal{P}X$ with respect to finite intersections.

In particular, we are also saying that $\emptyset$ (the empty union of subsets of X) and X (the empty intersection of subsets of X) are open in X.

(f) Closed sets in X satisfy the following fundamental properties

(Cl.1) $\mathcal{C}l(X)$ is stable in $\mathcal{P}X$ with respect to arbitrary intersections,

(Cl.2) $\mathcal{C}l(X)$ is stable in $\mathcal{P}X$ with respect to finite unions.

In particular, $\emptyset$ and X are closed in X.

(g)(*Bounded subsets*) The subset $A \subset \mathbb{R}^n$ is bounded (as defined in 3.1.3) if and only if, equivalently:

- A is contained in some open ball $B_n(a, r)$,
- A is contained in some open ball $B_n(0, r)$,
- A is contained in some closed ball $D_n(a, r)$,
- A is contained in some closed ball $D_n(0, r)$.

3.1.6 Interior, closure and boundary

We examine now a subset A of the euclidean space $X \subset \mathbb{R}^n$, as a subset of X – not as a space in itself.

The *interior* of A (in X), written as A° or $\mathrm{int}_X A$, is defined as the union of all the open subsets of X contained in A. It is thus the largest open subset of X contained in A, and coincides with A if and only if A is open in X. A point $a \in A^\circ$ is said to be *interior* to A in X.

The *closure* of A (in X), written as $\overline{A}$, or A^-, or $\mathrm{cl}_X A$, is defined as the intersection of all the closed subsets of X containing A. It is thus the smallest closed subset of X containing A, and coincides with A if and only if A is closed in X. A point $a \in \overline{A}$ is said to be *adherent to* A in X.

The *boundary* of A (in X), written as $\partial_X A$ or ∂A, is defined as the set of points of X that are adherent to A and to its complement

$$\partial A = \overline{A} \cap (X \setminus A)^-, \tag{3.14}$$

and is closed in X.

(The boundary of a subset should not be confused with the boundary of a surface, or a manifold: the latter is an absolute notion, not depending of an environment, that will be briefly outlined in 4.4.8.)

3.1.7 *Exercises and complements*

A is a subset of the euclidean space $X \subset \mathbb{R}^n$ and $a \in X$.

(a) Prove that:

(i) $a \in A^\circ \;\Leftrightarrow\;$ A is a nbd of a $\;\Leftrightarrow\; B_X(a,r) \subset A$, for some $r > 0$,

(ii) $a \in \overline{A} \;\Leftrightarrow\;$ every nbd of a meets A $\;\Leftrightarrow\;$ every $B_X(a,r)$ meets A,

(iii) $a \in \partial A \;\Leftrightarrow\;$ every nbd of a meets A and $X \setminus A$,

(iv) $C_X(\overline{A}) = (C_X A)^\circ, \qquad C_X(A^\circ) = (C_X A)^-, \qquad \partial A = \partial(C_X A)$,

(v) $X = A^\circ \cup \partial A \cup (C_X A)^\circ$ (a disjoint union),

(vi) $\overline{A} = A^\circ \cup \partial A = A \cup \partial A$,

(vii) A is open in X $\;\Leftrightarrow\; A \cap \partial_X A = \emptyset$,

(viii) A is closed in X $\;\Leftrightarrow\; A \supset \partial_X A$.

(b) *For the total space* $X = \mathbb{R}^n$ $(n > 0)$ compute interior, closure and boundary of the following subsets of X

$$B_n(a,r), \qquad D_n(a,r), \qquad S_n(a,r),$$

proving – in particular – that $D_n(a,r)$ is the closure of $B_n(a,r)$, while the sphere $S_n(a,r)$ is the boundary of both $B_n(a,r)$ and $D_n(a,r)$.

(c) *These properties can fail for a euclidean space* X *properly contained in* $\mathbb{R}^n$: in particular, $D_X(a,r)$ need not be the closure of $B_X(a,r)$. It is sufficient to give examples for $n = 1$.

(d) Letting $n \geqslant 2$, compute interior, closure and boundary in $\mathbb{R}^n$ of the subset

$$A = \{x \in \mathbb{R}^n \mid 0 < x_1 < 1,\ x_2 = ... = x_n = 0\}.$$

(e) Let I be an interval of the real line $\mathbb{R}$, and consider if it is open or closed in $\mathbb{R}$, *according to the present 'topological' definitions.* When speaking of an open or closed interval of $\mathbb{R}$, it will always be in this sense – unless differently specified.

(f) The closure operator $\mathrm{cl}\colon \mathcal{P}X \to \mathcal{P}X$ satisfies the following fundamental properties (for all $A, B \subset X$):

(cl.1) $A \subset \mathrm{cl}(A)$,

(cl.2) $\mathrm{cl}(A) = \mathrm{cl}(\mathrm{cl}(A))$,

(cl.3) $\mathrm{cl}(\emptyset) = \emptyset$, $\quad \mathrm{cl}(A \cup B) = \mathrm{cl}(A) \cup \mathrm{cl}(B)$.

The last property amounts to saying that closure preserves finite unions. Therefore, it is monotone: $A \subset B$ implies $\mathrm{cl}(A) \subset \mathrm{cl}(B)$.

The interior operator has dual properties, related to the previous ones by taking complements in X.

3.1.8 ***Theorem*** (Characterisations of continuity)

Let $a \in X \subset \mathbb{R}^n$ and $f \colon X \to \mathbb{R}$ be a mapping.

(a) The mapping f is continuous at a if and only if, for every neighbourhood N of $f(a)$ in $\mathbb{R}$, the preimage $f^{-1}(N)$ is a neighbourhood of a in X.

(b) The continuity of f on X is expressed by the following equivalent properties:

(i) f is continuous at every $a \in X$,
(ii) the preimage of any open subset of $\mathbb{R}$ is open in X,
(iii) the preimage of any closed subset of $\mathbb{R}$ is closed in X,
(iv) for every $A \subset X$: $\quad f(\overline{A}) \subset (f(A))^-$.

Proof (a) Suppose that f is continuous at a, as reviewed in 3.1.4.

If N is a neighbourhood of $f(a)$ in $\mathbb{R}$, it contains an open interval $N' = B_1(f(a), \varepsilon)$ of radius ε, and there is some $\delta > 0$ such that $f(B_n(a, \delta)) \subset N'$; therefore $B_n(a, \delta) \subset f^{-1}(N') \subset f^{-1}(N)$, and the latter is a neighbourhood of a in X. The converse is similarly proved.

(b) First, it is easy to see that (i) $\Leftrightarrow$ (ii). Suppose that f is continuous at every $a \in X$, and let V be an open subset of $\mathbb{R}$; if $a \in f^{-1}(V)$, then $f(a) \in V$, which is a neighbourhood of each of its points; applying the previous point, $f^{-1}(V)$ is a neighbourhood of a in X.

Conversely, assuming (ii), and taking $a \in X$, every neighbourhood N of $f(a)$ contains an open neighbourhood V, whose preimage is an open neighbourhood of a contained in $f^{-1}(N)$.

(ii) $\Rightarrow$ (iii). Preimages commute with complements.

(iii) $\Rightarrow$ (iv). We have to prove that $\overline{A} \subset f^{-1}(C)$, where $C = (f(A))^-$. In fact C is closed in $\mathbb{R}$, whence $f^{-1}(C)$ is closed in X; but $A \subset f^{-1}(fA) \subset f^{-1}(C)$, whence $\overline{A} \subset f^{-1}(C)$.

(iv) $\Rightarrow$ (i). Take an open neighbourhood N of $f(a)$ in $\mathbb{R}$; we have to prove that $f^{-1}(N)$ is a neighbourhood of a in X. We denote complements (in $\mathbb{R}$ and in X) by $(-)^*$ and let

$$A = f^{-1}(N^*) = (f^{-1}(N))^*,$$

so that, using the hypothesis (iv) and the fact that N is open:

$$f(\overline{A}) \subset (f(A))^- = (f(f^{-1}(N^*)))^- \subset (N^*)^- = N^*.$$

Now, $f(a) \notin f(\overline{A})$ and $a \notin \overline{A}$. Using 3.1.7(iv), $a \in \overline{A}^* = (f^{-1}(N))^\circ$ is interior to $f^{-1}(N)$.

□

3.1.9 Exercises and complements

(a) Let $f \colon X \to \mathbb{R}$ be a continuous mapping, on $X \subset \mathbb{R}^n$. For every $\lambda \in \mathbb{R}$, consider the following subsets of X

$$\begin{aligned} A = \{x \in X \mid f(x) < \lambda\}, \qquad B = \{x \in X \mid f(x) = \lambda\}, \\ C = \{x \in X \mid f(x) \leqslant \lambda\}, \end{aligned} \tag{3.15}$$

and show that some of them are open and others are closed (for every continuous f).

(b) Show that each of them *can* be open and closed, for some choice of f.

(c) The euclidean space $\mathbb{I} = [0, 1]$ is called the *standard interval*, while

$$\mathbb{I}^n = [0, 1]^n = \{x \in \mathbb{R}^n \mid 0 \leqslant x_i \leqslant 1, \text{ for all } i\} \tag{3.16}$$

is the *standard cube* of dimension n. Prove that it is closed in $\mathbb{R}^n$.

In particular, the standard square $\mathbb{I}^2 = [0, 1]^2$ is represented in the left figure below

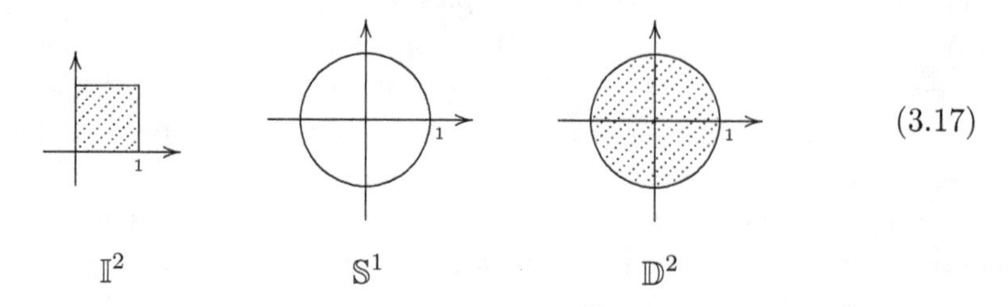

(3.17)

(d) We are particularly interested in the following subsets of $\mathbb{R}^n$, for $n > 0$:

$$\begin{aligned} &\mathbb{S}^{n-1} = S_n(0, 1) = \{x \in \mathbb{R}^n \mid \Sigma\, x_i^2 = 1\} \\ &\qquad \text{(the } \textit{standard sphere} \text{ of dimension } n - 1\text{)}, \\ &\mathbb{D}^n = D_n(0, 1) = \{x \in \mathbb{R}^n \mid \Sigma\, x_i^2 \leqslant 1\} \\ &\qquad \text{(the } \textit{standard disc}\text{, or closed ball, of dimension } n\text{)}. \end{aligned} \tag{3.18}$$

In particular, the standard circle $\mathbb{S}^1$ and the standard disc $\mathbb{D}^2$ are represented in the central and the right figure above.

We have already seen, in 3.1.7(b), that they are closed and $\mathbb{S}^{n-1}$ is the boundary of $\mathbb{D}^n$ in $\mathbb{R}^n$ (for $n > 0$). Moreover the interior of $\mathbb{D}^n$ is the open ball $\mathbb{D}^n \setminus \mathbb{S}^{n-1} = B_n(0, 1)$, and the interior of $\mathbb{S}^{n-1}$ is empty.

The superscripts of $\mathbb{S}^{n-1}$ and $\mathbb{D}^n$ do *not* refer to a cartesian power, but to 'topological dimension', which here is left at an intuitive level, as in 3.1.3(e). Formally, the expression 'standard sphere (or disc) of dimension k' should be taken as a whole, defined by the previous formula.

(e) For $n = 1$, $\mathbb{D}^1 = [-1, 1]$ is a closed interval and $\mathbb{S}^0 = \{-1, 1\}$ is the set of its endpoints.

We can also consider the singleton $\mathbb{R}^0 = \{*\}$, with $\mathbb{D}^0 = \{*\}$. (It may also be useful to let $\mathbb{S}^{-1} = \emptyset$.)

3.2 Topological spaces and continuous mappings

We now extend the basic topological notions, from the euclidean spaces to the general ones.

3.2.1 Main definitions

A *topological space* X is a set equipped with a *topology*, or *topological structure*: namely a set $\mathcal{O}(X) \subset \mathcal{P}X$ of subsets of X which satisfies the axioms (Op.1, 2) of 3.1.5(e): it is stable in $\mathcal{P}X$ under arbitrary unions and finite intersections.

We simply say a *space* when there is no risk of confusion (with vector spaces, affine spaces, Banach spaces, etc.)

The elements of $\mathcal{O}(X)$ are called *open (sub)sets* of X, and said to be *open in* X; in particular, the empty subset and the total subset are always open.

The underlying set (without any structure) can be written as $|X|$, but also here – as for algebraic structures – the distinction is left understood unless it is really needed. An element of X is often called a *point* (and we always mean $x \in |X|$, of course).

A subset $C \subset X$ is said to be *closed* (in X) if the complement $X \setminus C$ is open. In particular, the empty and the total subsets are always closed. (There can be other subsets of X that are both open and closed, depending on the space we are considering.)

Equivalently, we can define the topological structure of X by assigning a set $\mathcal{C}l(X) \subset \mathcal{P}X$ which satisfies the axioms (Cl.1, 2) of 3.1.5(f): it is stable in $\mathcal{P}X$ under arbitrary intersections and finite unions. Then its elements are called closed subsets, and the open subsets are – by definition – their complements in X.

Each set S has some topology. For instance, we can consider the *discrete topology* $D(S)$ where every subset is open (and closed). On the other hand, we can consider the *indiscrete*, or *chaotic*, *topology* $C(S)$ where the only open sets are $\emptyset$ and S (and the only closed sets are the same).

On each subset $X \subset \mathbb{R}^n$ we have already considered, in Section 3.1, an important topology, called the *euclidean topology* of X, where a subset U is open if, for every $a \in U$, it contains an open ball $B_X(a, r)$ *of* X, for some $r > 0$. With this topology, X is called a *euclidean space.*

Suppose now that X and Y are topological spaces. A mapping $f\colon X \to Y$ is said to be *continuous* (on X) if, for every open subset V of Y, its preimage $f^{-1}(V)$ is open in X.

Plainly, a composition of continuous mappings is continuous, and the identity mapping $\mathrm{id}\, X\colon X \to X$ of any space is continuous.

A *homeomorphism* $f\colon X \to Y$ is a continuous mapping that has a continuous inverse $g\colon Y \to X$. (We have already remarked that this is an 'isomorphism' of the structure we are considering.) Then the spaces X, Y are said to be *homeomorphic*, and we write $X \cong Y$.

A homeomorphism is thus the same as a bijective mapping $f\colon X \to Y$ which preserves open subsets, by images *and* preimages; it gives a bijective correspondence $\mathcal{O}(X) \to \mathcal{O}(Y)$.

As in the case of ordered sets, there is a crucial difference with algebraic structures: a bijective continuous mapping need not be invertible (within continuous mappings). For instance, the mapping

$$f\colon D(S) \to C(S), \tag{3.19}$$

whose underlying set-theoretical mapping is the identity of the set S, is obviously continuous (because every open subset of $C(S)$ is also open in $D(S)$), but its inverse is not – as soon as S has more than one point.

Exercises and complements. (a) The empty set and a singleton have a unique topology. The set $\{0, 1\} \subset \mathbb{R}$ has four topologies, two of which are homeomorphic; its euclidean topology is the discrete one.

In particular, $\{0, 1\}$ becomes the *Sierpinski space* when the singleton $\{1\}$ is open, and $\{0\}$ is not. Then, for any space X, a continuous mapping $X \to \{0, 1\}$ is the same as the characteristic function of an open set of X.

(b) A mapping $f\colon X \to Y$ between topological spaces is necessarily continuous in each of the following cases: the space X is discrete, the space Y is indiscrete, the mapping f is constant.

(c) (*Open and closed mappings*) A mapping $f\colon X \to Y$ between topological spaces is said to be *open* (resp. *closed*) if the image of every open subset (resp. closed subset) of X is open (resp. closed) in Y. Open mappings are stable under composition, as well as the closed ones.

(d) Let $f\colon X \to Y$ be a bijective mapping between topological spaces. Then its inverse f^{-1} is continuous if and only if f is open, if and only if f is closed.

(e) The linear structure of $\mathbb{R}^n$ on the real field gives the following mappings, for a vector $v \in \mathbb{R}^n$ and a scalar $\lambda \in \mathbb{R}^*$

$$\begin{aligned} t_v &: \mathbb{R}^n \to \mathbb{R}^n, \quad t_v(x) = x + v \quad \textit{(translation)}, \\ m_\lambda &: \mathbb{R}^n \to \mathbb{R}^n, \quad m_\lambda(x) = \lambda x \quad \textit{(homothety)}. \end{aligned} \tag{3.20}$$

All of them are homeomorphisms.

3.2.2 Comparing topologies

A topological structure on a set S is an element of $\mathcal{P}(\mathcal{P}S)$, and they are ordered by inclusion.

Suppose we have two topological structures X', X'' on the same underlying set S, and $\mathcal{O}(X') \subset \mathcal{O}(X'')$, i.e. every open set of the first structure is also open for the second. Then we say that X' has a *less fine*, or *coarser*, topology than X'', while X'' has a *finer* topology than X'.

For every set S, the indiscrete topology is the *coarsest* (or smallest), while the discrete topology is the *finest* (or largest).

Topological structures on S are stable under arbitrary intersections in $\mathcal{P}(\mathcal{P}S)$: given any family $(\mathcal{O}_i)_{i\in I}$ of topologies on S, the set $\mathcal{O} = \bigcap_i \mathcal{O}_i$ contains the subsets $U \subset S$ that are open in all the given topologies; if we take any family $(U_j)_{j\in J}$ of such subsets, their union is still open in all the given topologies, and therefore belongs to $\mathcal{O}$. Similarly, $\mathcal{O}$ is stable under finite intersections.

We conclude thus that the topological structures on S form a complete lattice. Let us note that the join $\mathcal{O}' \vee \mathcal{O}''$ of two topologies on S is the least topology containing both, and contains the set-theoretical union $\mathcal{O}' \cup \mathcal{O}''$, which is not a topology (generally): we have to 'stabilise' this collection of subsets of S under finite intersections and arbitrary unions.

Exercises and complements. (a) In a set S, every set $\mathcal{O}_0 \subset \mathcal{P}S$ of subsets *generates* a topology $\mathcal{O}$, namely the intersection of all topologies containing it.

Explicitly, $\mathcal{O}$ is the set of all unions of finite intersections of elements of $\mathcal{O}_0$.

(b) Given two topologies X', X'' on the same underlying set S, X' is less fine than X'' if and only if the identity $\mathrm{id}\, S$ gives a continuous mapping $X'' \to X'$.

For every topology X on S we have continuous mappings $D(S) \to X \to C(S)$, whose underlying mapping is always $\mathrm{id}\, S$.

(c) Let $f\colon X \to Y$ be a continuous mapping. Replacing the topology of X by any finer one, or the topology of Y by any less fine one, the mapping f remains continuous.

(d) Any set X can be equipped with the *cofinite topology*, whose closed subsets are the finite subsets and X. Thus a subset U is open in X if and only if it is empty or *cofinite* (i.e. the complement $X \setminus U$ is finite).

If the set X is infinite, this topology is not discrete. On $\mathbb{R}^n$ we get a topology coarser than the euclidean one.

3.2.3 Neighbourhoods and continuity at a point

Let X be a topological space. A *neighbourhood* of a point $a \in X$ is any subset $N \subset X$ which contains an open set U containing x. We write as $\mathcal{N}(a)$, or $\mathcal{N}_X(a)$, the set of all neighbourhoods of a in X.

These families of subsets satisfy the axioms we have already considered in 3.1.5(d), for every $a \in X$:

(Nb.1) a belongs to every element of $\mathcal{N}(a)$,

(Nb.2) every subset of X that contains an element of $\mathcal{N}(a)$ belongs to $\mathcal{N}(a)$,

(Nb.3) $\mathcal{N}(a)$ is stable in $\mathcal{P}X$ with respect to finite intersections,

(Nb.4) if $N \in \mathcal{N}(a)$, there is some subset $U \subset N$ such that $a \in U$ and, for every $b \in U$, $U \in \mathcal{N}(b)$.

A mapping $f\colon X \to Y$ between topological spaces is said to be *continuous at the point* $a \in X$ if it satisfies the equivalent conditions:

(i) for every $N \in \mathcal{N}_Y(f(a))$ there is some $M \in \mathcal{N}_X(a)$ such that $fM \subset N$,

(ii) for every $N \in \mathcal{N}_Y(f(a))$, $\;f^{-1}(N) \in \mathcal{N}_X(a)$.

Remarks and exercises. (a) We prove below, in Theorem 3.2.5, that the mapping f is continuous (on X) if and only if it is continuous at each point of X.

(b) Prove that a topological structure on a set X can equivalently be given by assigning a family $(\mathcal{N}(a))_{a \in X}$ of sets $\mathcal{N}(a) \subset \mathcal{P}X$, under the axioms (Nb.1 – 4). Then a subset $U \subset X$ is said to be open if $U \in \mathcal{N}(a)$, for every $a \in U$.

3.2.4 Interior, closure and boundary

We extend now Subsections 3.1.6 and 3.1.7, following the same pattern. X is a topological space and $A \subset X$.

The *interior* of A (in X), written as A° or $\mathrm{int}_X(A)$, is the union of all the open subsets of X contained in A. It is the largest open subset of X contained in A, and coincides with the latter if and only if A is open in X. A point $a \in A^\circ$ is said to be *interior* to A in X.

The *closure* of A (in X), written as $\overline{A}$ or A^- or $\mathrm{cl}_X(A)$, is the intersection of all the closed subsets of X containing A. It is the smallest closed subset of X containing A, and coincides with the latter if and only if A is closed in X. A point $a \in \overline{A}$ is said to be *adherent to* A in X.

The *boundary* of A (in X), written as ∂A or $\partial_X A$, is defined as the set of points of X that are adherent to A and to its complement

$$\partial A = \overline{A} \cap (X \setminus A)^{-}, \tag{3.21}$$

and is closed in X.

The subset $A \subset X$ is said to be *dense* (in X) if $\overline{A} = X$.

Exercises and complements. (a) For a subset A of the space X and a point $a \in X$:

(i) $a \in A^{\circ}$ if and only if A is a neighbourhood of a,

(ii) $a \in \overline{A}$ if and only if every neighbourhood of a meets A,

(iii) $a \in \partial A$ if and only if every neighbourhood of a meets A and $X \setminus A$,

(iv) $C_X(\overline{A}) = (C_X A)^{\circ}, \qquad C_X(A^{\circ}) = (C_X A)^{-}, \qquad \partial A = \partial(C_X A)$,

(v) $X = A^{\circ} \cup \partial A \cup (C_X A)^{\circ}$ (a disjoint union),

(vi) $\overline{A} = A^{\circ} \cup \partial A = A \cup \partial A$,

(vii) A is open in $X \;\Leftrightarrow\; A \cap \partial_X A = \emptyset$,

(viii) A is closed in $X \;\Leftrightarrow\; A \supset \partial_X A$.

(b) (*Kuratowski closure axioms*) A topology on the set X can be equivalently assigned by a closure operator $\mathrm{cl}\colon \mathcal{P}X \to \mathcal{P}X$, that satisfies the axioms (cl.1–3) of 3.1.7(f); they go back to a 1922 article of K. Kuratowski.

In this approach, the closed subsets are defined by the condition $\mathrm{cl}(A) = A$, or equivalently by $\mathrm{cl}(A) \subset A$.

We already noted that the closure preserves finite unions and is monotone. Moreover, for every family A_i $(i \in I)$ of subsets of X, we have:

$$\mathrm{cl}(\textstyle\bigcap_i A_i) \subset \bigcap_i \mathrm{cl}(A_i), \tag{3.22}$$

simply because $\mathrm{cl}(\bigcap_i A_i) \subset \mathrm{cl}(A_i)$, for every index i.

(c) The subset A of the space X is dense if and only if A meets every non-empty open set of X, if and only if A meets every neighbourhood of any point of X.

(d) Let A be a subset of the euclidean line $\mathbb{R}$, which is non-empty and upper bounded. Then $\sup A \in \overline{A}$. If A is closed in $\mathbb{R}$, $\sup A$ belongs to A, and is its maximum.

(e) The set $\mathbb{Q}$ of rational numbers is dense in $\mathbb{R}$. The same is true of its complement, the set $\mathbb{R} \setminus \mathbb{Q}$ of irrational numbers.

(f) (*Cluster points*) A point $a \in X$ is a *cluster point*, or *accumulation point*, of A in X if every neighbourhood of a in X meets A in some point other than a. Equivalently, a belongs to the closure of the subset $A \setminus \{a\}$ in X.

On the other hand, an *isolated point* of A is a point $a \in X$ that has a neighbourhood N in X such that $N \cap A = \{a\}$. *(Equivalently, the singleton $\{a\}$ is open in the *subspace* A, as defined in Section 3.4.)*

Plainly, the closure A in X is the disjoint union of the set A' of its cluster points (also called the *derived set* of A in X) and the set of its isolated points.

3.2.5 Theorem (Characterisations of continuity)

Let $f: X \to Y$ be a mapping between topological spaces. The continuity of f on X is expressed by the following equivalent properties:

(i) f is continuous at every $a \in X$,
(ii) the preimage of any open subset of Y is open in X,
(iii) the preimage of any closed subset of Y is closed in X,
(iv) for every $A \subset X$: $f(\overline{A}) \subset (f(A))^-$.

If f is continuous and A is dense in X, then $f(X) \subset (f(A))^-$.

Proof It is the same argument as in Theorem 3.1.8(b), replacing the domain and codomain of f with arbitrary spaces. □

3.2.6 Bases of open sets and local bases

We have seen, in Exercise 3.2.2(a), that in a set S, every set $\mathcal{O}_0 \subset \mathcal{P}S$ of subsets generates a topology: the set of all unions of finite intersections of elements of $\mathcal{O}_0$.

But we are interested in a more particular situation. Given a topological space X, a set $\mathcal{B} \subset \mathcal{O}(X)$ is said to be a *basis of open sets* for X, or a *basis of its topology*, if:

(i) every open set of X is a union of elements of $\mathcal{B}$.

Plainly, all the open sets of X form such a basis (no minimality condition is required), but there are often more 'economical' bases. For instance, the set of all open balls $B_n(a, r)$ (for $a \in \mathbb{R}^n$ and $r > 0$) is a basis of the euclidean topology of $\mathbb{R}^n$. Fixing any number $r_0 > 0$, the set of open balls $B_n(a, r)$ with $0 < r < r_0$ is also a basis.

For $n = 1$, the set of open intervals $]a, b[$ (for $a < b$) is a basis of the euclidean topology of $\mathbb{R}$; we can further require that $b - a < l$ (for a fixed real number $l > 0$).

Many conditions on open sets need only be checked on a basis. For instance, if $f: X \to Y$ is a mapping between topological spaces and $\mathcal{B}$ is a basis of the topology of Y, we only need to verify that the preimage of every $V \in \mathcal{B}$ is open in X, to conclude that f is continuous. In fact, preimages preserve arbitrary unions. (But, of course, we cannot expect that these preimages belong to a given basis of the topology of X.) Other examples are shown in the exercises below.

A similar perspective can simplify the use of neighbourhoods. In a topological space X, a *fundamental system of neighbourhoods* of the point a is any subset $\mathcal{N}_0 \subset \mathcal{N}_X(a)$ such that:

(ii) every $N \in \mathcal{N}_X(a)$ contains some element of $\mathcal{N}_0$,

or, equivalently, $\mathcal{N}_X(a)$ is the set of all $N \subset X$ that contain some element of $\mathcal{N}_0$.

For instance, in every topological space, the open neighbourhoods of any point form a fundamental system. In a euclidean space $X \subset \mathbb{R}^n$, the family of open balls $(B_X(a,r))_{r>0}$ of any point $a \in X$ forms a fundamental system of neighbourhoods of a. It is often convenient to use a fundamental system of *open* neighbourhoods of the point a, also called a *local basis at* a, as in the previous examples.

Now, if $f\colon X \to Y$ is a mapping between topological spaces, in order to verify that f is continuous at $a \in X$, it is sufficient to take a fundamental system of neighbourhoods of $f(a)$ in Y and prove that the preimage of each of them is a neighbourhood of a in X.

Exercises and complements. (a) For a set S, a set $\mathcal{B} \subset \mathcal{P}S$ is a basis of open sets for some topology (and indeed for the topology that it generates) if and only if $\mathcal{B}$ satisfies the following axioms:

(BO.1) the set S is a union of elements of $\mathcal{B}$,

(BO.2) an intersection $B \cap B'$ of elements of $\mathcal{B}$ is a union of elements of $\mathcal{B}$.

Equivalently, each finite intersection of elements of $\mathcal{B}$ is a union of elements of $\mathcal{B}$.

(b) Let X be a space, and $\mathcal{N}_0$ a fundamental system of neighbourhoods of the point a. For every $A \subset X$

$a \in A^\circ$ if and only if A contains some $N \in \mathcal{N}_0$,

$a \in \overline{A}$ if and only if every $N \in \mathcal{N}_0$ meets A,

$a \in \partial A$ if and only if every $N \in \mathcal{N}_0$ meets A and $C_X A$.

(c) In the same hypotheses, the set $\mathcal{N}_1 = \{N^\circ \mid N \in \mathcal{N}_0\}$ is a local basis at a.

(d) Let X be a space with a basis $\mathcal{B}$ of open sets. If $a \in X$, a subset $N \subset X$ is a neighbourhood of a if and only if there is some $U \in \mathcal{B}$ such that $a \in U \subset N$.

The set $\mathcal{B}(a)$ of all $U \in \mathcal{B}$ that contain the point a is a local basis at a. A subset $A \subset X$ is dense in X if and only if it meets every non-empty $U \in \mathcal{B}$.

(e) Let X be a space, and suppose we have for every $a \in X$ a local basis $\mathcal{B}(a)$. Then their union $\mathcal{B}$ is a basis of open sets for X.

(f) (*Nested systems*) Let $(N_k)_{k\in\mathbb{N}}$ be a fundamental system of neighbourhoods of the point a in the space X. Taking

$$N'_k = \textstyle\bigcap_{i \leqslant k} N_i,$$

we can replace it by a *nested system*, where $h \leqslant k$ implies $N'_h \supset N'_k$. Note also that $N'_k \subset N_k$ (for all k), and that a local basis is transformed into a nested local basis.

(g) The norm mapping $f\colon \mathbb{R}^n \to \mathbb{R}$, $f(x) = ||x||$ is continuous.

3.2.7 Topology and preorder

The similarity of the theories of preordered sets and topological spaces can easily be made precise.

Let X be an arbitrary topological space. For $a \in X$, we write as $\overline{a}$ the closure of the singleton $\{a\}$, also called the *closure of the point* a. The *specialisation preorder* of the space X is defined by

$$x \prec x' \;\Leftrightarrow\; x \in \overline{x'} \;\Leftrightarrow\; \overline{x} \subset \overline{x'}, \tag{3.23}$$

so that $\overline{a} = \{x \in X \mid x \prec a\}$.

Generally speaking, this preorder misses a large part of the information contained in the topology: for instance, in a space where all singletons are closed, this preorder is discrete: $x = x'$. Thus, on the set $\mathbb{R}$, the euclidean, the discrete and the cofinite topology give the same preorder.

However, this is no longer the case if we restrict in a convenient way the topologies we are considering.

An *Alexandrov topology*, named after Pavel S. Alexandrov, is a topology where the open sets are stable under arbitrary unions and arbitrary intersections; or, equivalently, the same is true of closed sets. A space X with such a topology is called here an *Alexandrov space*.

(This term can also refer to a different structure, not present in this book, and named after another Russian mathematician, Alexandr D. Alexandrov.)

In this case, for any subset $A \subset X$

$$\overline{A} \;=\; \bigcup_{a \in A} \overline{a}, \tag{3.24}$$

because this union is closed, and obviously the least closed subset containing A. Therefore, an Alexandrov topology is determined by the closure of its points, and also by the specialisation preorder.

The reader might now guess that preordered sets are 'the same as' Alexandrov spaces, and monotone mappings amount to continuous mappings between such spaces. All this is made precise in the exercises below.

(We are constructing a full embedding of the category of preordered sets into that of topological spaces: see Chapter 5.)

Exercises and complements. (a) We have already seen that any topological space X determines a preordered set $S(X)$, namely the same underlying set with the specialisation preorder. The other way round, if X is a preordered set, the *downward stable* subsets $C \subset X$, defined by the property:

- if $c \in C$ and $x \prec c$ in X, then $x \in C$,

are the closed sets of an Alexandrov space $A(X)$ on the same underlying set.

The open sets of $A(X)$ are the upward-stable subsets. The opposite preordered set X^{op} has the opposite Alexandrov topology, where open and closed subsets are interchanged.

(b) A continuous mapping $f\colon X \to Y$ between arbitrary topological spaces is monotone, as a mapping $S(X) \to S(Y)$.

(c) A monotone mapping $f\colon X \to Y$ between preordered sets is continuous, as a mapping $A(X) \to A(Y)$.

(d) If we restrict topological spaces to the Alexandrov ones, these two procedures are inverse to each other.

(e) Consider the Alexandrov topology $\mathbb{R}_A = A(\mathbb{R}, \leqslant)$ associated to the real line with the natural order; determine its closed and open sets.

(f) In an Alexandrov space X every subset A has an 'aperture', the intersection of all open sets containing it

$$\uparrow A = \{x \in X \mid a \prec x \text{ for some } a \in A\}, \tag{3.25}$$

which is the closure of A in the opposite Alexandrov topology.

In particular, for each $x \in X$, the set $\uparrow x$ of its upper bounds is the least open set containing x, and forms by itself *a local basis at x*.

3.2.8 Complements

(a) (*Initial topologies*) Suppose we have an indexed family of mappings

$$f_i\colon X \to Y_i \qquad (i \in I), \tag{3.26}$$

where X is a set and all Y_i are topological spaces. Trivially, all these mappings are made continuous if we put on X the discrete topology: this is the finest (or largest) topology having this effect.

We are interested in the less fine (or smallest) topology of X that makes all f_i continuous, called the *initial topology* of the family $(f_i)_{i\in I}$.

In this topology, for every open set U of some space Y_i, the preimage $f_i^{-1}(U)$ must be open in X. Our topology is generated by all these subsets of X, and consists of the unions of finite intersections of the preimages we have considered (by Exercise 3.2.2(a)).

When the family is reduced to a single mapping $f\colon X \to Y$, the initial topology on X just consists of the f-preimages of the open sets of Y, which are already stable under unions and finite intersections.

This procedure will be used to define subspaces and topological products.

(b) (*Final topologies*) Reversing all arrows, we have now an indexed family of mappings

$$f_i\colon Y_i \to X \qquad (i \in I), \tag{3.27}$$

where, again, X is a set and all Y_i are topological spaces. These mappings are made continuous if we put on X the indiscrete topology, which is the coarsest (or smallest) topology having this effect.

But we are interested in the finest (or largest) topology of X that makes all f_i continuous, called the *final topology* of the family $(f_i)_{i \in I}$.

This topology is easily described: a subset U is open in X if and only if, for every $i \in I$, the preimage $f_i^{-1}(U)$ is open in Y_i. In fact, if we fix an index i, the subsets of X whose preimage in Y is open form a topology $\mathcal{O}_i$ on X; the intersection of all $\mathcal{O}_i$ is a topology on X, and plainly the required one.

This procedure will be used to define quotient spaces and topological sums.

(c) For a space X, the homeomorphisms $X \to X$ form a subgroup of the symmetric group $\mathrm{Sym}(|X|)$ on the underlying set

$$\mathrm{Aut}(X) = \{f \colon X \to X \mid f \text{ is a homeomorphism}\}, \qquad (3.28)$$

which acts on the set $|X|$, as defined in 1.6.4 (and on the space X, as we will see).

We say that the points $x, y \in X$ are *topologically equivalent* in X if there is a homeomorphism $f \colon X \to X$ such that $f(x) = y$; in other words, x and y belong to the same orbit of the action of $\mathrm{Aut}(X)$ on X. We say that the space X is (topologically) *homogeneous* if all its points are topologically equivalent, i.e. if $\mathrm{Aut}(X)$ acts transitively on X.

This is the case of the euclidean space $\mathbb{R}^n$, as we can always use a translation $t_v(x) = x + v$ (see (3.20)) to take x to y (letting $v = y \dot{-} x$).

3.3 Sequences and countability axioms

After reviewing sequences of real numbers, convergence of a sequence is extended to any topological space. This property determines the topology, when the space satisfies a 'countability condition'.

3.3.1 Sequences of real numbers

A sequence in $\mathbb{R}$ is a mapping $x \colon \mathbb{N} \to \mathbb{R}$. It is often written as an indexed family $(x_k)_{k \in \mathbb{N}}$ of real numbers; we also write (x_k), leaving the rest understood.

We say that the sequence (x_k) *converges* to $\lambda \in \mathbb{R}$ if it satisfies the equivalent properties:

(i) for every $\varepsilon > 0$ there is some $h \in \mathbb{N}$ such that, for every $k \geqslant h$, $x_k \in B_1(\lambda, \varepsilon) =]\lambda - \varepsilon, \lambda + \varepsilon[$,

(ii) for every $N \in \mathcal{B}_1(\lambda)$ there is some $h \in \mathbb{N}$ such that, for every $k \geqslant h$, $x_k \in N$.

The second is often expressed saying that:

(iii) for every neighbourhood N of λ in $\mathbb{R}$, the sequence (x_k) lies *eventually* in N.

We also say that λ is the *limit* of the sequence, and we write $\lambda = \lim_k x_k$, or $\lambda = \lim x_k$, or $x_k \to \lambda$ (when $k \to \infty$). The limit, if it exists, is uniquely determined by the sequence: see Exercise 3.3.2(a).

The sequence $x = (x_k)$ is said to be *infinitesimal* if it converges to 0. This means that:

(iv) for every $\varepsilon > 0$ there is some $h \in \mathbb{N}$ such that, for every $k \geqslant h$, $|x_k| < \varepsilon$,

and is equivalent to saying that the sequence $(|x_k|)_k$ of its moduli is infinitesimal.

The sequence $x = (x_k)$ is said to be *bounded* if its image is contained in some bounded interval $]a, b[$, or equivalently if there exists some $r > 0$ such that $|x_k| < r$, for all $k \in \mathbb{N}$.

We use the same terminology for sequences $(x_k)_{k \geqslant K}$ that are defined on some subset $\uparrow K \subset \mathbb{N}$, like – for instance – the sequence $(1/k)_{k \geqslant 1}$. In fact, a sequence $(x_k)_{k \geqslant K}$ can always be transformed into a 'standard' sequence $(x_{k+K})_{k \in \mathbb{N}}$, so that – dealing with the general theory – we only have to consider standard sequences, defined on $\mathbb{N}$.

3.3.2 Exercises and complements

Let $x = (x_k)$, $y = (y_k)$ and $z = (z_k)$ be sequences of real numbers. The following points are not difficult to prove (and well known, in Calculus).

(a) The sequence (x_k) has at most one limit.

(b) Prove that $\lim_k 1/k = 0$.

(c) One should not confuse a sequence $x = (x_k)$ with its image $\mathrm{Im}\, x = \{x_k \mid k \in \mathbb{N}\}$, the set of its values. A sequence with image $\{-1, 1\}$ can converge to 1, or can converge to -1, or can have no limit.

(d) (*Comparison*) If x and y converge and $x_k \leqslant y_k$ (for all k) then $\lim x_k \leqslant \lim y_k$. (The hypothesis $x_k < y_k$ would give nothing more.)

(e) (*Squeeze Theorem*) If $x_k \leqslant z_k \leqslant y_k$ (for all k) and $\lim x_k = \lambda = \lim y_k$, then the sequence (z_k) also converges to λ.

(f) (*Bounded sequences*) If the sequence x is convergent, then it is bounded. If (x_k) is bounded and (y_k) is infinitesimal, then the sequence of products $(x_k y_k)_k$ is also infinitesimal.

(g) If x and y converge to λ and μ, respectively, then:

$$\lim(x_k + y_k) = \lambda + \mu, \qquad \lim(x_k . y_k) = \lambda\mu. \tag{3.29}$$

(h) If x converges to $\lambda > 0$, there exists some $K \geqslant 0$ such that $x_k > 0$ for all $k \geqslant K$, and the sequence $(1/x_k)_{k \geqslant K}$ converges to $1/\lambda$. If x converges to $\lambda \neq 0$, then $\lim 1/x_k = 1/\lambda$.

(i) The sequence (x_k) in $\mathbb{R}$ is said to be (weakly) *increasing* if $x_k \leqslant x_{k+1}$, for all k. Equivalently, this means that $h \leqslant k$ implies $x_h \leqslant x_k$.

Prove that an increasing sequence x converges if and only if it is upper bounded, and then it converges to $\sup(\operatorname{Im} x)$.

Dually (with respect to the ordering), a lower bounded (weakly) *decreasing* sequence converges to $\inf(\operatorname{Im} x)$, as in Exercise (b).

3.3.3 Sequences in a space

Let us consider now a topological space X.

A *sequence* in X is a mapping $x\colon \mathbb{N} \to X$; again, it will often be written as an indexed family $x = (x_k)_{k \in \mathbb{N}}$, or simply as (x_k).

We say that the sequence (x_k) *converges* to the point $a \in X$ if:

(i) for every neighbourhood N of a, the sequence lies *eventually* in N,

i.e. there exists some $h \in \mathbb{N}$ such that, for every $k \geqslant h$, $x_k \in N$.

One can equivalently use any fundamental system of neighbourhoods of a. We also say that a is *a limit* of the sequence.

We shall only use the symbol $\lim_k x_k$ when the limit-point is uniquely determined by the sequence. The exercises below show that this need not be the case. They also show that, independently of this uniqueness property, the convergence of sequences can determine the topological properties, under convenient hypotheses of 'countability' of the spaces involved, or of some of them.

3.3.4 Countability axioms

We recall that a set is said to be countable if it is finite or equipotent to $\mathbb{N}$ (see 1.7.8). 'Large' spaces can be more easily approached when their topology satisfies some 'countability condition', as we begin now to explore.

We say that the topological space X is *first-countable* if:

(i) every point of X has a countable fundamental system of neighbourhoods.

By Exercise 3.2.6(c), this is equivalent to saying that every point of X has a countable local basis. Then we can always use, at each point, a nested local basis $(N_k)_{k \in \mathbb{N}}$, where $h \leqslant k$ implies $N'_h \supset N'_k$ (by Exercise 3.2.6(f)).

We say that the topological space X is *second-countable* if:

(ii) the topology of X has a countable basis.

We say that the topological space X is *separable* if it has a countable dense subset.

Exercises and complements. (a) Every euclidean space $\mathbb{R}^n$ is first-countable.

(b) The euclidean line is second-countable and separable.

The same is true of every euclidean space $\mathbb{R}^n$, as one can prove now with little more work. (But this extension will follow automatically from general properties of product topologies.)

(c) Every second-countable space is also first-countable and separable. (But a direct approach in (a) can still be useful.)

3.3.5 Exercises and complements (Convergence of sequences)

These exercises are important and their results should be kept in mind; the solution is below.

(a) In an indiscrete space, each sequence converges to every point.

(b) In any space X, every sequence $(x_k)_k$ which is *eventually constant at* x converges (at least) to the point x. In a discrete space there are no other convergent sequences.

(c) A sequence $(x_k)_k$ in the euclidean space $\mathbb{R}^n$ can have at most one limit. (Spaces where this uniqueness property is always true will be studied in Section 3.6.)

(d) If a sequence $(a_k)_k$ of points of the subset $A \subset X$ converges to a point $x \in X$, then $x \in \overline{A}$.

(e) Conversely, if the point x has a *countable* fundamental system of neighbourhoods, and $x \in \overline{A}$, there is some sequence of points of A that converges to x in X.

(f) The last two points show that, if the space X is first-countable, then its closure operator (and its topology) can be reconstructed from the convergence of the sequences of X.

(g) Let $f\colon X \to Y$ be a mapping between topological spaces. If f is continuous, it preserves the convergence of sequences – which means that, if a sequence $(x_k)_k$ converges to x in X, then the sequence $(f(x_k))_k$ converges to $f(x)$ in Y. The converse certainly holds when the space X is first-countable.

(h) If $\mathbb{N}_{\mathrm{cof}}$ is the set of natural numbers with the cofinite topology (in Exercise 3.2.2(d)), a sequence eventually constant at x has no other limit, but the sequence $(k)_{k\in\mathbb{N}}$ converges to any point.

Solutions. (a) Here the only neighbourhood of a point is the total space.

(c) Suppose that, in $\mathbb{R}^n$, the sequence $(x_k)_k$ converges to two distinct points a_1 and a_2. Then $r = d(a_1, a_2) > 0$, and our points a_i have two *disjoint* neighbourhoods, for instance $N_i = B_n(a_i, r/2)$ (for $i = 1, 2$). The given sequence should eventually lie in both, which is a contradiction.

The existence of disjoint neighbourhoods of distinct points is precisely the 'separation property' of interest here – to be studied in Section 3.6.

(d) We have a sequence $(a_k)_k$ of points of A that converges to x in X. Therefore every neighbourhood of x meets A.

(e) Let $(N_k)_{k\in\mathbb{N}}$ be a nested local basis at x (which exists, by Exercise 3.2.6(f)). If $x \in \overline{A}$, each of them meets A, and we can choose a point $a_k \in N_k \cap A$. Now the sequence (a_k) converges to x, because any neighbourhood of x contains some N_h, and therefore, for $k \geqslant h$, all N_k and all points a_k.

(g) Suppose that f is continuous, and the sequence $(x_k)_k$ converges to x in X. If N is a neighbourhood of $f(x)$ in Y, the preimage $f^{-1}(N)$ is a neighbourhood of x in X, and eventually $x_k \in f^{-1}(N)$; thus $f(x_k) \in N$, eventually.

Conversely, suppose that X is first-countable and f preserves the convergence of all sequences of X. Taking an arbitrary subset $A \subset X$, it is sufficient to prove that $f(\overline{A}) \subset (f(A))^-$ (by Theorem 3.2.5). Using the previous exercises, for every $x \in \overline{A}$ we can choose a sequence $(a_k)_k$ of points of A that converges to x in X; then the sequence $(f(a_k))_k$ converges to $f(x)$ in Y, and $f(x) \in (f(A))^-$.

(h) Take a sequence eventually constant at x and some $y \neq x$ in $\mathbb{N}_{\mathrm{cof}}$. The set $N = \mathbb{N} \setminus \{y\}$ is an open neighbourhood of x, and our sequence cannot lie eventually in N.

Take now the sequence $(k)_{k\in\mathbb{N}}$ and any natural number x. Any neighbourhood U of x has a finite complement U^*. If $h = \max U^*$, then $x_k = k \in U$, for every $k > h$.

3.3.6 Exercises and complements (Real functions)

We now review some basic facts about the continuity of real-valued functions, which the reader may partially know from Calculus. The proof can be made simple by using the characterisation of continuity by convergent sequences, in 3.3.5(g), that applies when the domain is first-countable.

(a) The sum $s\colon \mathbb{R}^2 \to \mathbb{R}$ and the multiplication $m\colon \mathbb{R}^2 \to \mathbb{R}$ are continuous mappings.

(b) If $f, g\colon X \to \mathbb{R}$ are continuous mappings defined on any space X, so is their sum $f + g$ and their multiplication h

$$\begin{aligned} &(f+g)\colon X \to \mathbb{R}, \qquad (f+g)(x) = f(x) + g(x), \\ &h\colon X \to \mathbb{R}, \qquad h(x) = f(x).g(x). \end{aligned} \tag{3.30}$$

(c) For a fixed $\lambda \in \mathbb{R}$, the multiplication $m_\lambda(x) = \lambda x$ is a continuous mapping $\mathbb{R} \to \mathbb{R}$.

(d) The inversion $i(x) = x^{-1}$ is a continuous mapping $\mathbb{R}^* \to \mathbb{R}^*$.

If $f, g\colon X \to \mathbb{R}$ are continuous mappings and $g(x)$ is never 0, their quotient is a continuous function

$$f/g\colon X \to \mathbb{R}, \qquad (f/g)(x) = f(x)/g(x). \tag{3.31}$$

(e) Any polynomial function $f\colon \mathbb{R}^n \to \mathbb{R}$ in n variables is continuous.

(f) Any *rational function* $f/g\colon X \to \mathbb{R}$ in n variables is continuous.

Here $f, g\colon \mathbb{R}^n \to \mathbb{R}$ are polynomial functions and $X \subset \mathbb{R}^n$ is the complement of the set where the denominator g annihilates (an 'algebraic variety').

3.4 Subspaces and quotients

Every subset A of a space X becomes a subspace. Every set-theoretical quotient X/R becomes a topological quotient.

In contrast with algebraic structures, no stability condition is required of A and no compatibility condition is required of R.

3.4.1 Subspaces

Let X be a topological space. A subset $A \subset X$ becomes a topological space, and a *subspace* of X, when equipped with the initial topology for the inclusion mapping $j\colon A \to X$ (see 3.2.8(a)), which is the smallest making j continuous.

The open subsets of A are thus the *traces* $j^{-1}(U) = U \cap A$ of the open sets U of X on the subset A.

As a consequence, the closed subsets of A are the traces $j^{-1}(C) = C \cap A$ of the closed sets C of X (because preimages commute with complements). Similarly, the (open) neighbourhoods of a point $x \in A$ are the traces $N \cap A$ of the (open) neighbourhoods N of x in X.

We speak of an *open subspace* A of the space X when A is an open subset of X. Then the open subsets of A simply are the open subsets of X contained in A, and the inclusion mapping $A \to X$ is open.

Similarly, we speak of a *closed subspace* A of the space X when A is a closed subset of X. Then the closed subsets of A are the closed subsets of X contained in A, and the inclusion mapping $A \to X$ is closed.

(Of course, these two properties have nothing to do with the trivial fact that A is open and closed *in itself*, as any topological space.)

Every subset $A \subset \mathbb{R}^n$, equipped with the euclidean topology defined in 3.2.1, is a subspace of the euclidean $\mathbb{R}^n$. We generally refer to this topology of A, unless some other topology is specified.

3.4.2 Exercises and complements (Subspaces)

Let A be a subspace of the space X.

(a) If $a \in A$, a local basis $\mathcal{B}(a)$ of a in X produces a local basis for a in A, taking its traces on A. If X is first-countable, so is A.

(b) Similarly, a basis $\mathcal{B}$ of open sets of X gives a basis of A. If X is second-countable, so is A.

(c) We have a sequence (a_k) in A. If (a_k) *converges* to the point $a \in A$ in the space X, the same holds in A.

(d) For any subset S of A, $\mathrm{cl}_A(S) = \mathrm{cl}_X(S) \cap A$. If $f\colon X \to Y$ is a continuous mapping and A is dense in X, then $f(A)$ is dense in $f(X)$ (as a subspace of Y).

(e) (*Transitivity of the subspace relationship*) Take a subset $B \subset A$. Then the topology of B as a subspace of A coincides with its topology as a subspace of X. (In other words, if B is a subspace of A, which is a subspace of X, then B is a subspace of X.)

(f) (*Universal property of a subspace*) A mapping $h\colon X' \to A$ defined on a topological space is continuous if and only if so is its composite $jh\colon X' \to X$ with the inclusion $j\colon A \to X$.

Note. This universal property of substructures also holds in Algebra, where it is too obvious to be worth mentioning. For instance, a mapping from a group to a subgroup of G is a homomorphism if and only if it is with values in G.

(g) (*Restrictions of continuous mappings*) Take a continuous mapping $f\colon X \to Y$, and suppose that $f(A)$ is contained in the subspace B of Y. Then the following *restrictions* of f

$$\begin{aligned} &f_A\colon A \to Y, && f_A(x) = f(x) && (x \in A),\\ &f_{AB}\colon A \to B, && f_{AB}(x) = f(x) && (x \in A), \end{aligned} \tag{3.32}$$

are continuous. If f is a homeomorphism, its restriction $A \to f(A)$ is a homeomorphism.

3.4.3 Lemma and Definition (Continuity and covers)

Let $f\colon X \to Y$ be a mapping between topological spaces and let $(A_i)_{i\in I}$ be a cover *of the space X: this simply means that $X = \bigcup A_i$.*

(a) If $(A_i)_{i \in I}$ is an open cover *of the space X (which means that all A_i are open subspaces of X), the mapping $f\colon X \to Y$ is continuous if and only if all its restrictions $f_i\colon A_i \to Y$ are.*

(b) The same holds for a finite closed cover *$(A_i)_{i \in I}$, which means that all A_i are closed subspaces of X and the index-set I is finite.*

(c) Equivalently, we can start from a family of continuous mappings $f_i\colon A_i \to Y$ defined on the spaces of an open cover (or a finite closed cover) of the space X. Then this family admits a (continuous) pasting *$f\colon X \to Y$ (that extends all mappings f_i) if and only if it is consistent, that is:*

- if $x \in A_i \cap A_j$, then $f_i(x) = f_j(x)$ (for all $i, j \in I$).

Proof (a) If f is continuous, all its restrictions are. Conversely, if all restrictions $f_i\colon A_i \to Y$ are continuous, take an open set V of Y. Every preimage $f_i^{-1}(V)$ is open in the (open) subspace A_i, and therefore is also open in X. But $f_i^{-1}(V) = f^{-1}(V) \cap A_i$, so that

$$\bigcup_i f_i^{-1}(V) = \bigcup_i (f^{-1}(V) \cap A_i) = f^{-1}(V) \cap (\bigcup_i A_i) = f^{-1}(V)$$

is a union of open sets of X, and open itself.

(b) We can adapt the previous argument working with *closed* subsets, which are stable under finite unions.

(c) It is a rewriting of the previous points, taking into account that a consistent family of mappings (as above) can always be pasted into a mapping $f\colon X \to Y$. □

3.4.4 Exercises and complements (Euclidean subspaces)

A beginner should acquire a concrete knowledge of the euclidean topology, on simple subspaces of $\mathbb{R}^n$. The following exercises are worked out in Chapter 7, but personal work on these and other cases should not be spared.

(a) Draw the following subspaces of the euclidean plane $\mathbb{R}^2$:

- the closed square $[0,1]^2$,
- the open square $]0,1[^2$,
- the subspace $]0,1[\times[0,1]$, neither open nor closed,

and some simple open or closed subsets, in each of them.

(b) We have already introduced, in 3.1.9, the following closed subsets of $\mathbb{R}^n$, which are now viewed as closed subspaces:

- the standard sphere $\mathbb{S}^{n-1} = S_n(0,1)$,
- the standard closed ball $\mathbb{D}^n = D_n(0,1)$,

- the standard cube $\mathbb{I}^n = [0,1]^n$.

(c) The open arcs of $\mathbb{S}^1$ are a basis of its euclidean topology. It is often interesting to view $\mathbb{S}^1$ as a subspace of $\mathbb{C}$

$$\mathbb{S}^1 = \{z \in \mathbb{C} \mid |z| = 1\},$$

and a subgroup of the multiplicative group $\mathbb{C}^*$ (as in 2.7.6).

(d) (*The shape of the intervals, I*) Any non-degenerate interval of the euclidean line is homeomorphic to $]0,1[$, or $[0,1[$, or $[0,1]$.

More precisely, rational functions and their 'pastings' (in the sense of 3.4.3(c)) are sufficient to construct homeomorphisms (for all $a < b$ in $\mathbb{R}$):

(i) $\mathbb{R} \cong\ ]a,+\infty[\ \cong\ ]-\infty,b[\ \cong\ ]a,b[\ \cong\ ]0,1[$,

(ii) $[a,+\infty[\ \cong\ ]-\infty,b]\ \cong\ [a,b[\ \cong\ ]a,b]\ \cong\ [0,1[$,

(iii) $[a,b]\ \cong\ [0,1]$.

In particular, $\mathbb{R}$ is homeomorphic to any non-degenerate open interval, bounded or unbounded.

Proving that these three homeomorphism-types are *distinct* would now require some more work, that can be spared: the conclusion will follow examining the 'connected components' of an interval deprived of a point, in Exercise 4.1.7(b).

3.4.5 Topological quotients

Let X be a topological space and R an equivalence relation in X (more precisely, in its underlying set). The set-theoretical quotient $Y = X/R$ becomes a topological space, and a *topological quotient* of X, when equipped with the final topology for the canonical projection (see 3.2.8(b))

$$p \colon X \to X/R. \tag{3.33}$$

This is the finest topology of X/R that makes p continuous. Its open subsets are the subsets V of X/R whose preimage $p^{-1}(V)$ is open in X. Similarly, a subset of X/R is closed if and only if its preimage is closed in X. We also speak, for brevity, of a *topological quotient* $p \colon X \to X/R$.

All this is better understood saying that a subset $A \subset X$ is *saturated* for the relation R if it contains the equivalence class of each of its points. Any subset $A \subset X$ has an *associated saturated set*

$$\mathrm{sat}_R(A) = p^{-1}(p(A)), \tag{3.34}$$

which is the least R-saturated subset of X containing A, and the union of the equivalence classes of R that meet A.

Now, coming back to the topological quotient X/R, its open (resp. closed) sets can be described as the p-images of all the open (resp. closed) sets of X which are saturated for R.

In fact, if U is a saturated open set of X, then its image $V = p(U)$ in X/R is open, because $p^{-1}(V) = p^{-1}(p(U)) = U$ is open in X. Conversely, if V is open in X/R, then $U = p^{-1}(V)$ is open in X and obviously saturated for R; moreover $p(U) = p(p^{-1}(V)) = V$, as p is surjective. The same argument holds for closed subsets.

The equivalence relation R is said to be *open* if the canonical projection $p\colon X \to X/R$ is an open mapping (see 3.2.1). This means that the image of every open set of X is open in the quotient, and is equivalent to saying that the saturated of every open set of X is open. In this case the open subsets of X/R are the images of the open sets of X.

Similarly, the relation R is said to be *closed* if the projection p is a closed mapping; this means that the image of every closed set of X is closed in the quotient, or equivalently that the saturated of every closed set of X is closed. In this case the closed subsets of X/R are the images of the closed sets of X.

3.4.6 Exercises and complements (Quotients)

Consider a topological quotient $p\colon X \to X/R$.

(a) (*Universal property of a quotient*) A mapping $h\colon X/R \to Y$ with values in a topological space is continuous if and only if so is the composed mapping $hp\colon X \to X/R \to Y$.

(b) (*Induction of continuous mappings*) Take a continuous mapping $f\colon X \to Y$, and suppose that we have an equivalence relation S in Y, such that

$$x\,R\,x' \;\Rightarrow\; f(x)\,S\,f(x') \qquad \text{(for all } x, x' \in X). \tag{3.35}$$

Then the following mappings f' and f'', *induced* by f

$$\begin{aligned} &f'\colon X \to Y/S, && f'(x) = [f(x)]_S, \\ &f''\colon X/R \to Y/S, && f''[x]_R = [f(x)]_S, \end{aligned} \tag{3.36}$$

$$\begin{array}{ccc} X & \xrightarrow{\;f\;} & Y \\ {\scriptstyle p}\downarrow & \searrow^{f'} & \downarrow{\scriptstyle q} \\ X/R & \xrightarrow[\;f''\;]{} & Y/S \end{array}$$

are continuous.

If f is a homeomorphism, and xRx' is equivalent to $f(x)Sf(x')$, then the mapping f'' is also a homeomorphism.

3.4.7 The canonical factorisation

From the corresponding set-theoretical result (in 1.2.5), we readily have that a continuous mapping $f\colon X \to Y$ between topological spaces has a *canonical factorisation*

$$\begin{array}{ccc} X & \xrightarrow{f} & Y \\ {\scriptstyle p}\downarrow & & \uparrow{\scriptstyle m} \\ X/R_f & \xrightarrow[g]{} & \operatorname{Im} f \end{array} \qquad f = mgp, \qquad (3.37)$$

where:

- p is the canonical projection onto the topological quotient X/R_f, modulo the equivalence relation associated to f,
- m is the inclusion of the subspace $\operatorname{Im} f = f(X)$ into the codomain Y,
- $g\colon X/R_f \to \operatorname{Im} f$ is the (bijective) mapping defined by $g[x] = f(x)$ (for all $x \in X$).

The continuity of g follows from the universal properties of a subobject and a quotient (Exercises 3.4.2(f) and 3.4.6(a)): the mapping $f = mgp$ is continuous, whence gp is also, and g as well.

But of course g need not be a homeomorphism: it is sufficient to take as f a bijective continuous mapping that is not a homeomorphism, as in (3.19). One can find more interesting examples, where the domain and codomain are both euclidean spaces: see Exercise 3.4.8(c).

Actually, the mapping g is a homeomorphism if and only if it is an open mapping, if and only if it is a closed mapping. The first condition amounts to saying that, for every open subset U of X which is saturated for R_f, the image $f(U)$ is open in $\operatorname{Im} f$. The second condition requires that any saturated closed subset of X has a closed image in $\operatorname{Im} f$.

We say that $f\colon X \to Y$ is a *topological embedding* if it is an injective mapping and X has the initial topology for f: the open sets of X are (precisely) the preimages $f^{-1}(V)$ of the open sets of Y. Then f restricts to a homeomorphism $X \to f(X)$ between X and a subspace of Y, and we may want to 'identify' X with its image in Y, so that it becomes a subspace. Plainly, topological embeddings are stable under composition. Any injective continuous mapping that is open (or closed) is always a topological embedding.

On the other hand, we say that $f\colon X \to Y$ is a *topological projection* if it is a surjective mapping where Y has the final topology for f: the open sets of Y are (precisely) the subsets V whose preimage $f^{-1}(V)$ is open in X. Then f induces a homeomorphism $X/R_f \to Y$ between a quotient of X and the space Y, and we may want to 'identify' Y with this quotient of X. Topological projections are stable under composition. Any surjective continuous mapping which is open (or closed) is always a topological projection.

3.4.8 Comments, exercises and complements

(a) (*Transitivity of the quotient relationship*) This property is more complex than transitivity of subspaces (in 3.4.2(e)), because set-theoretically 'a quotient of a quotient' of the set X can only be viewed as a quotient of X up to a canonical bijection.

Essentially, we only have to reconsider what we said at the end of 3.4.7: topological projections are stable under composition. Composing two topological quotients we get a topological projection, which is the 'same as a quotient', up to homeomorphism.

This can be analysed in a more concrete way. We have a topological space X and a topological quotient $p\colon X \to X/R = Y$. Then we have a second topological quotient $q\colon Y \to Y/S$. Letting R' be the equivalence relation of X defined by:

$$xR'x' \quad \Leftrightarrow \quad qp(x) = qp(x') \quad \Leftrightarrow \quad p(x)\, S\, p(x') \text{ in } Y, \tag{3.38}$$

the composed mapping qp induces a homeomorphism g

$$\begin{array}{ccc} X & \xrightarrow{\;p\;} & Y \\ {\scriptstyle p'}\downarrow & & \downarrow{\scriptstyle q} \\ X/R' & \xrightarrow[g]{} & Y/S \end{array} \qquad \begin{array}{l} g\colon X/R' \to (X/R)/S, \\ g[x]_{R'} = [\,[x]_R\,]_S. \end{array} \tag{3.39}$$

(b) The usual parametrisation of the standard circle $\mathbb{S}^1$

$$f\colon \mathbb{R} \to \mathbb{R}^2, \qquad f(t) = (\cos t, \sin t), \tag{3.40}$$

induces a homeomorphism $g\colon \mathbb{R}/R_f \to \mathbb{S}^1$. The mapping f is also called the *exponential* mapping, with reference to the complex exponential e^{it} (see 2.7.9).

**Note.* We have already seen this homeomorphism g as an isomorphism

$\mathbb{R}/(2\pi\mathbb{Z}) \to \mathbb{S}^1$ of commutative groups, in (2.150). Combining both aspects, the algebraic and the topological one, we have an isomorphism of topological groups: see 4.6.1.*

(c) This mapping f can be restricted to a continuous bijective mapping $[0, 2\pi[\ \to \mathbb{S}^1$, that is not a homeomorphism. Note that the domain and codomain are both euclidean spaces.

3.4.9 Retracts

Suppose we have two consecutive continuous mappings j, p with

$$A \xrightarrow{j} X \xrightarrow{p} A \qquad\qquad pj = \mathrm{id}\, A. \tag{3.41}$$

In this situation j is a topological embedding and p is a topological projection (see Exercise (a), below), so that A is at the same time a subspace and a quotient of X (up to homeomorphism). The mapping j is called a *section*, and p is called a *retraction* (of topological spaces).

The space A is called a *retract* of X. In fact, this term is commonly used when A is a subspace of X and j is its inclusion; but we can always rewrite (3.41) in this form, replacing A with its (homeomorphic) image $j(A)$ in X.

Intuitively, the fact that a subspace $A \subset X$ is a retract has a clear meaning: we can continuously deform X into A, without moving the points of A.

As an important example, the sphere $\mathbb{S}^n$ is a retract of the 'pierced space' $X = \mathbb{R}^{n+1} \setminus \{0\}$, by the *normalisation* mapping N (see 3.2.6(g))

$$\mathbb{S}^n \xrightarrow{j} X \xrightarrow{N} \mathbb{S}^n \qquad N(x) = x/||x|| \qquad (Nj = \mathrm{id}\,). \tag{3.42}$$

Now, $\mathbb{S}^n$ is *not* a retract of $\mathbb{R}^{n+1}$, nor of the closed ball $\mathbb{D}^{n+1}$. This is intuitively clear (in low dimension): trying to 'spread' $\mathbb{D}^{n+1}$ on the sphere, without moving the points of the latter, we would break continuity somewhere inside the open ball; which is why we have taken out a point, in X.

However, to prove that no such retraction exists (not just that N cannot be continuously prolonged in the origin!) requires higher tools, of Algebraic Topology. (For $n = 0$ we only need the Intermediate Value Theorem, or Section 4.1 on connected spaces.)

Exercises and complements. (a) Prove that, in (3.41), j is a topological embedding and p is a topological projection.

(b) Any point of a non-empty space is a retract of the latter. The empty space is only a retract of itself.

(c) Any bounded closed interval $[a, b]$ is a retract of the euclidean line.

*(d) A retract of a 'good' space is closed in the latter, as we will see in Exercise 3.6.7(b).

3.5 Topological products and sums

A finite product of topological spaces has an obvious definition. On the other hand, the structure of an infinite product is less obvious, and was defined by Andrey Tychonoff in 1935 [Ty2].

This is indeed the 'good' definition, determined by the universal property of a product. Its relationship with compactness, in the famous Tychonoff's Theorem (see 4.2.4), solved one of the main open problems in Topology.

3.5.1 Finite topological products

Let us begin by considering the binary *product* $X = X_1 \times X_2$ of two topological spaces. This is the cartesian product of the underlying sets, equipped with the initial topology for the projections $p_i \colon X \to X_i$ $(i = 1, 2)$, the coarsest that makes both of them continuous.

There is a canonical basis $\mathcal{B}$ of the product topology, whose elements are the products $U_1 \times U_2$, where each U_i is open in X_i. First, this is indeed a basis of open sets on X, because it contains the total space and is closed under binary intersections (more than we need)

$$(U_1 \times U_2) \cap (V_1 \times V_2) = (U_1 \cap V_1) \times (U_2 \cap V_2). \tag{3.43}$$

Second, the topology defined by this basis makes both projections continuous, because $p_1^{-1}(U_1) = U_1 \times X_2$, and symmetrically. Third, each topology on X that makes both projections continuous is finer than the one we are considering, because it must contain as open sets all the sets

$$p_1^{-1}(U_1) \cap p_2^{-1}(U_2) = (U_1 \times X_2) \cap (X_1 \times U_2) = U_1 \times U_2,$$

where each U_i is open in X_i.

All this is readily extended to a finite *topological product* $X = \prod_i X_i$ $(i = 1, ..., n)$. The canonical basis $\mathcal{B}$ of the product topology consists of all products $\prod_i U_i$, where each U_i is an open subset of the space X_i. A general open set of X is a union of elements of $\mathcal{B}$. This basis happens to be stable under finite intersections.

Exercises and complements. We have a finite topological product $X = \prod_i X_i$.

(a) Suppose that, in each space X_i, the point $a_i \in X_i$ has a local basis $\mathcal{B}_i(a_i)$. Then the set of all products $U = \prod_i U_i$, where each U_i belongs to $\mathcal{B}_i(a_i)$, forms a local basis $\mathcal{B}(a)$, at the point $a = (a_1, ..., a_n)$ of the space X. If each space X_i is first-countable, so is X.

(b) Suppose that each space X_i has a basis of open sets $\mathcal{B}_i$. Then the products $\prod_i U_i$, where each U_i belongs to $\mathcal{B}_i$, form a basis $\mathcal{B}$ for the product topology of $X = \prod_i X_i$. If each space X_i is second-countable, so is X.

(c) (*An important fact*) The euclidean topology of $\mathbb{R}^n$ coincides with the n-th cartesian power of the euclidean topology of $\mathbb{R}$.

3.5.2 Cartesian products

We define now the *topological product* $X = \prod_i X_i$ of a family of spaces $(X_i)_{i\in I}$ indexed by an arbitrary set I.

Again, it is the cartesian product of the underlying sets, equipped with the initial topology for the projections $p_i \colon X \to X_i$ $(i \in I)$. Now, the elements of the canonical basis $\mathcal{B}$ are the products $\prod_i U_i$, where each U_i is open in X_i, and *coincides with* X_i *except for a finite subset of* I: the family (U_i) is 'quasi total'.

In fact, the preimage of an open set U_j of X_j with respect to p_j is

$$p_j^{-1}(U_j) = \prod_i U_i, \qquad \text{where } U_i = X_i \text{ for } i \neq j, \tag{3.44}$$

and the finite intersections of these sets are precisely the elements of $\mathcal{B}$. These elements will be called *canonical open sets* of the product topology.

As usual, the *universal property* says that: for every space Y and every family of continuous mappings $f_i \colon Y \to X_i$, there exists a unique continuous mapping $f \colon Y \to X$ such that

$$\begin{array}{ccc} Y & \overset{f}{\dashrightarrow} & X \\ & \searrow_{f_i} & \downarrow{\scriptstyle p_i} \\ & & X_i \end{array} \qquad p_i f = f_i \ \ (\text{for } i \in I). \tag{3.45}$$

Indeed, the mapping f is already determined at the level of sets, as $f(y) = (f_i(y))_{i\in I}$ (for $y \in Y$), and we only have to verify its continuity. The preimage of the open set of formula (3.44) is $f^{-1}p_j^{-1}(U_j) = f_j^{-1}(U_j)$, which is open in Y, so that the preimage of any open set in $\mathcal{B}$ is a finite intersection of open sets of Y.

Again, this mapping f is often written as (f_i). If we have a family $(f_i \colon X_i \to Y_i)$ of continuous mappings, we write as

$$f = \prod_i f_i \colon \prod_i X_i \to \prod_i Y_i, \qquad f(x_i)_{i\in I} = (f_i(x_i))_{i\in I}, \tag{3.46}$$

the continuous mapping that makes the following diagrams commutative

$$\begin{array}{ccc} \prod_i X_i & \xrightarrow{f} & \prod_i Y_i \\ {\scriptstyle p_i}\downarrow & & \downarrow{\scriptstyle q_i} \\ X_i & \xrightarrow[f_i]{} & Y_i \end{array} \qquad q_i f = f_i p_i \quad (\text{for } i \in I). \tag{3.47}$$

(It is determined by the universal property of $\prod_i Y_i$, with respect to its projections q_i.)

3.5.3 Exercises and complements (Products)

We have a topological product $X = \prod_{i \in I} X_i$.

(a) Suppose that, in each space X_i, the point $a_i \in X_i$ has a local basis $\mathcal{B}_i(a_i)$. Then the set of all quasi total products $U = \prod_i U_i$, where each U_i belongs to $\mathcal{B}_i(a_i)$ or is X_i, forms a local basis $\mathcal{B}(a)$, at the point $a = (a_i)$ of the space X.

(b) Suppose that each space X_i has a basis of open sets $\mathcal{B}_i$. Then the quasi total products $\prod_i U_i$, where each U_i belongs to $\mathcal{B}_i$ or is X_i, form a basis $\mathcal{B}$ for the product topology of $X = \prod_i X_i$.

(c) A product $C = \prod_i C_i$ of closed subsets $C_i \subset X_i$ is closed in X.

(d) A sequence $(x_k)_k$ in X converges to a point $a = (a_i)$ if and only if, for every $i \in X$, the sequence $(p_i x_k)_k$ converges to $p_i a = a_i$ in X_i.

(e) The topology of the power space

$$\mathrm{Map}(I, X) = \prod_{i \in I} X \tag{3.48}$$

is called the *topology of pointwise convergence.*

(f) An infinite product $\prod_i X_i$ of discrete spaces having at least two points (each of them) is not discrete.

(g) Let $X = \prod_i X_i$ be a product of preordered sets. Compare the Alexandrov topology of X with the topological product of the Alexandrov spaces $A(X_i)$ (see 3.2.7). *Hints*: one should distinguish finite products from the infinite ones.

(h) A product

$$f = \prod_i f_i \colon \prod_i X_i \to \prod_i Y_i \tag{3.49}$$

of open surjective mappings is open.

It is sufficient to suppose that all mappings f_i are open, and *nearly* all of them are surjective: in other words, surjectivity can fail for a finite subset of indices i. Thus, a finite product of open mappings is always open.

*(i) (*Box topology*) On a product $X = \prod_i X_i$ of topological spaces, it can also be useful to consider the *box topology*, where a basis of open sets is given by all the products $\prod_i U_i$ of open sets of the factors. For an infinite set of indices, this can be strictly finer than the product topology: then the cartesian projections stay continuous, but no longer satisfy the universal property.

3.5.4 ***Exercises and complements*** (Products and subspaces)

We have a topological product $X = \prod_{i \in I} X_i$.

(a) If A_i is a subspace of X_i (for $i \in I$), then the topological product $A = \Pi A_i$ is a subspace of $X = \Pi X_i$. If each A_i is a retract of X_i, then A is a retract of X.

(b) Suppose that each factor X_j contains a point a_j. Then the factor X_i can be embedded in X, by the following topological embedding

$$\begin{aligned} &m_i \colon X_i \to X, \qquad && m_i(x) = (x_j)_{j \in I}, \\ &x_i = x, && x_j = a_j \ \text{ for } j \neq i. \end{aligned} \tag{3.50}$$

(c) (*A warning*) We have thus proved that each factor X_i can be embedded in X *when all factors are non-empty.* Similarly, the projection $p_i \colon X \to X_i$ is surjective *when all factors are non-empty.*

These two facts will be used to prove that some properties of the product space X carry on to all its factors, *provided* no empty factors are present (see Theorems 3.6.4(c), 4.1.4(c), 4.2.4). This general remark is less pedantic than it may seem: important results are proved by verifying that an 'unwanted' subset is empty, and this only works if the empty set is not forgotten in our study.

(d) A continuous mapping $f \colon X \times Y \to Z$ restricts, for every $y \in Y$, to a continuous mappimg $f_y \colon X \to Z$, $f_y(x) = f(x, y)$. If the space Y is discrete, the converse holds: the continuity of all mappings f_y gives the continuity of f.

3.5.5 ***Exercises and complements*** (Products and quotients)

We have a topological product $X = \prod_i X_i$, and an equivalence relation R_i in each factor X_i. We denote by R the obvious equivalence relation in X

$$(x_i)_{i \in I} \, R \, (y_i)_{i \in I} \ \Leftrightarrow \ (\text{for every } i \in I, \ x_i \, R_i \, y_i). \tag{3.51}$$

(a) The canonical bijection

$$f \colon X/R \to \textstyle\prod_i (X_i/R_i), \qquad f[(x_i)_{i \in I}] = ([x_i])_{i \in I}, \tag{3.52}$$

is continuous.

(b) If all the equivalence relations R_i are open, also R is open and f is a homeomorphism.

*(c) When the previous hypothesis fails, the conclusion can fail as well, even for a binary product $X_1 \times X_2$, a trivial equivalence relation R_1 (the equality in X_1) and a closed equivalence relation R_2.

An example can be found in an exercise of [Bou2], Section I.5. The (difficult) proof is based on the fact that the product by a Hausdorff space X_1 which is not locally compact need not preserve quotients.

3.5.6 Topological sums

We define now the *topological sum* $X = \sum_{i \in I} X_i$ of a family of spaces $(X_i)_{i \in I}$ indexed by a set I.

It is the sum of the underlying sets (see 1.2.3(d)), equipped with the final topology for the injections $u_i \colon X_i \to X$ $(i \in I)$. In other words, viewing each u_i as an inclusion, a subset $U \subset X$ is open if and only if all its traces $U_i = U \cap X_i$ are open in the original spaces, if and only if U is a union of open subsets of the latter.

Each space X_i is an open subspace of X; it is also closed, as its complement is obviously open.

The *universal property* says that: for every pair $(Y, (f_i \colon X_i \to Y)_{i \in I})$ formed of space Y and a family of continuous mappings $f_i \colon X_i \to Y$, there exists precisely one continuous mapping $f \colon X \to Y$ such that

$$\begin{array}{ccc} X & \xrightarrow{f} & Y \\ {\scriptstyle u_i}\uparrow & \nearrow{\scriptstyle f_i} & \\ X_i & & \end{array} \qquad fu_i = f_i \ \text{(for } i \in I). \tag{3.53}$$

The map f can be written as $[f_i]$, and called the map $X \to Y$ of *co-components* f_i (for $i \in I$).

3.5.7 Exercises and complements

We have a topological sum $X = \sum_{i \in I} X_i$. The following facts are obvious or easy, and left to the reader.

(a) If each space X_i is first-countable, so is X.

(b) If I is countable and each space X_i is second-countable, so is X.

(c) If I is countable and each space X_i is separable, so is X.

(d) A subset $C \subset X$ is closed if and only if all its traces $C_i = C \cap X_i$ are closed in the original spaces.

(d) A sequence $(x_k)_k$ in X converges to a point $a \in X_i$ if and only if the sequence lies eventually in the space X_i (for all k bigger than some $h \in \mathbb{N}$), and converges to a in X_i (when so restricted).

(e) Suppose that the space X' is the disjoint union of a family (X_i) of subspaces. The canonical mapping $v\colon \Sigma_i X_i \to X'$ whose co-components are the inclusions $v_i\colon X_i \to X'$ is obviously bijective. It is a homeomorphism if and only if all X_i are open in X'.

(f) Letting $X_i = Y$ (for all $i \in I$), the space $\Sigma_{i\in I} Y$ is canonically isomorphic to the product $Y \times I_d$ of Y with the discrete space on the set of indices.

In particular, the sum $\mathbb{R} + \mathbb{R}$ of two euclidean lines can be embedded as the subspace of $\mathbb{R}^2$ formed by two parallel lines, and the sum $\Sigma_{i\in\mathbb{R}} \mathbb{R}$ is homeomorphic to the product $\mathbb{R} \times \mathbb{R}_d$ (a topology of the plane).

3.5.8 Actions of groups on spaces

Actions of groups on sets have been considered in 1.6.4.

We say that the group G *acts* on the topological space X, and that X is a *G-space*, when we have assigned a (left) action of G on the underlying set $|X|$, which is continuous, in the sense that each operator $g \in G$ gives a continuous mapping

$$g_X\colon X \to X, \qquad x \mapsto gx. \tag{3.54}$$

Then g_X is inverse to $(g^{-1})_X$, and all these mappings are homeomorphisms.

Defining such an action is the same as giving a group-homomorphism $G \to \mathrm{Aut}(X)$, from G to the group of homeomorphisms $X \to X$ (see 3.2.8(c)).

The *orbit space* of the action is the orbit set X/G, equipped with the quotient topology. Its elements are the orbits

$$[x] = Gx = \{gx \mid g \in G\},$$

i.e. the equivalence classes of the congruence modulo G. The saturated of any subset $A \subset X$ for this congruence is: $\mathrm{sat}_G(A) = \bigcup_g gA$.

Exercises and complements. (a) In a G-space X, the projection $p\colon X \to X/G$ is always open: the congruence modulo G in X is an open equivalence relation.

(b) We have described, in 3.2.1(e), an action of the multiplicative group $G = \mathbb{R}^*$ on the space $\mathbb{R}^n$, by homotheties

$$m_\lambda \colon \mathbb{R}^n \to \mathbb{R}^n, \qquad m_\lambda(x) = \lambda x \qquad (\lambda \in \mathbb{R}^*). \tag{3.55}$$

Its orbits are the singleton $\{0\}$ and the lines Gx through the origin, without the latter.

(This action, restricted to the subspace $\mathbb{R}^n \setminus \{0\}$, will be used to define the projective real space of dimension $n-1$, in Section 4.4.)

3.6 Limits of mappings and Hausdorff spaces

After reviewing the convergence of real functions, we extend it to mappings between topological spaces, which are not required to be continuous.

Then we deal with the Hausdorff property, which ensures the uniqueness of limits – besides being crucial in other aspects. It is named after Felix Hausdorff, one of the founders of Topology, who introduced in 1914 the topological spaces that satisfy this condition [Hd], giving the first formal definition of a topological space (in this stronger form).

3.6.1 Limits of real functions

Suppose we have a function $f\colon A \to \mathbb{R}$ defined on a subset A of $\mathbb{R}^n$, and a point $a \in \mathbb{R}^n$ (which can belong to A or not).

The limit of f at the point a makes sense when a is *adherent to* A, which means that every open ball $B_n(a, r)$ meets A: we want to be able to compute $f(x)$ on points x that are *arbitrarily close* to a.

If this is the case, we say that the real number λ is the *limit* of the function f *at the point* a if, equivalently:

(i) for every $\varepsilon > 0$ there is some $\delta > 0$ such that, if $x \in A$ and $d(a, x) < \delta$, then $|f(x) - \lambda| < \varepsilon$,

(ii) for every $N \in \mathcal{B}_1(\lambda)$ there is some $N' \in \mathcal{B}_n(a)$ such that $f(N' \cap A)$ is contained in N.

We also say that $f(x)$ *tends to* $\lambda \in \mathbb{R}$ *as the variable* x *tends to* a in A. The limit-value λ is written as $\lim_a f$, or more usually as

$$\lim_{x \to a} f(x). \tag{3.56}$$

In the previous hypotheses, one can prove that the function f has at most one limit, at the point a. The reader can easily give a proof similar to that of Exercise 3.3.2(a), for limits of sequences. Or wait until a more general result is proved, in Exercise 3.6.3(a).

In Calculus, if $a \in A$, one often excludes the value $f(a)$ from any influence on the limit of f, considering the restriction of f to the subset $A \setminus \{a\}$. Accordingly, the point a is required to be a *cluster point* of A, i.e. adherent to $A \setminus \{a\}$: see 3.2.4(f).

This gives some difference in terminology.

3.6.2 Limits of a mapping

More generally, let $f\colon A \to Y$ be a mapping defined on a subset A of a space X, with values in a space Y. Let $a \in \overline{A}$.

We say that the point $b \in Y$ is *a limit* of the mapping f *at the point* a if:

(i) for every $N \in \mathcal{N}_Y(b)$ there is some $N' \in \mathcal{N}_X(a)$ such that $f(N' \cap A)$ is contained in N.

We also say that $f(x)$ *tends to* b, in Y, *as the variable* x *tends to* a, in A.

The limit of f at a need not be unique: for instance, if the space Y is indiscrete, any mapping $f\colon A \to Y$ admits any point $b \in Y$ as a limit at any $a \in \overline{A}$.

When our limit is unique, it can be written as $\lim_a f$, or as

$$\lim_{x \to a} f(x). \tag{3.57}$$

We will see in Section 4.3 how limits 'at infinity' can be dealt with, by suitable compactifications of euclidean spaces. Or, more generally, by 'limits at filters', briefly described in 4.3.5.

Exercises and complements. (a) (*Limits and continuity*) In the previous situation, if $a \in A$, the mapping f is continuous at a if and only if $f(a)$ is a limit of f at a.

On the other hand, if $a \notin A$, the point b is a limit of f at a if and only if the following extension of the mapping f

$$g\colon A \cup \{a\} \to Y, \qquad g(x) = f(x) \text{ for } x \neq a, \quad g(a) = b, \tag{3.58}$$

is continuous at a.

(b) Use a homeomorphism $\varphi\colon [0, r[\to [0, +\infty[$ (Exercise 3.4.4(d)) to prove that the open ball $B_n(0, r)$ is homeomorphic to $\mathbb{R}^n$ (for $r > 0$).

(c) $\mathbb{R}^n$ is homeomorphic to any open cube $\prod_i]a_i - r, a_i + r[$ (for $r > 0$).

3.6.3 Hausdorff spaces

As an important property related to limits, a topological space X is said to be *Hausdorff*, or a *T_2-space*, if it satisfies the Hausdorff separation axiom:

(T_2) two distinct points $x, y \in X$ have disjoint neighbourhoods.

Then, in each pair of local bases $\mathcal{B}(x)$, $\mathcal{B}(y)$ of two distinct points x, y we can find two disjoint elements, $U \in \mathcal{B}(x)$ and $V \in \mathcal{B}(y)$.

In a Hausdorff space, a sequence can only converge to one point, by the same obvious argument already used in Exercise 3.3.5(c): the sequence cannot eventually lie in two disjoint sets.

Exercises and complements. (a) A mapping $f: A \to Y$ defined on a subset A of a space X, with values in a Hausdorff space Y, can have at most one limit at a point a adherent to A in X.

(b) Every discrete space is Hausdorff. An indiscrete space with at least two points is not Hausdorff. The Sierpinski space (see 3.2.1(a)) is not Hausdorff. An infinite space with the cofinite topology is not Hausdorff.

3.6.4 Theorem

(a) (Comparison) *Any topology finer than a Hausdorff topology is Hausdorff.*

(b) (Subspaces) *A subspace of a Hausdorff space is Hausdorff.*

(c) (Products) *A product $X = \Pi_i X_i$ of Hausdorff spaces is Hausdorff.*

Conversely (or 'nearly so'), if all the factors X_i are non empty and X is Hausdorff, so is each space X_i.

Proof The first two points are obvious. The third is easy, and can be proved as an exercise.

Suppose that all the spaces X_i are Hausdorff, and take two distinct points $x = (x_i)$ and $y = (y_i)$ in X. Then there exists an index $i \in I$ such that $x_i \neq y_i$, and we have two disjoint neighbourhoods U, V of these points in X_i. Their preimages $p_i^{-1}(U)$ and $p_i^{-1}(V)$ are disjoint neighbourhoods of x, y in X.

Conversely, if X is Hausdorff and all factors X_i are non-empty, then each X_i is homeomorphic to a subspace of X, by Exercise 3.5.4(b), and Hausdorff too. □

3.6.5 Exercises and complements

(a) The space $\mathbb{R}^n$ is Hausdorff. Any euclidean space $X \subset \mathbb{R}^n$ is also.

(b) A quotient X/R of an arbitrary topological space X is Hausdorff if and only if:

- for every pair $[x] \neq [y]$ of points of X/R there exist two disjoint neighbourhoods of x and y in X, which are saturated for R.

(c) (*Hausdorff orbit spaces*) If X is a Hausdorff space and G is a finite group acting on X, the orbit space $Y = X/G$ is Hausdorff.

(d) Give an example of a Hausdorff quotient of a non-Hausdorff space, and an example of a non-Hausdorff quotient of a Hausdorff space. For the second example one can use the action of an infinite group.

(e) A sum $X = \Sigma_i X_i$ of topological spaces is Hausdorff if and only if all the spaces X_i are Hausdorff.

3.6.6 Proposition (The diagonal of a Hausdorff space)

A topological space X is Hausdorff if and only if the diagonal of X

$$\Delta_X = \{(x, x) \mid x \in X\}, \tag{3.59}$$

is a closed subset of the cartesian product $X \times X$.

Proof Let $x_1 \neq x_2$ in X, so that $(x_1, x_2) \notin \Delta_X$. If Δ_X is closed, the point (x_1, x_2) has a neighbourhood N in $X \times X$ that does not meet the diagonal; then N contains a product $N_1 \times N_2$ of neighbourhoods of x_1 and x_2, which are disjoint.

The converse is proved by the same argument, backwards. □

3.6.7 The equaliser

The *equaliser* of two continuous mappings $f, g\colon X \to Y$ is the subspace of X on which they coincide

$$\mathrm{Eq}(f, g) = \{x \in X \mid f(x) = g(x)\} \subset X. \tag{3.60}$$

The equaliser is defined in any category, by a morphism $E \to X$ satisfying a universal property: see 6.1.3.

Exercises and complements. (a) In the previous situation, suppose that the space Y is Hausdorff. Then $\mathrm{Eq}(f, g)$ is closed in X. If the mappings f, g coincide on a subset A that is dense in X, then they are the same.

As a well known consequence, if two continuous functions $f, g\colon \mathbb{R} \to \mathbb{R}$ coincide on all rational numbers, they coincide everywhere.

(b) Any retract of a Hausdorff space is closed in the latter.

3.6.8 Other separation axioms

Many separation properties are important in Topology. Basically, they deal with separating points, or subsets of a topological space X, by means of neighbourhoods. By a *neighbourhood of a subset* A of X we mean any subset N containing an open set that contains A; equivalently, N is a neighbourhood of each point of A (in X).

We only mention some of these properties.

(i) X is a *T_0-space* if, for $x \neq y$ in X, there is a neighbourhood of one of them that does not contain the other point. Equivalently, $x \neq y$ implies $\mathcal{N}(x) \neq \mathcal{N}(y)$: *the topology of X distinguishes the points.*

(ii) X is a *T_1-space* if, for $x \neq y$ in X, there is a neighbourhood of each of them that does not contain the other point. Equivalently, all singletons of X are closed.

(iii) We have already defined T_2-spaces, or Hausdorff spaces.

(iv) X is a *T_3-space* (or a *regular Hausdorff* space) if all its points are closed, and for any closed set C and any point $x \notin C$, there exist two disjoint open sets U, V such that $C \subset U$ and $x \in V$.

(v) X is a *T_4-space* (or a *normal Hausdorff* space) if all its points are closed, and for any pair of disjoint closed sets C, D, there exist two disjoint open sets U, V such that $C \subset U$ and $D \subset V$.

Exercises and complements. The following exercises are obvious or easy, except the last, for which a reference is given.

(a) If a space satisfies one of the properties (i)–(v), it also satisfies the previous ones. A discrete space satisfies all these properties; a chaotic space with at least two points none of them.

(b) The Sierpinski space is T_0, and not T_1. A cofinite topology on an infinite set is T_1, and not T_2. On any set, the cofinite topology is the coarsest T_1-topology.

*(c) Every euclidean space, more generally every metric space, is T_4 ([Bou3], IX.4.1, Proposition 2).

4
Topological structures, II

We explore now two fundamental features of topological spaces: connectedness and compactness.

Then we deal briefly with metric spaces and algebro-topological structures, like topological groups and topological vector spaces.

4.1 Connected spaces

Connectedness is a quite intuitive property of topological spaces: loosely speaking, a connected space is 'made of one piece'. Every space is the disjoint union of its connected components.

For instance, the real line $\mathbb{R}$, all its intervals and all $\mathbb{R}^n$ are connected. The euclidean space $\mathbb{R} \setminus \{0\}$ is not, and has two connected components, while $\mathbb{R} \setminus \{0, 1\}$ has three of them. The connectedness of the intervals of the real line gives (an extended version of) the Intermediate Value Theorem (in 4.1.5(b)), a well-known theorem of elementary Calculus.

Formally, this property has a simple definition: in a connected space the only subsets which are both open and closed are the 'unavoidable' ones, namely the empty subset and the total one.

4.1.1 Definition

A topological space X is said to be *connected* if it satisfies the following properties, which are plainly equivalent:

(i) the only subsets of X that are open and closed are the empty subset and the total one,

(ii) if X is a disjoint union of two open subsets, one of them is empty,

(ii$'$) if X is a disjoint union of two open subsets, one of them is total,

(iii) if X is a disjoint union of two closed subsets, one of them is empty,

(iii′) if X is a disjoint union of two closed subsets, one of them is total.

If this is not true we say that X is *disconnected*. When we use these terms for a *subset* A of a space X, we refer to the topology of A as a subspace of X (unless differently specified).

Exercises and complements. The following points are straightforward.

(a) The empty space and each singleton are connected. Every discrete space with at least two points is disconnected. Every indiscrete space is connected. On the two-point set $\{0, 1\}$, the Sierpinski topology (in Exercise 3.2.1(a)) is connected.

(b) *Note.* Some authors exclude the empty space from the connected ones. In a particular context, one choice or the other can be preferable. Here we follow the definition above.

(c) In any space, the union of two disjoint, non-empty open subsets (or closed subsets) is a disconnected subspace.

(d) A sum $X_1 + X_2$ of non-empty topological spaces is disconnected.

(e) The Alexandrov topology on a totally ordered set is connected. In particular, the Alexandrov space $A(\mathbb{R}, \leqslant)$ is connected.

(f) A space X is connected if and only if every continuous mapping $f\colon X \to D\{0, 1\}$ with values in the two-point discrete space is constant.

(g) (*Connected induction*) If X is a connected space and A is a subset, the following steps (similar to induction on natural numbers) imply that A is the whole space:

(*i*) (*The initial step*) A is not empty,
(*ii*) (*First inductive step*) A is open in X,
(*iii*) (*Second inductive step*) A is closed in X.

4.1.2 Theorem (Subspaces and connectedness)

Let X be a space.

(a) If $A = \bigcup_i A_i$ is a union of connected subsets of X and $\bigcap_i A_i \neq \emptyset$, then A is connected.

(b) If A is a connected subset of X, its closure $\overline{A}$ is also connected.

Note. In (a) one can replace the hypothesis $\bigcap_i A_i \neq \emptyset$ with a weaker one: there exists an index $j \in I$ such that each intersection $A_i \cap A_j$ is non-empty. (The proof only needs minor adaptations, or one can apply (a) to each subset $B_i = A_i \cup A_j$ and then to their family.)

Proof (a) Let $x_0 \in \bigcap_i A_i$, and suppose that I is non-empty (otherwise $A = \emptyset$ is connected). Recall that the topology of A_i as a subspace of X is also induced by A (by transitivity of subspaces, in 3.4.2(e)).

Suppose that A is a disjoint union $U \cup V$ of open subsets. Then $A_i =$

$(A_i \cap U) \cup (A_i \cap V)$ is a disjoint union of open subsets of A_i, and in each case one of them is the whole A_i. Fixing a particular index $j \in I$, we can suppose that $A_j = A_j \cap U$, so that $x_0 \in U$. Then each $A_i \cap U$ is non-empty (it contains x_0), whence $A_i = A_i \cap U$, for every i, and finally $A = U$.

(b) Suppose that $\overline{A}$ is a disjoint union $C \cup D$ of closed subsets of $\overline{A}$, which are also closed in X. Then A is the disjoint union of their traces, that are closed in A, and must coincide with one of them: say that $A \subset C$. It follows that $\overline{A}$ is also contained in C. □

4.1.3 Theorem and Definition (Connected components)

(a) A topological space X can be decomposed as a disjoint union of its connected components, *the maximal connected subspaces. For every $x \in X$ we write as* $\mathrm{Cc}(x)$, *or* $\mathrm{Cc}_X(x)$, *the connected component of X containing the point x.*

(b) The connected components of a space X are closed in X. If all of them are also open in X, the space is homeomorphic to their topological sum. This certainly holds when there is a finite number of connected components. It also holds if X is an Alexandrov space.

(c) A space is said to be totally disconnected *if all its connected components are singletons, and it has at least two points.*

Note. The connected components of a space are components of points: the empty space has none, while a non-empty connected space has precisely one. This is a main cleavage between the empty space and the other connected spaces.

Proof (a) Consider the relation $x \sim_C y$ defined by the existence of a connected subspace A of X that contains x and y. This relation is reflexive, because every singleton is connected; it is trivially symmetric; it is also transitive: if $x, y \in A$ and $y, z \in B$ (connected subspaces), then $A \cup B$ is connected by Theorem 4.1.2(a).

X is thus the disjoint union of the equivalence classes of this relation.

The equivalence class $\mathrm{Cc}(x)$ of the point x is the union of all the connected subsets of X that contain this point. It is connected (applying again Theorem 4.1.2(a)), and obviously a maximal connected subset of X.

(b) We know that the closure of a connected subspace is connected (Theorem 4.1.2(b)). The second assertion follows from Exercise 3.5.7(e). Finally, a connected component is the complement of the union of the others; this subset is certainly closed when there is a finite number of components, and also when X is an Alexandrov space. □

4.1.4 Theorem

(a) (Comparison) *Any topology less fine than a connected one is connected.*

(b) (Continuous images) *If $f\colon X \to Y$ is a surjective continuous mapping and the space X is connected, so is Y. In particular, every quotient and every retract of a connected space is connected.*

(c) (Products) *A product $X = \prod_i X_i$ of connected spaces is connected. Conversely, if the factors X_i are non empty and X is connected, so is each space X_i.*

Proof (a) Obvious.

(b) Let Y be a disjoint union $U \cup V$ of open subsets. Then

$$X = f^{-1}(U) \cup f^{-1}(V)$$

is also a disjoint union of open subsets, and one of them is empty; therefore U or V is empty.

(c) If all the factors X_i are non-empty, the projections $p_i\colon X \to X_i$ are surjective, and the connectedness of X implies that of X_i.

Conversely, we begin by proving that a product $X \times Y$ of connected spaces is connected. We suppose that there is a point (a, b) in $X \times Y$, and we show that any other point (a', b') belongs to the same connected component.

In fact the subspaces $X \times \{b\}$ and $\{a'\} \times Y$ are homeomorphic to X and Y, respectively (by Exercise 3.5.4(b)), and connected. But they meet at (a', b), whence their union is connected (by Theorem 4.1.2(a)) and contains (a', b').

By induction, any finite product of connected spaces is connected.

Consider now an arbitrary product $X = \prod_i X_i$ of connected spaces, and take a point $a = (a_i) \in X$ (if any); we want to prove that its connected component $C_X(a)$ in X is the total space. For every finite subset $J \subset I$, let

$$A_J = \{(x_i) \in X \mid x_i = a_i \text{ for } i \in I \setminus J\},$$

where the only 'free' coordinates are those indexed by J.

The subspace A_J is homeomorphic to the finite product $\prod_{i \in J} X_i$ (an obvious extension of what we have already used when J is a singleton), and connected.

Moreover $a \in A_J$, whence $C_X(a)$ contains $A = \bigcup A_J$, the subset of all points x which differ from a on a finite number of indices.

Finally, we prove that A is dense in X, so that $C_X(a) = X$ (by Theorem 4.1.3(b)). In fact, a non-empty canonical open set $U = \prod_i U_i$ of X has $U_i = X_i$ for $i \in I \setminus J$ (for a finite set $J \subset I$), and meets A_J in the subset

$$\{(x_i) \in X \mid x_i = a_i \text{ for } i \in I \setminus J, \text{ and } x_i \in U_i \text{ for } i \in J\},$$

that is not empty; thus $U \cap A \neq \emptyset$. □

4.1.5 Theorem (Euclidean spaces and connectedness)

(a) A subspace A of the euclidean line $\mathbb{R}$ is connected if and only if it is an interval. In particular, $\mathbb{R}$ is connected.

(b) (*Intermediate Value Theorem*) The image of a continuous mapping $f\colon X \to \mathbb{R}$ defined on a connected space is an interval of the line.

(c) All the euclidean spaces $\mathbb{R}^n$ are connected.

(d) The euclidean space $\mathbb{Q}$ of rational numbers is totally disconnected; its connected components – the singletons – are closed and not open. The non-degenerate intervals of $\mathbb{Q}$ are also totally disconnected.

Proof (a) If $A \subset \mathbb{R}$ is connected, we prove that it satisfies the characterisation of intervals, in Proposition 1.7.3. Let $a, b \in A$ and $x \in \mathbb{R}$, with $a < x < b$. Then $x \in A$, otherwise the pierced line

$$\mathbb{R} \setminus \{x\} = \,]-\infty, x[\, \cup \,]x, +\infty[$$

would give, by intersection with A, a disjoint union $A = U \cup V$ of open sets of A, both non-empty as $a \in U$ and $b \in V$.

Conversely, suppose that A is an interval. If it is not connected, it is a disjoint union $A = C \cup D$ of non-empty closed subsets of A. Taking $a \in C$ and $b \in D$, we can suppose that $a < b$ and we get a disjoint union $[a, b] = C' \cup D'$ of closed subsets of $[a, b]$, which are also closed in the line.

Taking $x = \sup C'$, it follows that $]x, b] \subset D'$. But C' and D' are closed in $\mathbb{R}$, whence $x \in C'$ (by Exercise 3.2.4(d)) and $x \in [x, b] \subset D'$, a contradiction.

(b), (c) From the previous point and Theorem 4.1.4.

(d) If a subspace $A \subset \mathbb{Q}$ is connected, then it is a connected subspace of the real line, and therefore a real interval; being contained in $\mathbb{Q}$, it is degenerate: either empty or a singleton. Thus all the connected components of $\mathbb{Q}$ are singletons, and none is open in the euclidean topology of $\mathbb{Q}$. □

4.1.6 Path-connected spaces

There is a stronger notion of connectedness, based on (continuous) paths, which we briefly sketch here.

By definition, a *path* in the space X is a continuous mapping

$$a \colon \mathbb{I} \to X, \tag{4.1}$$

defined on the standard interval $\mathbb{I} = [0, 1]$ (with euclidean topology). Its *endpoints* are $x = a(0)$ and $y = a(1)$, and we say that a is a path in X, *from x to y*.

We say that the space X is *path-connected*, or *pathwise connected*, if any two points of X are linked by a path in X. A subset of X is said to be *path-connected* if it is, as a subspace. (Also here, some authors exclude the empty space.)

Any space X is the disjoint union of its *path-components*, which are the maximal path-connected subsets. In fact, we have the relation $x \sim_P y$ defined by the existence of a path in X, from x to y. It is an equivalence relation, as (easily) verified in Exercise (a), below. Its equivalence classes form the components we are interested in.

The empty space has no path-components, while every other pathwise connected space has one, and only one.

Exercises and complements. (a) The relation $x \sim_P y$ defined above is an equivalence relation.

(b) The euclidean spaces $\mathbb{R}^n$ are path-connected.

(c) The same is true of every *convex* subset $A \subset \mathbb{R}^n$, which means that: if $x, y \in A$, the line segment from x to y is contained in A.

More generally, the same is true of every subset $A \subset \mathbb{R}^n$ which is *starred* with respect to some $x \in A$; this means that: if $y \in A$, the line segment from x to y is contained in A.

(d) The path-component of x in X is contained in the connected component $\mathrm{Cc}(x)$. Every path-connected space is connected.

The converse is not true: an example can be found in the Solutions; it also shows that the path-components need not be closed.

(e) Path-connected spaces are stable under continuous image.

(f) Path-connected spaces are stable under cartesian products.

4.1.7 Exercises and complements

(a) (*Distinguishing spaces by connectedness*) Homeomorphic spaces have the same number (or cardinal) of connected components, and the same number (or cardinal) of path-components. This can be used to prove that some spaces are not homeomorphic.

(b) (*The shape of the intervals, II*) Using connectedness in a more refined way, we can complete the analysis of the topological type of the intervals of the euclidean line, in Exercise 3.4.4(d), showing that the homeomorphism-types of the intervals $]0,1[$, $[0,1[$ and $[0,1]$ are distinct.

(c) Consider the following subspace of the euclidean space $\mathbb{R}^n$

$$A_n = \{(x_i) \in \mathbb{R}^n \mid x_i \neq 0 \text{ for at most one index } i\}, \qquad (4.2)$$

namely the union of the n cartesian axes x_i (which meet at the origin). Prove that all the spaces A_n, for $n \in \mathbb{N}$, have distinct homeomorphism-type.

(d) In a preordered set X, consider the equivalence relation $\approx$ generated by the relation $x \prec y$. This means that $x \approx y$ when there is a finite sequence $x_0, x_1, ..., x_n$ of pairwise *comparable* points (i.e. $x_{i-1} \prec x_i$ or $x_i \prec x_{i-1}$), from $x_0 = x$ to $x_n = y$.

Prove that the equivalence classes of X for this relation are the same as the connected components of the Alexandrov space $A(X)$.

(e) If, in a preordered set X, each pair of points has an upper bound, then $A(X)$ is connected. In particular, this applies to $A(\mathbb{R}^n, \leqslant)$.

4.1.8 Local connectedness

A space X is said to be *locally connected* if every point has a local basis formed of connected open sets. It is *locally path-connected* if every point has a local basis formed of path-connected open sets. The second property implies the first.

Exercises and complements. (a) The space $\mathbb{R}^n$ is locally path-connected.

(b) A locally path-connected space that is connected is also path-connected.

*(c) A connected space need not be locally connected. *Hints:* use the subspace $B \subset \mathbb{R}^2$ constructed in the solution of Exercise 4.1.6(d).

*4.1.9 *Exercises and complements* (Invertible real functions)

(a) Let $f\colon I \to \mathbb{R}$ be a continuous function, defined on a real interval. If f is injective, then it is either strictly increasing or strictly decreasing.

Hints. This is a well known result of elementary Calculus, which can be given a brief, elegant proof using the connectedness of the subset

$$A = \{(x_1, x_2) \in I \times I \mid x_1 < x_2\} \subset \mathbb{R}^2.$$

(b) Let $f\colon I \to \mathbb{R}$ be an injective function defined on an interval of $\mathbb{R}$. Then

f gives a homeomorphism $I \to f(I)$ between intervals of the line, and an isomorphism or an anti-isomorphism of ordered sets.

(c) (*Real roots*) We can now deduce that, for each $n \geqslant 1$, the power function $f\colon I \to I$, $f(x) = x^n$ is a homeomorphism of the interval $I = [0, +\infty[$ on itself; its inverse is the n-th real root function.

4.2 Compact spaces

Compactness is another crucial property of topology, which – historically – went through various formulations, before reaching the conclusive one.

The primitive notion, now called a 'sequentially compact' space, requires that every sequence of points in the space contains a subsequence that has a limit. This notion works well in the euclidean spaces, but fails to do so in more complex spaces.

A general notion of compactness, based on the open covers of a space, came forth in the beginning of the 20th century. The results of the Moscow school of Topology, in particular Tychonoff's Theorem in 1930–35 [Ty1, Ty2], achieved showing that it was the good notion.

The compactness of the intervals $[a, b]$ of the real line gives another basic result of elementary Calculus, the Extreme Value Theorem (in 4.2.6(b)), in an extended version.

Compact Hausdorff spaces are particularly important, as shown by Theorem 4.2.7: they form a category with some characters of 'algebraicity'.

(Bourbaki and others use the term 'compact' as including the Hausdorff property; a compact space in the present sense is then called 'quasi compact'.)

4.2.1 Main definitions

Let X be a topological space.

We say that X is *compact* if every open cover $(U_i)_{i \in I}$ of X has a finite *subcover*: this means that there exists a finite subset $J \subset I$ such that $X = \bigcup_{i \in J} U_i$.

(Open covers of X can equivalently be treated as indexed families of open sets of X, or sets $\mathcal{U} \subset \mathcal{O}(X)$ of open sets.)

A subset $A \subset X$ is said to be *compact* in X if it is, as a subspace. It is said to be *relatively compact* in X if its closure in X is compact.

Exercises and complements. (a) Every finite space is compact. A discrete space is compact if and only if it is finite. Every indiscrete space is compact.

(b) The compact euclidean spaces will be characterised in Exercise 4.2.6(a), by the Heine–Borel Theorem. But the reader can easily see now that the euclidean line is not compact, and the intervals $]a, b[$ and $[a, b[$ are neither (for $a < b$).

(c) The compactness of a space X can equivalently be assessed using open covers formed of elements of a given basis $\mathcal{B}$ of open sets.

(d) A space X is compact if and only if every family $(N_x)_{x \in X}$ where each N_x is an open neighbourhood of the point x, has a finite subcover.

(e) A subspace $K \subset X$ is compact if and only if every family $(U_i)_{i \in I}$ of open sets of X which *covers* K (in the sense that $K \subset \bigcup_{i \in I} U_i$) has a finite subcover (of K).

(f) Every space with the cofinite topology is compact.

4.2.2 **Theorem**

(a) (Comparison) *Any topology less fine than a compact one is compact.*

(b) (Continuous images) *If $f: X \to Y$ is a surjective continuous mapping and the space X is compact, so is Y.*

In particular, every quotient and every retract of a compact space is compact.

Proof The first point is obvious.

For the second, take an open cover $(V_i)_{i \in I}$ of Y. Then the preimages $U_i = f^{-1}(V_i)$ form an open cover of X, from which we can extract a cover $(U_i)_{i \in J}$ indexed by a finite subset J of I. Thus $X = \bigcup_{i \in J} U_i$ and

$$Y = f(X) = \bigcup_{i \in J} f(U_i) = \bigcup_{i \in J} V_i.$$

□

4.2.3 **Theorem** (Subspaces and compactness)

(a) A closed subspace C of a compact space K is compact.

(b) A finite union $A = \bigcup_j K_j$ of compact subspaces of a space X is compact.

(c) A compact subspace K of a Hausdorff space X is a closed subset of X. Moreover, for every $y \notin K$ we can find two disjoint open sets U, V of X such that $K \subset U$ and $y \in V$.

(d) Every compact Hausdorff space is T_3 (see 3.6.8).

Proof We use the characterisation 4.2.1(e) of compact subspaces.

(a) Take a family $(U_i)_{i \in I}$ of open sets of K that covers C. Adding the open set $U = K \setminus C$, we obtain a family that covers the whole space K. There is thus a finite subfamily $(U_i)_{i\in J}$ which, together with U, covers K. But then the family $(U_i)_{i\in J}$ covers C.

(b) From a family $(U_i)_{i\in I}$ of open sets of X that covers A we can extract a finite subfamily that covers K_j; all of them – together – form a finite subcover of A.

(c) Let us fix a point $y \notin K$. For every $x \in K$ there exist two disjoint open sets, U_x and V_x, with $x \in U_x$ and $y \in V_x$. The open sets U_x cover K, and there exists a finite set of points $x_1, ..., x_n \in K$ such that $K \subset U = \bigcup_i U_{x_i}$.

Then the finite intersection $V = \bigcap_i V_{x_i}$ is an open neighbourhood of y that does not meet U

$$V \cap (\bigcup_i U_{x_i}) = \bigcup_i (V \cap U_{x_i}) \subset \bigcup_i (V_{x_i} \cap U_{x_i}) = \emptyset.$$

We have also proved that every point $y \notin K$ is interior to $X \setminus K$.

(d) It is a straightforward consequence of (a) and (c). If the space X is compact Hausdorff, a closed set K is compact and any point $y \notin K$ can be separated from K, as above. □

4.2.4 Tychonoff's Theorem

A product $X = \prod_{i\in I} X_i$ of compact spaces is compact. Conversely, if the factors X_i are non empty and X is compact, so is each space X_i.

Proof We only write the (non-obvious part of the) proof for a finite set of indices I. For an infinite cartesian product, an interested reader can see [Ke], Chapter 5, Theorem 13, or [Bou2], Section I.9.5 (and the preliminary parts required by these proofs).

If all the factors X_i are non empty, the projections $p_i \colon X \to X_i$ are surjective, and the compactness of X implies that of X_i.

For the converse and a finite set of indices, we only have to consider a product $Z = X \times Y$ of two compact spaces. Using Exercise 4.2.1(c), we start from an open cover $(U_i \times V_i)_{i\in I}$ of $X \times Y$ formed of canonical open sets of the product; then (U_i) and (V_i) are open covers of X and Y, respectively.

Fixing a point $x \in X$, the space $\{x\} \times Y$ is homeomorphic to Y and compact. There is thus a finite subset $J(x) \subset I$ such that the family $(U_i \times V_i)_{i\in J(x)}$ covers $\{x\} \times Y$.

For every $x \in X$, $U(x) = \bigcap_{i\in J(x)} U_i$ is an open neighbourhood of x in X and

$$U(x) \times Y \subset \bigcup_{i\in J(x)} U_i \times V_i. \tag{4.3}$$

Now, the family of all $U(x)$ covers X, and there is a finite family $x_1, ..., x_m$ of points of X such that

$$X = U(x_1) \cup ... \cup U(x_m). \tag{4.4}$$

We have now a finite subset $J = J(x_1) \cup ... \cup J(x_m) \subset I$ such that, applying (4.3)

$$X \times Y \subset (U(x_1) \times Y) \cup ... \cup (U(x_m) \times Y) \subset \bigcup_{i \in J} U_i \times V_i. \tag{4.5}$$

This gives a finite subcover of $(U_i \times V_i)_{i \in I}$. □

4.2.5 Theorem (Heine–Borel)

A bounded closed interval $[a, b]$ of the euclidean line is compact.

Note. The theorem is named after Eduard Heine and Émile Borel, but came out of the research of many authors in the 19th century, before the present (general) definition of compactness.

Proof The proof can be quite simple, as we already know that the interval $X = [a, b]$ is a connected space. Let $\mathcal{U}$ be an open cover of X, and

$$A = \{x \in X \mid [a, x] \text{ is covered by a finite subset of } \mathcal{U}\}. \tag{4.6}$$

We want to prove that $A = X$. Plainly $a \in A$, that is not empty.

Now, if $x \in X$, there is some $r > 0$ such that $]x-r, x+r[\cap X$ is contained in an open $U \in \mathcal{U}$.

If $x \in A$, the interval $[a, x]$ is covered by a finite subset of the cover $\mathcal{U}$, and $]x - r, x + r[\cap X \subset A$; therefore A is open in X.

On the other hand, if $x \notin A$, no point of $]x - r, x + r[\cap X$ can belong to A; therefore $X \setminus A$ is also open in X. Finally, $A = X$.

Note. If X is the (non-compact) interval $[a, b[$ (for $a < b$), we conclude as above that $A = X$; but this only says that every interval $[a, x] \subset [a, b[$ is covered by a finite subset of $\mathcal{U}$, which is of little consequence for $[a, b[$.

□

4.2.6 Exercises and complements

The following results are important, and should be kept in mind. Their proof is easy, after the previous theorems, and can be found below, unless it is trivial.

(a) (*Compact euclidean spaces*) A subspace of the euclidean space $\mathbb{R}^n$ is compact if and only if it is closed in $\mathbb{R}^n$ and bounded.

(b) (*Extreme Value Theorem*) A continuous function $f\colon K \to \mathbb{R}$ defined on a compact non-empty space has a minimum value $m = \min(\operatorname{Im} f)$ and a maximum value $M = \max(\operatorname{Im} f)$.

(c) (*Corollary*) The image of a continuous function $f\colon X \to \mathbb{R}$ defined on a connected compact non-empty space is a bounded closed interval $[m, M]$, whose endpoints are the minimum and maximum values. In particular this holds for $X = [a, b]$, in the euclidean line.

(d) (*A warning*) To prevent a frequent error, let us stress the fact that compactness *is an absolute property of a space*, independent of the superspaces where we may view it.

Thus, a subspace $A \subset \mathbb{Q}$ is compact if and only if it is a compact space, if and only if it is bounded and *closed in* $\mathbb{R}$.

For instance, the subset $\{1/n \mid n \in \mathbb{N}^*\} \cup \{0\}$ is compact, but a non-degenerate interval $[a, b] \cap \mathbb{Q}$ of the 'rational line' (with $a < b$ in $\mathbb{Q}$) cannot be compact, because its closure in $\mathbb{R}$ contains (infinitely many) irrational points.

(e) Every subspace of a space X with the cofinite topology has the cofinite topology, and is compact (by Exercise 4.2.1(f)). It is only closed in X if it is a finite subset, or the total one.

Solutions. (a) A closed bounded subspace K of $\mathbb{R}^n$ is contained in some disc $D_n(0, r)$ and in some cube $[-r, r]^n$, which is compact by Tychonoff's and Heine–Borel Theorems; being closed there, it is compact.

Conversely, if K is a compact subspace of $\mathbb{R}^n$, then it is closed. It is also bounded, because the open balls $B_n(0, r)$ centred at the origin (for instance) cover K, which is contained in a finite union of them, namely the largest.

(b) $K' = \operatorname{Im} f$ is a compact non-empty space, closed in $\mathbb{R}$ and bounded. It follows that $M = \sup K'$ belongs to K' (by 3.2.4(d)), and $M = \max K'$. Similarly $m = \inf K'$ is the least value of f.

(c) A straightforward consequence of the previous point, together with the Intermediate Value Theorem, in 4.1.5(b).

4.2.7 **Theorem** (Compact Hausdorff spaces)

Let $f\colon X \to Y$ be a continuous mapping with values in a Hausdorff space, and $g\colon X/R_f \to \operatorname{Im} f$ the associated bijective continuous mapping in the canonical factorisation of f.

(a) If X is compact, f is a closed mapping and the induced mapping g is a homeomorphism.

(b) If X is compact and f is bijective, then it is a homeomorphism.

(c) If there is a compact subspace K of X such that $f(K) = f(X)$, then the induced mapping $g\colon X/R_f \to \operatorname{Im} f$ is a homeomorphism.

Note. The reader will note that the structure of compact Hausdorff spaces (with their continuous mappings) has some formal similarity with algebraic structures (with their homomorphisms): a bijective 'distinguished' mapping has always a distinguished inverse.

This will become clearer in Chapter 6: the category of compact Hausdorff spaces can be viewed as a category of 'generalised algebras', see 6.6.5.

Proof (a) Follows from the previous results: a closed subset C of X is compact (by 4.2.3(a)), therefore its image $f(C)$ is a compact space (by 4.2.2(b)), and a closed subset of the Hausdorff space Y (by 4.2.3(c)).

(b) A straightforward consequence.

(c) Since $f(K) = f(X)$, the projection $p\colon X \to X/R_f$ has $p(K) = X/R_f$. Therefore the space X/R_f is compact (by Theorem 4.2.2(b)), and g is a homeomorphism by the previous point. □

4.2.8 Local compactness

We say that a space X is *locally compact* if every point has a compact neighbourhood.

(This notion is undisputed when X is Hausdorff, also because of the next theorem; in the general case, there are various stronger properties that can bear this name, according to different authors.)

Exercises. (a) Every compact space and every discrete space are (obviously) locally compact. The space $\mathbb{R}^n$ is locally compact.

(b) The euclidean space $\mathbb{Q}$ is not locally compact: no point has a compact neighbourhood.

4.2.9 Theorem

Let X be a locally compact Hausdorff space. Then every point has a fundamental system of compact neighbourhoods.

Proof Take a point $x \in X$ and a compact neighbourhood K of x in X. If U is any open neighbourhood of x in X, we can suppose $U \subset K^\circ \subset K$.

The closed set $C = K \setminus U$ does not contain x. Since K is T_3 (by Theorem 4.2.3(d)), there exist disjoint open sets V, W of K with $x \in V$ and $C \subset W$. Then $x \in V \subset (K \setminus W) \subset U$, and $K \setminus W$ is compact, being closed in K.

But V is a neighbourhood of x in K, whence $V \cap K^\circ$ is a neighbourhood of x in K°, and also in X, so that $K \setminus W$ is a compact neighbourhood of x in X, contained in U.

□

4.3 Limits at infinity and compactifications

Classically, a compactification of a space is the embedding of a space in a compact one, as a dense subspace. (We will see in 6.6.5 that there are useful generalisations: the universal Stone–Čech compactification of a space need not be an embedding.)

Adding 'points at infinity' is a classical way of compactifying spaces. This allows us to deal with limits at infinity. More formally, limits on 'filters' will be briefly examined at the end of the section.

Filters were introduced by Henri Cartan [Car] in 1937; they are extensively studied and used in Bourbaki [Bou2]. Alternatively, one can use Moore–Smith 'nets', which are generalised sequences defined on 'directed' ordered sets [Ke]. Filters are less intuitive, but have the advantage of not requiring auxiliary sets of indices.

4.3.1 Extending limits

In Calculus we meet expressions like

$$\lim_{n\to\infty} 1/n = 0, \qquad \lim_{x\to+\infty} 1/x = 0, \tag{4.7}$$

$$\lim_{x\to+\infty} (-x) = -\infty, \qquad \lim_{x\to 0} 1/x = \infty. \tag{4.8}$$

If we want to interpret them at the light of our definition, in 3.6.2, as

- a limit of a mapping $f\colon A \to Y$, at a point a adherent to A in a space X,

we have to introduce extensions of the euclidean spaces $\mathbb{N}$ and $\mathbb{R}$, by adding 'points at infinity'.

This leads to compactifications of our spaces, i.e. topological embeddings in compact spaces. We examine some of them, in an informal way, before dealing with the general theory.

(a) The set $\mathbb{N}$ of natural numbers, as a subspace of the real line, is a discrete euclidean space. It has an interesting *one-point compactification*, the set

$$\mathbb{N}^\bullet = \mathbb{N} \cup \{\infty\} \qquad (\infty \notin \mathbb{N}), \tag{4.9}$$

where the open sets are the subsets of $\mathbb{N}$ together with the cofinite subsets of $\mathbb{N}^\bullet$ that contain the added point ∞.

(The reader can easily verify that this is indeed a topology on the extended set, and a compact one; or wait to see how this comes out of general results – the Alexandrov compactification of the discrete space $\mathbb{N}$.)

This space allows us to view a limit of a sequence $x\colon \mathbb{N} \to Y$ (as defined in 3.3.3) as a limit at the point ∞ of a mapping defined on the subset $\mathbb{N}$ of the space $\mathbb{N}^\bullet$ (as defined in 3.6.2): it is sufficient to note that ∞ is adherent to $\mathbb{N}$ in $\mathbb{N}^\bullet$ (all its neighbourhoods in $\mathbb{N}^\bullet$ meet $\mathbb{N}$), and that a local basis of ∞ is formed by the open sets $\uparrow k \cup \{\infty\}$.

Coming back to the first expression in (4.7), we note that the sequence $x_n = 1/n$ is defined for $n \in \mathbb{N}^*$; this makes no problem, because ∞ is also adherent to $\mathbb{N}^*$ (and to any unbounded subset of $\mathbb{N}$) in $\mathbb{N}^\bullet$.

(b) Directing our attention to larger spaces, the interval $[0,1]$ is obviously a compactification of $]0,1[$, obtained by adding two points. But $]0,1[$ is homeomorphic to the real line, and we are also interested in compactifying the latter by adding two (distinct) new points.

We have already considered the *extended real line*

$$\overline{\mathbb{R}} = \{-\infty\} \cup \mathbb{R} \cup \{+\infty\}, \tag{4.10}$$

as a totally ordered set. We put on this set a topology that makes it homeomorphic to the compact interval $[0,1]$: a basis of open sets is formed by all the intervals (of the ordered set $\overline{\mathbb{R}}$) of the following kinds

$$[-\infty, a[, \qquad]a, b[, \qquad]b, +\infty] \qquad\qquad (\text{for } a < b \text{ in } \mathbb{R}), \tag{4.11}$$

which are even stable under binary intersections, if we add the empty set. The real line is dense in $\overline{\mathbb{R}}$.

Now, the expression

$$\lim_{x \to +\infty} f(x) = \lambda \in \mathbb{R}$$

makes sense (according to 3.6.2) for each function $f \colon A \to \mathbb{R}$ defined on a non-empty, upper unbounded subset $A \subset \mathbb{R}$ (so that $+\infty$ is adherent to A in the extended line).

Moreover, we can view f as a mapping $f \colon A \to \overline{\mathbb{R}}$, covering expressions like

$$\lim_{x \to +\infty} f(x) = -\infty, \qquad \lim_{x \to +\infty} f(x) = +\infty, \ ...$$

as the first limit in (4.8).

(c) To cover the second expression in (4.8) we should consider the function $f(x) = 1/x$ as a mapping $\mathbb{R}^* \to \mathbb{R}^\bullet$, using a one-point compactification of $\mathbb{R}$.

We can visualise $\mathbb{R}^\bullet$ as the topological quotient of the extended line $\overline{\mathbb{R}}$, by identifying the added points $\pm\infty$; or also as a space homeomorphic to the circle $\mathbb{S}^1$, where the real line is embedded as the complement of a point.

More formally, we consider the set

$$\mathbb{R}^\bullet = \mathbb{R} \cup \{\infty\}, \tag{4.12}$$

with the topology whose open sets are the open sets of $\mathbb{R}$ together with the complements in $\mathbb{R}^\bullet$ of the compact subsets of $\mathbb{R}$. This is indeed a topology (as we prove below), and $\mathbb{R}$ is a dense subset.

A local basis of ∞ in this space is the family of the complements of the compact intervals $[-r, r]$ $(r > 0)$. The last expression in (4.8), according to the general notion of limit for a function defined on $\mathbb{R}^*$, means thus:

- for every $r > 0$ there is some $\varepsilon > 0$ such that, if $0 < |x| < \varepsilon$, then $|1/x| > r$.

(d) Some rudiments of complex analysis are sufficient to see that the following limits in the complex variable z

$$\lim_{z \to \infty} 1/z = 0, \qquad \lim_{z \to 0} 1/z = \infty, \tag{4.13}$$

are covered by the one-point compactification of the euclidean space $\mathbb{C}$, where a local basis of ∞ is formed by the complements of the compact discs $D_{\mathbb{C}}(0, r)$, for $r > 0$ (or also $r \in \mathbb{N}^*$).

4.3.2 The Alexandrov compactification

In this chapter, a *compactification* of a space X is a topological embedding of X into a compact space K.

It is certainly of interest that X be dense in K, as we can always replace the latter with the closure K' of X in K, which is compact – and get rid of a part of K with 'no topological influence' on X. Yet the formal theory works better if we do not assume this condition.

We begin by studying the *Alexandrov compactification* $X^\bullet$, or *one-point compactification*, of a topological space X (named again after Pavel S. Alexandrov, see 3.2.7).

It is a beautiful construction, that requires some non-trivial work on the properties of compactness, without being really difficult. An interested reader might like to work it out, after the hints we have seen in 4.3.1. Or begin to read the following, and go on independently, as soon as possible.

(a) We start from a general topological space X, and define a space

$$X^\bullet = X \cup \{\infty\}, \tag{4.14}$$

by adding a point $\infty \notin X$. Its topology $\mathcal{O}$ contains open sets of two (disjoint) kinds:

(i) the open sets U of X,

(ii) the complements $V_K = X^\bullet \setminus K$ of the compact closed subsets K of X.

(If X is Hausdorff, a compact subset of X is automatically closed in X; but we are not assuming this, here.)

(b) In fact, the set $\mathcal{O}$ is stable under arbitrary unions and finite intersections.

For unions, we remark that:

- a union of sets of type (i) is of the same type,

- a non-empty union of sets of type (ii) is of the same type, because every non-empty intersection of closed compact subsets is a closed subset of each of them, and therefore a closed compact subset of X,

- a 'mixed union' $U \cup V_K$ is an open subset of type (ii), because its complement in $X^\bullet$

$$(X^\bullet \setminus U) \cap K = (X \setminus U) \cap K,$$

is a closed subset of K, and therefore compact closed in X.

For finite intersections (in $X^\bullet$):

- a non-empty finite intersection of sets of type (i) is of the same type,

- a finite intersection of sets of type (ii) is of the same type (by 4.2.3(b)),

- a 'mixed intersection' $U \cap V_K = U \cap (X \setminus K)$ is an open of type (i).

(c) The space $X^\bullet$ is compact.

Let $\mathcal{U}$ be an open cover of this space. The added point ∞ belongs to some element of $\mathcal{U}$, which is thus an open set $V_K = X^\bullet \setminus K$ of type (ii). The compact closed subset K of X is covered by the remaining elements of $\mathcal{U}$; as K is also a compact subspace of $X^\bullet$, it is covered by a finite subset of $\mathcal{U}$; adding V_K we have a finite subcover of $\mathcal{U}$.

(d) The original space X is an open subspace of $X^\bullet$.

If X is not compact, then it is dense in $X^\bullet$; otherwise the added point is isolated, i.e. open in $X^\bullet$ (and the only interest of adding it is to avoid an 'exception').

In fact, the traces of the open sets of type (i) and (ii) on the subset X give all the open sets of X, and nothing else. A non-empty open set U of X meets X in itself; a (necessarily non-empty) open set $V_K = X^\bullet \setminus K$ meets X in $X \setminus K$, which is open in X, and can only be empty if X is compact.

(e) The construction is particularly interesting when the space X is Hausdorff and locally compact: then $X^\bullet$ is Hausdorff, as (easily) proved in the theorem below.

Moreover, $X^\bullet$ is determined by being *the* compact Hausdorff space containing X, as a subspace, and one more point (up to the name of the latter, of course).

4.3.3 **Theorem**

(a) If the space X is Hausdorff and locally compact, then its Alexandrov compactification $X^\bullet$ is Hausdorff.

(b) Let A be a compact Hausdorff space, and $a \in A$. Then the (open)

subspace $X = A \setminus \{a\}$ *is Hausdorff locally compact, and* A *has the same topology as the one-point compactification of* X*, with the addition of the point* $a = \infty$.

Proof (a) Two distinct points of X already have disjoint open neighbourhoods in X, and it is sufficient to 'separate' each point $x \in X$ from ∞.

By hypothesis, there is a compact subspace K of X that contains an open set U containing x. Since X is Hausdorff, K is closed in X and we have found two disjoint neighbourhoods, as required: $x \in U$ and $\infty \in V_K = X^\bullet \setminus K$.

(b) The subspace X is obviously Hausdorff, and open in A.

For every point $x \in X$, there are disjoint open neighbourhoods U, V of x and a in A; therefore U is open in X and contained in $K = A \setminus V$, which is closed in A and therefore a compact (closed) subset of X. We have proved that X is locally compact.

Comparing the two topologies, let V be an open set of A. If $a \notin V$, then V is open in X. If $a \in V$, then $K = A \setminus V$ is closed in K and a compact subset of X, whence $V = V_K$ is also an open set of the one-point compactification.

Conversely, if U is open in X it is also open in A. Finally, a subset $V_K = A \setminus K$, where K is compact in X, is a complement of a closed subset of A and open in A. □

4.3.4 Remarks, exercises and complements

(a) We are interested in spaces up to homeomorphism, and each space A homeomorphic to $X^\bullet$ will also be called an Alexandrov compactification of X – specifying how X is embedded in A, or which point $a \in A$ plays the role of the 'added point', when all this is not obvious.

But let us note that 'different' spaces (in the sense of non-homeomorphic) can have the 'same' Alexandrov compactification.

For instance, the compact interval $A = [0, 2]$ can be viewed as the one-point compactification of $[0, 2[$ (with added point 2), but *also* of the disconnected euclidean space $[0, 1[\cup]1, 2]$ (with added point 1). In fact, we have already remarked that the space $[0, 2]$ is not topologically homogeneous (in Exercise 4.1.7(b)): taking out different points can lead to non-homeomorphic spaces.

(b) The discrete space $\mathbb{N}$ is Hausdorff, and locally compact (because all points are open). The one-point compactification $\mathbb{N}^\bullet$ is the space we have introduced in 4.3.1(a), as related to limits of sequences in any topological space.

(c) The euclidean space $\mathbb{R}^n$ is Hausdorff, locally compact. The one-point compactification $(\mathbb{R}^n)^\bullet$ has been considered for $n = 1$, in 4.3.1(c).

Prove that $(\mathbb{R}^n)^\bullet$ is homeomorphic to the sphere $\mathbb{S}^n$, applying Theorem 4.3.3. *Hints*: one can use a 'stereographic projection', and some geometric intuition to spare lengthy computations.

(d) In particular, the Alexandrov compactification of $\mathbb{C}$, already considered in 4.3.1(d), is homeomorphic to the sphere $\mathbb{S}^2$.

*(e) The space $\mathbb{S}^n$ is topologically homogeneous, as any any point $x \in \mathbb{S}^n$ can be taken to $e_1 = (1, 0, ..., 0)$ by a transformation of coordinates that preserves distances. (More formally, the orthogonal group $\mathrm{O}(n+1, \mathbb{R})$ acts transitively on $\mathbb{S}^n$, see (2.109)). Therefore $\mathbb{S}^n$ can only be the Alexandrov compactification of a space homeomorphic to $\mathbb{R}^n$.

(Here we are not saying that the natural number n is determined by these types of homeomorphism: this is true, but requires the Theorem of Topological Dimension, mentioned in 3.1.3(e).)

(f) The euclidean space $\mathbb{R}^n$ is homeomorphic to the open ball $B_n(0, 1)$ (by Exercise 3.6.2(b)). This gives a compactification $\mathbb{R}^n \to \mathbb{D}^n$, whose points at infinity form the whole sphere $\mathbb{S}^{n-1} \subset \mathbb{D}^n$.

*4.3.5 *Limits at a filter*

We have seen that, for a mapping $f\colon A \to Y$ defined on a subspace $A \subset X$, with values in a space Y, we can consider whether f has a limit at a point a adherent to A in X.

Instead of referring to the 'topological environment' X of A, we only need to consider the set $\mathcal{F}$ of the traces $N' = N \cap A$ of the neighbourhoods N of a in X. In fact, definition 3.6.2 can be rewritten saying that the point $b \in Y$ is *a limit* of the function f *along the filter* $\mathcal{F}$ if:

(i) for every $N \in \mathcal{N}_Y(b)$ there is some $N' \in \mathcal{F}$ such that $f(N') \subset N$.

We are thus lead to abstracting the notion of a 'filter', as a family of subsets of a space on which a mapping can converge.

A *filter* on a set X is a set $\mathcal{F} \subset \mathcal{P}X$ such that:

(Fl.1) every subset of X that contains an element of $\mathcal{F}$ belongs to $\mathcal{F}$,

(Fl.2) $\mathcal{F}$ is stable under finite intersections,

(Fl.3) the empty set does not belong to $\mathcal{F}$.

The neighbourhoods of a point a in the space X form a filter of X, and their traces on a subset A form a filter of the latter, provided that a is adherent to A. A filter on A can thus simulate a point that would only belong to some extension of A.

For instance, on the set $\mathbb{N}$ of natural numbers, the filter $\mathcal{F}(\mathbb{N})$ of cofinite subsets (called the *Fréchet filter* of $\mathbb{N}$) can be used to rewrite the definition of a limit of a sequence $f\colon \mathbb{N} \to X$, as in (i).

In a space X, a point a is said to be a *limit* of a filter $\mathcal{F}$ if the latter contains the filter $\mathcal{N}_X(a)$ of neighbourhoods of a in X. This amounts to saying that the inclusion $A \to X$ has limit a along the filter $\mathcal{F}$, in the sense defined above.

We already mentioned [Bou2] as a good reference for this topic.

4.4 Projective spaces and compact surfaces

Other compactifications of the euclidean spaces give important spaces, like the projective spaces, at the basis of Projective Geometry [Be].

In dimension two, compactifying the plane gives various enticing surfaces, like the Möbius band, the torus, and the Klein bottle, briefly sketched here.

The study and classification of compact surfaces is a classical topic of Algebraic Topology, dealt with in Massey's book [Mas1].

4.4.1 Analysing the spheres

Before studying the projective spaces, it will be useful to say something more about the geometry of the spheres

$$\mathbb{S}^n = \{x \in \mathbb{R}^{n+1} \mid ||x|| = 1\} \subset \mathbb{D}^{n+1} \subset \mathbb{R}^{n+1}. \tag{4.15}$$

As already said in 3.1.1(c), we view the spaces $\mathbb{R}^n$ as a nested family $\mathbb{R} \subset \mathbb{R}^2 \subset ...$, identifying $\mathbb{R}^n$ with the hyperplane $\{x \in \mathbb{R}^{n+1} \mid x_{n+1} = 0\}$.

As a consequence, the standard spheres also form a nested family

$$\{-1, 1\} = \mathbb{S}^0 \subset \mathbb{S}^1 \subset \mathbb{S}^2 \subset ... \tag{4.16}$$

The sphere $\mathbb{S}^n$ is the union of two compact hemispheres, which meet at $\mathbb{S}^{n-1}$ in the hyperplane $x_{n+1} = 0$, as shown below (for $n = 2$)

$$\begin{aligned} &\mathbb{D}^n_+ = \{x \in \mathbb{S}^n \mid x_{n+1} \geqslant 0\}, && \mathbb{D}^n_- = \{x \in \mathbb{S}^n \mid x_{n+1} \leqslant 0\}, \\ &\mathbb{D}^n_+ \cup \mathbb{D}^n_- = \mathbb{S}^n, && \mathbb{D}^n_+ \cap \mathbb{D}^n_- = \mathbb{S}^{n-1}, \end{aligned} \tag{4.17}$$

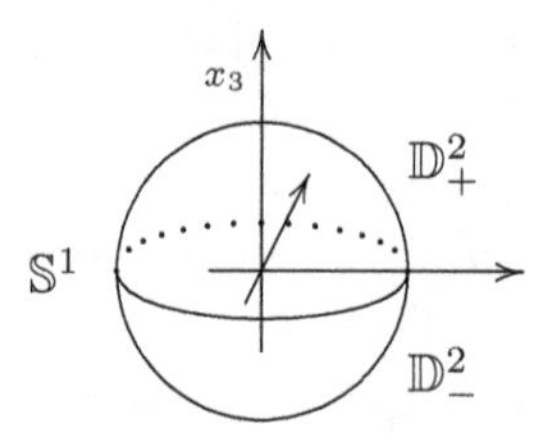

The notation $\mathbb{D}^n_+$ is justified by the fact that the orthogonal projection onto this hyperplane

$$p\colon \mathbb{R}^{n+1} \to \mathbb{R}^n, \qquad p(x_1, ..., x_{n+1}) = (x_1, ..., x_n), \tag{4.18}$$

restricts to a continuous bijection $\mathbb{D}^n_+ \to \mathbb{D}^n$ between compact Hausdorff spaces, which is a homeomorphism. The same holds for $\mathbb{D}^n_-$.

Remarks. (a) We have already remarked, in 3.4.9, that $\mathbb{S}^n$ is a retract of the pierced space $X = \mathbb{R}^{n+1} \setminus \{0\}$, by the normalisation mapping $N(x) = x/||x||$. But $\mathbb{D}^n_+$ is a retract of $\mathbb{S}^n$, with retraction

$$N'(x_1, ..., x_n, x_{n+1}) = (x_1, ..., x_n, |x_{n+1}|). \tag{4.19}$$

(b) Composing retracts, $\mathbb{D}^n_+$ (homeomorphic to $\mathbb{D}^n$) is also a retract of X

$$\mathbb{D}^n_+ \xrightarrow{j'} \mathbb{S}^n \xrightarrow{j} X \xrightarrow{N} \mathbb{S}^n \xrightarrow{N'} \mathbb{D}^n_+ \tag{4.20}$$

$$N'Njj' = \mathrm{id}\, \mathbb{D}^n_+.$$

(c) We know that the pierced line $\mathbb{R}^* = \mathbb{R} \setminus \{0\}$ has two connected components. In higher dimension, the pierced space $\mathbb{R}^k \setminus \{0\}$ is path-connected for $k \geqslant 2$.

This fact, geometrically obvious, can be proved analytically considering that the points $e_1 = (1, 0, ..., 0)$ and $-e_1 = (-1, 0, ..., 0)$ generate two starred subsets of X (see 4.1.6(c)) that cover X, and whose intersection contains $e_2 = (0, 1, 0, ..., 0)$. (This argument needs two coordinates, at least.)

4.4.2 Theorem and Definition (The projective spaces)

(a) On the pierced space $X = \mathbb{R}^{n+1} \setminus \{0\}$ there is a (continuous) action of the multiplicative group $G = \mathbb{R}^*$ of non-zero scalars, by homotheties

$$\lambda(x_1, ..., x_{n+1}) = (\lambda x_1, ..., \lambda x_{n+1}) \qquad (x \in X, \lambda \in \mathbb{R}^*), \tag{4.21}$$

coming from the linear structure of $\mathbb{R}^{n+1}$ on the real field.

(b) The projective real space *is defined as the orbit space of this action*

$$\mathrm{P}^n\mathbb{R} = (\mathbb{R}^{n+1} \setminus \{0\})/G \qquad (n \in \mathbb{N}). \tag{4.22}$$

It is a path-connected compact Hausdorff space. Its points, the orbits of the action, are lines of $\mathbb{R}^{n+1}$ *through the origin, pierced at this point*

$$[x] = [x_1, ..., x_{n+1}] = Gx \qquad (x \neq 0), \tag{4.23}$$

so that two points of X *are identified when they are allineated with the origin. The point* $[x]$ *has* homogeneous coordinates $x_1, ..., x_{n+1}$, *defined up to a multiplicative factor* $\lambda \neq 0$.

$\mathrm{P}^0\mathbb{R}$ *is a singleton.*

(c) The inclusion $\mathbb{S}^n \subset \mathbb{R}^{n+1} \setminus \{0\}$ *induces a homeomorphism*

$$\mathbb{S}^n/T \;\to\; X/G = \mathrm{P}^n\mathbb{R}, \tag{4.24}$$

where the subgroup $T = \{1, -1\} \subset G$ *acts on* $\mathbb{S}^n$ *by* antipodism: *each class* $[a] = \{a, -a\}$ *is formed of two* antipodal *points of* $\mathbb{S}^n$ *(symmetric with respect to the origin).*

Note. T is the 0-dimensional sphere, and the subgroup $U_2 \subset \mathbb{S}^1$ of (2.140).

Proof To study the topology of $\mathrm{P}^n\mathbb{R}$, as a quotient X/G of the pierced space $X = \mathbb{R}^{n+1} \setminus \{0\}$, we form a commutative diagram of continuous mappings

$$\begin{array}{ccccc} \mathbb{S}^n & \xrightarrow{\;j\;} & X & \xrightarrow{\;N\;} & \mathbb{S}^n \\ {\scriptstyle p'}\downarrow & & \downarrow{\scriptstyle p} & & \downarrow{\scriptstyle p'} \\ \mathbb{S}^n/T & \xrightarrow[f]{} & X/G & \xrightarrow[g]{} & \mathbb{S}^n/T \end{array} \qquad \begin{array}{l} Nj = \mathrm{id}\,\mathbb{S}^n, \\ \\ gf = \mathrm{id}\,(\mathbb{S}^n/T). \end{array} \tag{4.25}$$

The upper row presents $\mathbb{S}^n$ as a retract of X, by the normalisation mapping N (in 3.4.9). The congruence modulo G on X restricts to the congruence modulo the subgroup $T = \{1, -1\}$ on $\mathbb{S}^n$, and we have continuous mappings f and g, induced by j and N, in the lower row (Exercise 3.4.6(b)); it is again a retraction pair, because $(gf)p' = p'(Nj) = p'$, and p' is surjective.

The injective mapping f is also surjective, because every orbit Gx of X meets $\mathbb{S}^n$ (in two antipodal points), whence f is bijective. Thus f and g are inverse to each other, and homeomorphisms.

The space $\mathbb{S}^n/T$ is compact Hausdorff, as a quotient of a compact Hausdorff space by a finite group (Exercise 3.6.5(c)), and $\mathrm{P}^n\mathbb{R}$ is also.

$\mathrm{P}^0\mathbb{R}$ is a singleton; all the higher $\mathrm{P}^n\mathbb{R}$ are also path-connected, because the space $\mathbb{R}^{n+1} \setminus \{0\}$ is path-connected for $n \geqslant 1$ (as remarked in 4.4.2(c)). □

4.4.3 Theorem (The projective spaces, Part II)

(a) The inclusions $\mathbb{D}^n_+ \subset \mathbb{S}^n \subset \mathbb{R}^{n+1} \setminus \{0\}$ *induce homeomorphisms*

$$\mathbb{D}^n_+/\!\sim \;\to\; \mathbb{S}^n/T \;\to\; \mathrm{P}^n\mathbb{R}, \tag{4.26}$$

where $\sim$ *is the relation of antipodism* $x \sim -x$ restricted *to the sphere* $\mathbb{S}^{n-1} = \mathbb{D}^n_+ \cap \mathbb{R}^n$, *and discrete elsewhere (in* $\mathbb{D}^n_+$*).*

(b) One can similarly present $\mathrm{P}^n\mathbb{R}$ *as the topological quotient* $\mathbb{D}^n/\!\sim$ *of the compact disc* $\mathbb{D}^n$ *modulo the relation of antipodism* $x \sim -x$ restricted *to* $\mathbb{S}^{n-1} \subset \mathbb{D}^n$.

(c) $\mathrm{P}^n\mathbb{R}$ *is a compactification of* $\mathbb{R}^n$*, by the following mapping*

$$u\colon \mathbb{R}^n \to \mathrm{P}^n\mathbb{R}, \qquad u(x_1, ..., x_n) = [x_1, ..., x_n, 1], \tag{4.27}$$

which is an open topological embedding with a dense image.

For $n = 0$*, this mapping is a homeomorphism between singletons. For* $n > 0$ *it is not surjective, and the complement of the image (the space of the added points) is homeomorphic to* $\mathrm{P}^{n-1}\mathbb{R}$.

Proof (a) We proceed as in (4.25), using, using the fact that $\mathbb{D}^n_+$ is a retract of $\mathbb{S}^n$ (see (4.19)) that meets all its T-orbits

$$\begin{array}{ccccc} \mathbb{D}^n_+ & \xrightarrow{j'} & \mathbb{S}^n & \xrightarrow{N'} & \mathbb{D}^n_+ \\ {\scriptstyle p''}\downarrow & & \downarrow{\scriptstyle p'} & & \downarrow{\scriptstyle p''} \\ \mathbb{D}^n_+/\!\sim & \xrightarrow[f']{} & \mathbb{S}^n/T & \xrightarrow[g']{} & \mathbb{D}^n_+/\!\sim \end{array} \qquad \begin{array}{l} N'j' = \mathrm{id}\,, \\ \\ g'f' = \mathrm{id}\,. \end{array} \tag{4.28}$$

(b) Obvious, using the canonical homeomorphism $\mathbb{D}^n_+ \to \mathbb{D}^n$ (from (4.18)), which leaves $\mathbb{S}^{n-1}$ fixed and does not change the non-trivial part of the equivalence relation $\sim$.

(c) For $X = \mathbb{R}^{n+1} \setminus \{0\}$, we consider the commutative diagram

$$\begin{array}{ccccc} \mathbb{R}^n & \xrightarrow{v} & X & \xrightarrow{N} & \mathbb{S}^n \\ & {\scriptstyle u}\searrow & \downarrow{\scriptstyle p} & & \downarrow{\scriptstyle p'} \\ & & X/G & \xrightarrow[g]{} & \mathbb{S}^n/T \end{array} \tag{4.29}$$

where $v(x_1, ..., x_n) = (x_1, ..., x_n, 1)$, and $u = pv\colon \mathbb{R}^n \to \mathrm{P}^n\mathbb{R}$ is an injective map. We want to prove that it is open – a topological embedding of an open subspace.

The composite

$$Nv\colon \mathbb{R}^n \to \mathbb{S}^n, \qquad Nv(x_1, ..., x_n) = N(x_1, ..., x_n, 1),$$

gives a continuous bijective mapping h onto its image, the upper open hemisphere of $\mathbb{S}^n$ (open in the sphere)

$$h\colon \mathbb{R}^n \to \mathbb{D}^n_+ \setminus \mathbb{S}^{n-1} = \{x \in \mathbb{S}^n \mid x_{n+1} > 0\},$$
$$h(x_1, ..., x_n) = (\lambda x_1, ..., \lambda x_n, \lambda), \qquad \lambda = 1/(x_1^2 + ... + x_n^2 + 1)^{1/2}.$$

But h is a homeomorphism, with inverse

$$k\colon \mathbb{D}^n_+ \setminus \mathbb{S}^{n-1} \to \mathbb{R}^n, \qquad k(y_1, ..., y_n, \lambda) = \lambda^{-1}(y_1, ..., y_n).$$

Thus Nv is an open mapping; the mapping p' is also open, as any projection on an orbit space (see 3.5.8). It follows that the composite $gu = p'Nv$ is also open, and u as well (because g is a homeomorphism).

We have already seen that the image of Nv is dense in $\mathbb{S}^n$, whence the image of $p'Nv$ is dense in the projective space. Finally, the complement of $\operatorname{Im} u$ in X is:

$$X \setminus \operatorname{Im} u = p(\mathbb{R}^n \setminus \{0\}) = \mathrm{P}^{n-1}\mathbb{R}. \tag{4.30}$$

□

4.4.4 Nesting the projective spaces

The nesting of the spaces $\mathbb{R}^n$ gives a nesting of the corresponding pierced spaces

$$\mathbb{R} \setminus \{0\} \subset \mathbb{R}^2 \setminus \{0\} \subset ... \subset \mathbb{R}^{n+1} \setminus \{0\} \subset ... \tag{4.31}$$

and the action of the group $G = \mathbb{R}^*$ on these spaces is consistent with these inclusions.

We have thus a sequence of injective continuous mappings

$$\mathrm{P}^0\mathbb{R} \to \mathrm{P}^1\mathbb{R} \to ... \to \mathrm{P}^n\mathbb{R} \to ... \tag{4.32}$$

which are topological embeddings (as their domain is compact and their codomain is Hausdorff).

We have seen, in (4.30), that $\mathrm{P}^{n-1}\mathbb{R}$ is the set of points at infinity for the compactification $u\colon \mathbb{R}^n \to \mathrm{P}^n\mathbb{R}$.

We have already remarked that $\mathrm{P}^0\mathbb{R}$ is a singleton. The *projective real line* $\mathrm{P}^1\mathbb{R}$ is homeomorphic to $\mathbb{S}^1$, even though the projection $\mathbb{S}^1 \to \mathbb{S}^1/T$ is – obviously – not injective.

In fact, we already know that $\mathrm{P}^1\mathbb{R} \setminus \mathrm{P}^0\mathbb{R}$ is homeomorphic to $\mathbb{R}$, whence $\mathrm{P}^1\mathbb{R}$ is the one-point compactification of $\mathbb{R}$, that is homeomorphic to $\mathbb{S}^1$.

As a more intuitive way of viewing this, the orbit space $\mathbb{S}^1/T$ 'wraps' the circle on itself, twice; this gives back $\mathbb{S}^1$.

Exercises and complements. (a) The last remark can be made precise, viewing $\mathbb{S}^1$ in the complex field (or using polar coordinates in $\mathbb{R}^2$).

(b) More generally, 'wrapping $\mathbb{S}^1$ on itself n times' (for $n \geqslant 1$) we still get a space homeomorphic to $\mathbb{S}^1$.

4.4.5 Compactifications of the plane

We have already seen three compactifications of the plane $\mathbb{R}^2$:

(a) the Alexandrov compactification $(\mathbb{R}^2)^\bullet \cong \mathbb{S}^2$, with one point at infinity,

(b) the projective compactification $\mathrm{P}^2\mathbb{R}$, with a projective line $\mathrm{P}^1\mathbb{R} \cong \mathbb{S}^1$ at infinity,

(c) the disc-compactification $\mathbb{R}^2 \to \mathbb{D}^2$ (see 4.3.4(f)), which again has the circle $\mathbb{S}^1 \subset \mathbb{D}^2$ at infinity.

Now, $\mathbb{R}^2$ is homeomorphic to the open square $U =]0,1[^2$, as we have seen in Exercise 3.6.2(c), and we can easily find other compactifications of U, and therefore of the plane.

The standard square $\mathbb{I}^2$ is an obvious compactification of U, with its boundary $\partial\mathbb{I}^2$ (in $\mathbb{R}^2$) added 'at infinity'. (This corresponds to the last example, above.)

Let R be an equivalence relation of $\mathbb{I}^2$ that is *trivial out of the boundary*: each equivalence class of R which does not meet $\partial\mathbb{I}^2$ is a singleton. Then the composed mapping

$$f = pj \colon U \to \mathbb{I}^2 \to \mathbb{I}^2/R \tag{4.33}$$

is continuous and injective. It is actually an open topological embedding: if V is open in U, then it is open in $\mathbb{I}^2$ and saturated for R, so that $f(U) = pj(U)$ is open in the quotient. We have thus proved that $f \colon U \to \mathbb{I}^2/R$ is a compactification of U, with $p(\partial\mathbb{I}^2)$ added at infinity. We also note that f has a dense image (because U is dense in $\mathbb{I}^2$).

The compactifications of $\mathbb{R}^2$ recalled above correspond to the compactifications of U derived from the equivalence relation R that:

(a$'$) identifies all the points of $\partial\mathbb{I}^2$,

(b$'$) identifies the points of $\partial\mathbb{I}^2$ which are symmetric with respect to the centre $(1/2, 1/2)$ of the square,

(c$'$) is the discrete equivalence relation $x = y$ on $\partial\mathbb{I}^2$ (and $\mathbb{I}^2$).

We end this section exploring other compactifications of this kind.

4.4.6 The torus

The *torus* was already presented in 3.1.3, as the 'surface of a life buoy'

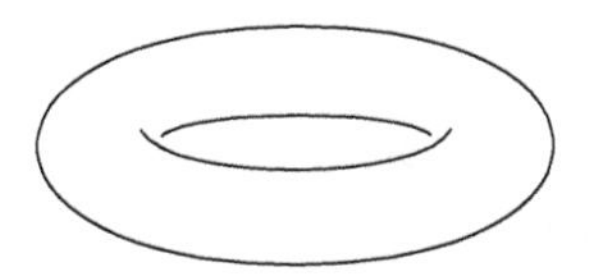

It can be defined as the subspace $X \subset \mathbb{R}^3$ produced by the rotation around the 'vertical' axis z of a circle C

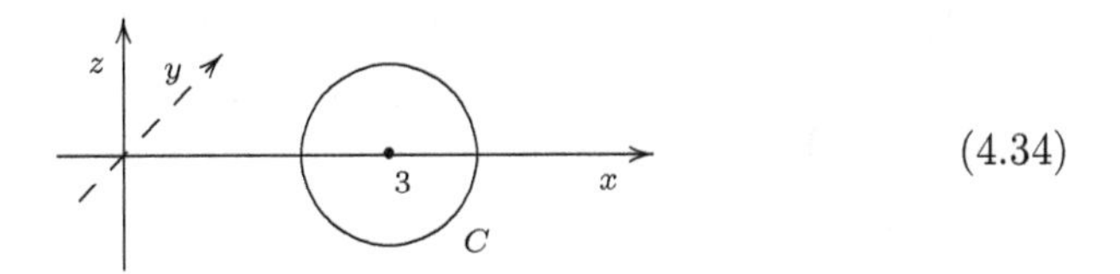

(4.34)

$$C = \{(x, y, z) \in \mathbb{R}^3 \mid y = 0 \text{ and } (x-3)^2 + z^2 = 1\},$$

that lies in the plane $y = 0$, with centre $(3, 0, 0)$ and radius 1. (The rotation axis and the circle are coplanar and do not meet.)

The circle has parametric equations (for $0 \leqslant t \leqslant 2\pi$, or $t \in \mathbb{R}$)

$$x = 3 + \cos t, \qquad y = 0, \qquad z = \sin t.$$

Our surface is thus the image of the following continuous mapping

$$f\colon \mathbb{R}^2 \to \mathbb{R}^3,$$
$$f(s, t) = ((3 + \cos t)\cos s,\ (3 + \cos t)\sin s,\ \sin t), \tag{4.35}$$

which can be restricted to the compact square $K = [0, 2\pi]^2$ (or to the standard square $\mathbb{I}^2$, with a different parametrisation).

The associated equivalence relation R_f in $\mathbb{R}^2$ is

$$f(s, t) = f(s', t') \;\Leftrightarrow\; s - s' \in 2\pi\mathbb{Z} \text{ and } t - t' \in 2\pi\mathbb{Z},$$

namely the congruence modulo the subgroup $(2\pi\mathbb{Z})^2$.

Theorem 4.2.7(c) says that f induces a homeomorphism $\mathbb{R}^2/R_f \to X$, and $K/R_f \to X$. But the equivalence relation R_f is trivial in the interior $K^\circ =]0, 2\pi[^2$ of the square, and we conclude – as in 4.4.5 – that the torus is a compactification of the open square, or equivalently of the plane.

Furthermore, the congruence modulo $2\pi\mathbb{Z}$ is open in $\mathbb{R}$, and the quotient $\mathbb{R}^2/R_f$ is homeomorphic to the product $\mathbb{S}^1 \times \mathbb{S}^1$ of two circles, by 3.5.5(b).

Finally, we can also define the torus as the topological product $\mathbb{S}^1 \times \mathbb{S}^1$, and more generally the *n-dimensional torus* $\mathbb{T}^n$ as the cartesian power $(\mathbb{S}^1)^n$.

*4.4.7 *Other compactifications of the plane*

There are many others compactifications obtained as in 4.4.5, from equivalence relations R of $\mathbb{I}^2$ that are trivial outside of the boundary.

We briefly describe the following ones, which are important in Topology (and can be studied with the tools of Algebraic Topology).

(i) Identifying two parallel edges of $\mathbb{I}^2$, 'respecting their direction', gives a compact cylinder, homeomorphic to the product $\mathbb{S}^1 \times \mathbb{I}$.

(ii) Identifying two parallel edges 'reversing their direction', produces a *Möbius band*, or *Möbius strip*.

These two spaces can be realised in $\mathbb{R}^3$, and a 'model' of each of them can be obtained, gluing differently a rectangular strip of paper, as shown below (see also [Mas1], Figure 1.1, and [Hat], page 2)

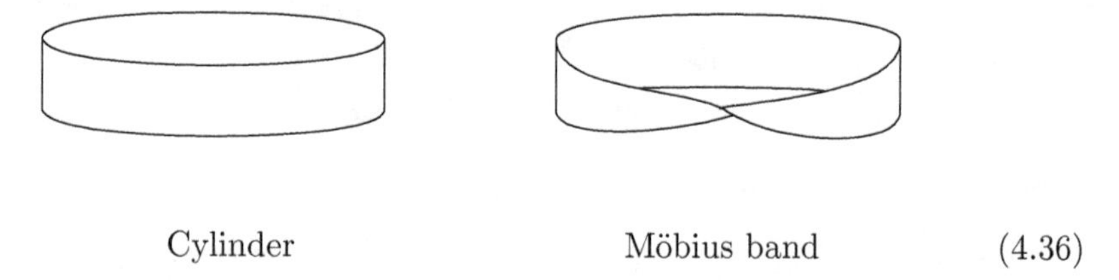

Cylinder Möbius band (4.36)

(iii) Identifying two parallel edges 'respecting their direction', and the other two in the same way, gives a *torus*, as we have seen above (in 4.4.6).

(iv) Identifying two parallel edges 'respecting their direction', and the other two 'reversing their direction', gives a *Klein bottle*. This surface cannot be embedded in $\mathbb{R}^3$, but has a clear model in $\mathbb{R}^4$, that can be effectively drawn in the plane: see [Hat], page 19.

(v) Identifying two parallel edges 'reversing directions', and the other two in the same way, we come back to the equivalence relation described in 4.4.5(b′), which gives the projective plane. A model of the latter can be realised in $\mathbb{R}^4$, but is not easily represented.

*4.4.8 *Hints at surfaces*

A *surface* is a topological space locally homeomorphic to the euclidean plane (which means that every point has a neighbourhood homeomorphic to $\mathbb{R}^2$); we also require that the space be Hausdorff and second-countable (it is not a consequence).

Compact connected surfaces are important. The main examples we have seen are the sphere, the torus, the projective plane, the Klein bottle.

A *surface with a boundary* also has points with a neighbourhood homeomorphic to a closed half-plane. This is the case of the compact cylinder and the Möbius band: their boundary forms two circles or one circle, respectively. (Note that this meaning of 'boundary' is absolute, different from the boundary of a subset *in* a space.)

A brilliant result of Algebraic Topology classifies, up to homeomorphism, the compact connected surfaces, proving that they form a countable family (see [Mas1]).

According to this classification theorem, every compact connected surface can be obtained 'connecting' a sphere, a finite number of tori, possibly a projective plane or a Klein bottle, by an operation called 'connected sum' of surfaces.

The compact connected 'surfaces with a boundary' can be obtained from the previous ones, removing a finite number of 'open discs' with a disjoint closure in the space.

4.5 Metric spaces

A metric space has a distance function, that generalises the euclidean metric of $\mathbb{R}^n$. The associated topology is defined as in Section 3.1; but the distance gives a richer information, which allows us to define uniform continuity and completeness.

We give here a brief presentation of these aspects, with some hints at more general structures where all this makes sense: uniform spaces.

A good reference for further study is [Bou2].

4.5.0 Main definitions

A *metric space* X is a set equipped with a *distance*, or *metric*, written as d, or d_X when useful. For every pair $(x, y) \in X^2$ the distance $d(x, y)$ *between* x *and* y satisfies the following axioms (for all $x, y, z \in X$):

(M.0) $d(x, y) \in [0, +\infty[$ and $d(x, x) = 0$,

(M.1) $d(x, y) + d(y, z) \geqslant d(x, z)$ (*triangle inequality*),

(M.2) $d(x, y) = d(y, x)$ (*symmetry*),

(M.3) if $d(x, y) = 0$ then $x = y$ (*separation*).

The main example is the set $\mathbb{R}^n$ with the euclidean distance

$$d(x,y) = \sqrt{\Sigma_i\,(x_i - y_i)^2} \qquad (x, y \in \mathbb{R}^n), \tag{4.37}$$

that we have already examined, in Section 3.1. Every subset $A \subset \mathbb{R}^n$, with the restricted euclidean metric, is also an example.

Let X be a metric space, with distance d, and $a \in X$. The basic terminology of the euclidean metric is extended to the present, general case.

The *open ball* of X, of centre a and radius $r > 0$ is defined as the subset

$$B_X(a,r) = \{x \in X \mid d(x,a) < r\}. \tag{4.38}$$

We write as $\mathcal{B}_X(a) \subset \mathcal{P}X$ the collection of all the subsets $B_X(a,r)$, for $r > 0$.

A subset $U \subset X$ is said to be an *open subset* (of X) if:

- for every $a \in U$ there is some $r > 0$ such that $B_X(a,r) \subset U$.

These open sets define a topology on X, called the *topology associated* to the metric of X, or the *metric topology* of X. A metric space X is viewed as a topological space, with the associated topology; all topological properties of X, or in X, make reference to this topology.

In particular, a sequence (x_k) converges to x in X if (as defined in 3.3.3):

- for every $\varepsilon > 0$ there exists some $h \in \mathbb{N}$ such that, for every $k \geqslant h$, $d(x_k, x) < \varepsilon$.

Exercises and complements. X is a metric space, with distance d.

(a) Every open ball $B_X(a,r)$ is open in the associated topology. The set $\mathcal{B}_X(a)$ is a local basis at a.

(b) A topological space is said to be *metrisable* if its topology is associated to a metric. Every discrete space is metrisable: one can always use the *discrete metric* D on the underlying set, where $D(x,y) = 0$ if $x = y$ and $D(x,y) = 1$ otherwise. Every indiscrete space with at least two points is not metrisable (because of the following point).

(c) The metric space X is always Hausdorff and first-countable. Therefore, each sequence in X converges to at most one point, and the topology of X is determined by the convergence of sequences: if $A \subset X$, then $x \in \overline{A}$ if and only if there exists a sequence (a_k) in A that converges to x in X.

(d) Two metrics on a set are said to be *topologically equivalent* if they have the same associated topology.

For every $\lambda > 0$, the metric d is topologically equivalent to λd. The euclidean metric and the discrete metric on $\mathbb{R}^n$ are not topologically equivalent, for $n > 0$. The euclidean metric on $\mathbb{Z}$ gives the discrete topology, and is topologically equivalent to the discrete metric.

(e) (*Bounded subsets*) The diameter diam(A) of a subset $A \subset X$ is defined as in (3.6):

$$\mathrm{diam}(A) = \sup\{d(x,y) \mid x, y \in A\} \in \overline{\mathbb{R}}, \tag{4.39}$$

and A is bounded if $\text{diam}(A) < +\infty$, or equivalently if A is contained in some open ball $B_X(a,r)$.

(f) For $a, b, a', b' \in X$ we have: $\mid d(a,b) - d(a',b') \mid \leqslant d(a,a') + d(b,b')$.

(g) There are many useful ways of weakening the axioms (M.0–3). In particular, a *semimetric space* need not satisfy the separation axiom (M.3); the associated *semimetric topology* need not be Hausdorff: it is if and only if (M.3) holds.

Other generalisations will be considered in Section 6.5.

4.5.1 Mappings of metric spaces

Let $f\colon X \to Y$ be a mapping between metric spaces. We can consider various ways in which f can behave, with respect to the metric structure.

(i) Saying that f is *continuous* means (of course) that it is for the metric topologies of X and Y, or equivalently that:

- for every $a \in X$ and every $\varepsilon > 0$ there is some $\delta > 0$ such that $f(B_X(a,\delta)) \subset B_Y(f(a),\varepsilon)$.

Since these spaces are first-countable, the continuity of f amounts to preserving the limit of every convergent sequence of X (see 3.3.5(g)).

(ii) We say that f is *uniformly continuous* if, equivalently:

- for every $\varepsilon > 0$ there is some $\delta > 0$ such that, for every $a \in X$: $f(B_X(a,\delta)) \subset B_Y(f(a),\varepsilon)$,

- for every $\varepsilon > 0$ there is some $\delta > 0$ such that, if $d(x,x') < \delta$ then $d(f(x),f(x')) < \varepsilon$.

Comparing (i) and (ii), one should note that in the second case δ only depends on ε, and must be chosen independently of a (or 'uniformly' with respect to a).

(iii) We say that f is a *Lipschitz mapping*, with *Lipschitz constant* the number $L > 0$, if

$$d(f(x),f(x')) \leqslant L.d(x,x') \qquad (\text{for } x,x' \in X). \tag{4.40}$$

(iv) We say that f is a *weak contraction* if it admits 1 as a Lipschitz constant, i.e.

$$d(f(x),f(x')) \leqslant d(x,x') \qquad (\text{for } x,x' \in X). \tag{4.41}$$

Each of these four kinds of mappings is stable under composition, and includes all the identities of metric spaces. Each of these properties implies the previous ones. Each of them gives a different way of evaluating the 'type' of a metric space.

In particular, the mapping $f\colon X \to Y$ is an *isometry* if $d(f(x),f(x')) = d(x,x')$, for all $x,x' \in X$. Then it is obviously injective.

4.5.2 Invertible mappings of metric spaces

We examine the invertible mappings between two metric spaces X, Y, within the four kinds (i)–(iv) considered above. Firstly (and obviously)

(i) the mapping $f\colon X \to Y$ is said to be a *homeomorphism* if it is for the metric topologies of X and Y.

This is a weak property, in the theory of metric spaces: we already know that the euclidean line is homeomorphic to the open interval $]0, 1[$, and we will see that these metric spaces behave differently, with respect to crucial properties. Continuous mappings and homeomorphisms are adequate to study *metrisable* topological spaces, rather then spaces with a given metric.

(ii) A *uniform isomorphism* is a bijective mapping $f\colon X \to Y$ that is uniformly continuous, together with the inverse mapping.

This stronger property will be seen to preserve completeness, a crucial feature of metric spaces. Yet, the fact of being a bounded metric space is not invariant under uniform isomorphism, as a simple example will show, in Exercise (a).

(iii) A *Lipschitz isomorphism* is a bijective mapping $f\colon X \to Y$ that is Lipschitz, together with the inverse mapping.

This is equivalent to saying that f is surjective and there exist two constants $L, M > 0$ such that

$$L.d(x, x') \leqslant d(f(x), f(x')) \leqslant M.d(x, x') \qquad (\text{for } x, x' \in X). \tag{4.42}$$

It is perhaps the best way of comparing metric spaces. It is easy to see that each non-degenerate bounded open interval $]a, b[$ is *Lipschitz isomorphic* to any other, but is not to any unbounded interval.

(iv) A *bijective isometry*, or *isometric isomorphism*, is a bijective mapping $f\colon X \to Y$ that is a weak contraction, together with the inverse mapping.

This is the same as a surjective mapping such that

$$d(f(x), f(x')) = d(x, x') \qquad (\text{for } x, x' \in X), \tag{4.43}$$

because injectivity is a consequence. Then the spaces X, Y are said to be *isometric*, the strongest way of saying that X, Y are 'essentially the same metric space'. Two intervals $]a, b[$ and $]a', b'[$ are isometric if they have the same length (by a translation), and only in this case.

Exercises and complements. (a) The discrete space $\mathbb{Z}$ is unbounded for the euclidean metric d, but bounded for the discrete metric D. Prove that the metric spaces $(\mathbb{Z}, d)$ and $(\mathbb{Z}, D)$ are uniformly homeomorphic; this shows that boundedness is not invariant up to uniform isomorphism.

(b) (*Comparing metrics*) Suppose we have two metrics d, d' on the same set S, and call X, X' the corresponding metric spaces. Consider the mapping $f\colon X \to X'$ given by $\mathrm{id}\, S$. Then f is homeomorphism if and only if d, d' are topologically equivalent.

But the more interesting case is when f is a Lipschitz isomorphism, which amounts to the existence of two constants $L, M > 0$ such that

$$L.d(x,x') \leqslant d'(x,x') \leqslant M.d(x,x'). \tag{4.44}$$

In this case we say that the metrics d, d' are *Lipschitz-equivalent.*

4.5.3 Subspaces and finite products

A subset A of a metric space X inherits a restricted metric, which we write in the same way. Then the inclusion $A \to X$ is an isometry, the open balls of A are the traces on A of the open balls of X

$$B_A(a,r) = B_X(a,r) \cap A \qquad (a \in A), \tag{4.45}$$

and the metric topology of A is the same as its topology as a subspace of X.

Suppose now that we have a *finite* family $(X_i)_{i \in I}$ of metric spaces, with metrics $(d_i)_{i \in I}$. On the cartesian product X of their underlying sets we put the l_∞*-metric*

$$d_\infty(x,y) = \max{}_{i \in I}\, d_i(x_i, y_i), \tag{4.46}$$

discussed in the exercises below.

Exercises and complements. (a) The formula (4.46) does define a distance on the finite product X.

(b) The metric space X, with the cartesian projections $p_i \colon X \to X_i$, satisfies the universal property of the product for the four kinds of mappings in 4.5.1(i)–(iv); doing so for the weak contractions strictly determines the metric.

(c) The topology associated to the l_∞-metric is the product topology of the spaces X_i. In particular the l_∞-metric of the space $\mathbb{R}^n = \mathbb{R} \times \dots \times \mathbb{R}$ is topologically equivalent to the euclidean metric, also called the l_2-metric $d_2(x,y)$ (see the next exercise). These metrics are even Lipschitz-equivalent:

$$d_\infty(x,y) \leqslant d_2(x,y) \leqslant \sqrt{n}.d_\infty(x,y). \tag{4.47}$$

(d) More generally, on the finite product X one can consider the l_p-metric,* depending on a parameter $p \in [1, \infty]$. After the case $p = \infty$, the other instances are defined as

$$\begin{gathered} d_p(x,y) = (\textstyle\sum_i (d_i(x_i,y_i))^p)^{1/p} \qquad (1 \leqslant p < \infty), \\ d_1(x,y) = \textstyle\sum_i d_i(x_i,y_i). \end{gathered} \tag{4.48}$$

All these metrics are Lipschitz equivalent: if $1 \leqslant p \leqslant q \leqslant \infty$

$$d_\infty(x,y) \leqslant d_q(x,y) \leqslant d_p(x,y) \leqslant d_1(x,y) \leqslant n.d_\infty(x,y). \tag{4.49}$$

In particular, we have the equivalent l_p-metrics of the cartesian power $\mathbb{R}^n$, produced by the standard metric $d(x,y) = |x - y|$ on $\mathbb{R}$.

4.5.4 *Completeness*

Loosely speaking, a metric space is complete if a sequence of points which become arbitrarily close to each other always converges. For instance, the euclidean line $\mathbb{R}$ is complete, while the rational line is not. The definition is based on Cauchy sequences, named after Augustin-Louis Cauchy, a French mathematician working in the first half of the 19th century.

Let $x = (x_k)$ be a sequence in the metric space X. We say that (x_k) is a *Cauchy sequence*, if

(i) for every $\varepsilon > 0$ there is some $h \in \mathbb{N}$ such that, for every $m, n \geqslant h$, $d(x_m, x_n) < \varepsilon$.

This means that $d(x_m, x_n) \to 0$ when $(m, n) \to \infty$: we are viewing $d(x_m, x_n)$ as a mapping $\mathbb{N} \times \mathbb{N} \to \mathbb{R}$, and the limit at infinity is taken with respect to the one-point compactification $X^\bullet$ of the discrete space $X = \mathbb{N} \times \mathbb{N}$: in this space the neighbourhoods of ∞ are the complements of the finite subsets of X; the complements of the subsets $\{(m, n) \in X \mid m, n < h\}$ form a local basis.

Every convergent sequence is Cauchy (see Exercise (a)). The metric space X is said to be *complete* if the converse holds: every Cauchy sequence in X converges to a point.

Exercises and complements. (a) In a metric space, every convergent sequence is Cauchy.

(b) In a metric space, every Cauchy sequence is bounded (i.e. so is its image).

(c) A uniformly continuous mapping of metric spaces preserves Cauchy sequences.

(d) A uniform isomorphism of metric spaces preserves completeness.

(e) A finite product of complete metric spaces is complete.

(f) A closed subspace A of a complete metric space X is complete.

(g) A complete metric subspace A of a metric space X is closed in X.

4.5.5 *Exercises and complements* (Sequences of real numbers)

We have already seen various results on the limits of sequences of real numbers, in Section 3.3. The following are also important; the solution is below, and is easy except for point (a).

(a) The euclidean metric space $\mathbb{R}$ is complete; all its finite powers $\mathbb{R}^n$ are also.

Hints: starting from a Cauchy sequence of real numbers, one can construct an increasing one, which converges.

(b) A non-degenerate bounded open interval $]a, b[$ is not complete. Homeomorphisms of metric spaces do not preserve completeness.

(c) In a metric space X, a sequence (x_k) converges to x if and only if the sequence of real numbers $(d(x, x_k))_k$ converges to 0 in $\mathbb{R}$ (or in the real interval $[0, +\infty[$).

(d) In a metric space X we have sequences (x_k) and (y_k) that converge to x and y, respectively. Then the sequence $(d(x_k, y_k))_k$ converges to $d(x, y)$ in $\mathbb{R}$. (In other words, the mapping $d\colon X \times X \to \mathbb{R}$ is continuous.)

Solutions. (a) It is sufficient to consider $\mathbb{R}$, by 4.5.4(e). Let (x_k) be a Cauchy sequence in $\mathbb{R}$.

We know from 3.3.2 that its image $A = \{x_k \mid k \in \mathbb{N}\}$ is bounded in $\mathbb{R}$, obviously non-empty, and we let $a = \sup A$.

We take the sequence $y_k = \inf\{x_n \mid n \geqslant k\}$, which is weakly increasing (because $\{x_n \mid n \geqslant k\} \supset \{x_n \mid n \geqslant k+1\}$) and upper bounded by a; then the sequence (y_k) converges to its least upper bound $c = \sup(y_k)$ (by Exercise 3.3.2(i)), and we use the Cauchy property of (x_k) to prove that also the latter converges to c.

For an $\varepsilon > 0$ there is some $h \in \mathbb{N}$ such that the following three facts hold true (because of the Cauchy property of (x_k), the convergence of (y_k) and the inf-property of the real number y_h):

(i) for every $m, n \geqslant h$, $d(x_m, x_n) < \varepsilon/3$,

(ii) for every $k \geqslant h$, $d(y_k, c) < \varepsilon/3$,

(iii) there is some $n \geqslant h$ such that $x_n \in [y_h, y_h + \varepsilon/3[$.

Then, for every $m \geqslant h$, using a natural number $n \geqslant h$ that satisfies (iii):

$$d(x_m, c) \leqslant d(x_m, x_n) + d(x_n, y_h) + d(y_h, c) < \varepsilon.$$

(b) The sequence $x_k = a + 1/k$ lies in $]a, b[$, for $k > 1/(b - a)$. It converges to a in $\mathbb{R}$, whence it is a Cauchy sequence of $\mathbb{R}$ and $]a, b[$, that has no limit in the latter.

(c) Obvious.

(d) It is a straightforward consequence of Exercise 4.5.0(f):

$$|d(x, y) - d(x_k, y_k)| \leqslant d(x, x_k) + d(y, y_k).$$

4.5.6 Theorem and Definition (The metric completion)

Every metric space X has an isometric embedding in a complete metric space

$$\eta\colon X \to \hat{X}, \tag{4.50}$$

which satisfies the following universal property*:*

- for every uniformly continuous mapping $f\colon X \to Y$ with values in a complete metric space, there is a unique uniformly continuous mapping $g\colon \hat{X} \to Y$ such that $g\eta = f$.

The image $\eta(X)$ is dense in $\hat{X}$. If f has Lipschitz constant L, the same holds of g. If f is an isometry, so is g.

The space $\hat{X}$ *that solves the universal problem for weak contractions is determined up to isometric isomorphism, and is called the (metric)* completion *of* X.

Proof (a) Let $C \subset \mathrm{Map}(\mathbb{N}, X)$ be the set of all Cauchy sequences of X. The general idea is to identify two elements of C when they are 'infinitesimally near', and prove that the quotient set becomes a complete metric space.

The proof is rather long, but not difficult; the reader can try to write it, using the properties of sequences of real numbers, in the Exercises of 3.3.2.

(b) We already have the set C, whose elements will be written as $x = (x_n)$, $y = (y_n)$, etc. We want to make C into a semimetric space (see Exercise 4.5.0(g)).

We define the mapping

$$D\colon C \times C \to \mathbb{R}, \qquad D(x, y) = \lim_{n\to\infty} d(x_n, y_n). \tag{4.51}$$

This is legitimate: applying Exercise 4.5.0(f), we deduce that $(d(x_n, y_n))$ is a Cauchy sequence of real numbers

$$0 \leqslant |d(x_m, y_m) - d(x_n, y_n)| \leqslant d(x_m, x_n) + d(y_m, y_n) \to 0,$$

and converges in $\mathbb{R}$ (that is complete).

The mapping D satisfies the first three axioms (M.0–2) of distances. The triangle inequality follows again from the properties of sequences of real numbers of 3.3.2

$$\begin{aligned} D(x, y) + D(y, z) &= \lim d(x_n, y_n) + \lim d(y_n, z_n) \\ &= \lim(d(x_n, y_n) + d(y_n, z_n)) \geqslant \lim d(x_n, z_n) = D(x, z). \end{aligned}$$

The set C is thus a semimetric space.

(c) We take the quotient $\hat{X} = C/{\sim}$, where $x \sim y$ if $D(x, y) = 0$. This relation is plainly reflexive and symmetric; it is also transitive because of the triangle inequality: $0 \leqslant D(x, z) \leqslant D(x, y) + D(y, z)$.

The semimetric D is invariant under this equivalence relation: if $x \sim x'$ and $y \sim y'$

$$|D(x, y) - D(x', y')\,| \leqslant D(x, x') + D(y, y') = 0.$$

Therefore, writing as $\hat{x}$ the equivalence class of $x \in C$ in $\hat{X}$, we have a semimetric in $\hat{X}$

$$d(\hat{x}, \hat{y}) = D(x, y), \tag{4.52}$$

which is actually a metric: $d(\hat{x}, \hat{y}) = 0$ means that $x \sim y$ and $\hat{x} = \hat{y}$.

(d) We define the mapping $\eta\colon X \to \hat{X}$ sending the point $a \in X$ to the equivalence class $\eta(a)$ of the constant sequence $(a)_{n\in\mathbb{N}}$; in fact this sequence converges to a in X, and is a Cauchy sequence.

The mapping η is an isometry (necessarily injective): if $a, b \in X$

$$d(\eta(a), \eta(b)) = \lim_{n\to\infty} d_X(a, b) = d_X(a, b).$$

We identify a with $\eta(a)$, so that X becomes a metric subspace of $\hat{X}$. It is dense in the latter, because a point $\hat{x} \in \hat{X}$ comes from a Cauchy sequence $x = (x_n)$ in X, and it is sufficient to prove that the sequence $(\eta(x_n))_n$ converges to $\hat{x}$ in $\hat{X}$. Indeed, the sequence of real numbers λ_n

$$\lambda_n = d(\eta(x_n), \hat{x}) = \lim_{k\to\infty} d_X(x_n, x_k),$$

converges to 0 in $\mathbb{R}$ (because x is a Cauchy sequence).

(e) To verify the universal property of our construction, we take a uniformly continuous mapping $f\colon X \to Y$ with values in a complete metric space.

A Cauchy sequence $x = (x_n) \in C$ is taken by f to a Cauchy sequence $(f(x_n))_n$ in Y (by Exercise 4.5.4(c)), that converges. If $x \sim x'$ in C, we have:

- for every $\varepsilon > 0$ there is some $\delta > 0$ such that, if $d_X(x, x') < \delta$, then $d_Y(f(x), f(x')) < \varepsilon$,

- for this $\delta > 0$ there is some $h \in \mathbb{N}$ such that, if $n \geqslant h$, then $d_X(x_n, x'_n) < \delta$.

This proves that $d_Y(f(x_n), f(x'_n)) \to 0$, and thus $\lim f(x_n) = \lim f(x'_n)$ (Exercise 4.5.5(d)).

We now define a mapping $g\colon \hat{X} \to Y$ letting $g(\hat{x}) = \lim f(x_n)$, legitimate by the previous argument, and uniformly continuous:

$$\begin{aligned} d_Y(g(\hat{x}), g(\hat{x'})) \leqslant\ & d_Y(g(\hat{x}), f(x_n)) + d_Y(f(x_n), f(x'_n)) \\ & + d_Y(f(x'_n), g(\hat{x'})). \end{aligned}$$

In fact, for every $\varepsilon > 0$ we have some $\delta > 0$ such that $d_X(x, x') < \delta$ implies $d_Y(f(x), f(x')) < \varepsilon/3$; choosing n sufficiently large, we deduce that $d_Y(g(\hat{x}), g(\hat{x'})) < \varepsilon$ when $d_X(x, x') < \delta$.

Now $g\eta = f$, and the uniqueness of g for this property follows from Exercise 3.6.7(a), because $\eta(X)$ is dense in $\hat{X}$ and Y is Hausdorff.

(f) If the previous $f\colon X \to Y$ has Lipschitz constant L, then

$$\begin{aligned} d_Y(g(\hat{x}), g(\hat{x'})) &= \lim d_Y(f(x_n), f(x'_n)) \\ &\leqslant L.\lim d_X(x_n, x'_n) = L.d(\hat{x}, \hat{x'}). \end{aligned} \tag{4.53}$$

As this holds for $L = 1$, the solution of the universal problem for weak contractions is determined up to isometric isomorphism.

Finally, if f is an isometry, the inequality (4.53) becomes an equality for $L = 1$, and g too is an isometry.

□

4.5.7 Comments

It is easy to prove that, if X is a dense subspace of a complete metric space Y, the inclusion $\vartheta \colon X \to Y$ is the metric completion of X, up to isometric isomorphism.

In fact, taking the metric completion $\eta \colon X \to \hat{X}$, the isometry ϑ can be written as $\vartheta = g\eta$, by a unique isometry $g \colon \hat{X} \to Y$, which is easily seen to be surjective: its image is a complete metric subspace of Y, and therefore a closed one (by 4.5.4(g)); but $g(\hat{X})$ contains X and is dense in Y.

In particular, the euclidean metric space $\mathbb{R}$ is the metric completion of the euclidean metric space $\mathbb{Q}$, because the latter is a dense metric subspace of $\mathbb{R}$.

Here *we should avoid entering into a circular argument*: we cannot *define* $\mathbb{R}$ as the metric completion of $\mathbb{Q}$, since we need $\mathbb{R}$ to define metric spaces and establish the completion theorem.

*However, before introducing metric spaces, it is possible to construct $\mathbb{R}$ from $\mathbb{Q}$, by a procedure similar to that of the completion theorem: one would define the Cauchy sequences (x_k) of rational numbers by the property:

- for every *rational number* $\varepsilon > 0$ there is some $h \in \mathbb{N}$ such that, if $m, n \geqslant h$, then $|x_m - x_n| < \varepsilon$.

These sequences form a unital subring C of the ring $\mathbb{Q}^{\mathbb{N}} = \mathrm{Map}(\mathbb{N}, \mathbb{Q})$; the infinitesimal sequences form an ideal I. The quotient ring C/I is the real field.*

4.5.8 *Compact metric spaces

Compactness has strong relations with metric spaces. Here we mention some of them, without proofs. These can be found in [Bou3], Section IX.2.9 or [Ke].

(a) A metric space is compact if and only if it is sequentially compact.

A space X is said to be *sequentially compact* if every sequence $x = (x_k)$ in X has a subsequence that admits a limit in X; a *subsequence* of x is obtained as a composite $xf \colon \mathbb{N} \to X$, where $f \colon \mathbb{N} \to \mathbb{N}$ is strictly increasing.

(Note that the uniqueness of limits is of no relevance here.)

(b) A metric space is compact if and only if it is complete and totally bounded.

A metric space X is said to be *totally bounded* if, for every $\varepsilon > 0$, it is covered by a finite family of open balls $B_X(a_i, \varepsilon)$, with $a_1, ..., a_n \in X$.

(c) (*Heine–Cantor Theorem*) A continuous mapping $f\colon X \to Y$ between metric spaces, defined on a compact one, is uniformly continuous.

(d) (*Lebesgue Number Lemma*) Let X be a compact metric space and $(U_i)_{i\in I}$ an open cover of X. Then there exists a number $\varepsilon > 0$ such that every subset $A \subset X$ with $\operatorname{diam}(A) < \varepsilon$ is contained in some element U_i of the cover.

Such a number is called a *Lebesgue number of* $\mathcal{U}$.

4.5.9 *A synopsis of uniform spaces

Uniform continuity can be defined on metric spaces, but can also be defined for topological abelian groups (see 4.6.1), where – for every neighbourhood N of 0 – we can say that two points x, y are *near of order* N if $x - y \in N$.

We may want to find the 'natural domain' of uniform continuity, which should extend the previous situations and be independent of auxiliary structures like the real field. This leads us to a beautiful theory, established by André Weil in 1937.

The following sketch can give some idea of the theory, and might prompt someone to study (part of) it; a good reference is Bourbaki's [Bou2], which contains all the results mentioned below. Some further hints at the uniform structure of algebro-topological objects will be given in the next section.

(a) A *uniform space* is a set X equipped with a non-empty collection $\mathcal{E}$ of sets $E \subset X\times X$ (called *entourages*) that satisfies the following axioms.

(Un.1) Every $E \in \mathcal{E}$ contains the diagonal $\Delta_X = \{(x,x)|x \in X\}$ of X.

(Un.2) If $E \in \mathcal{E}$ and $E \subset F \subset X\times X$, then $F \in \mathcal{E}$.

(Un.3) If $E, F \in \mathcal{E}$, then $E \cap F \in \mathcal{E}$.

(Un.4) If $E \in \mathcal{E}$, there is some $F \in \mathcal{E}$ such that $F\circ F \subset E$, where $F\circ F$ is a composite of binary relations:

$$F\circ F = \{(x,z) \in X\times X \mid \exists\, y \in X\colon (x,y),(y,z) \in F\}. \tag{4.54}$$

(Un.5) If $E \in \mathcal{E}$, then $E^\sharp \in \mathcal{E}$, where $E^\sharp = \{(x,y) \mid (y,x) \in E\}$.

If $(x,y) \in E \in \mathcal{E}$, we say that x and y are *E-close*. If all pairs of points in a subset A of X are E-close, the subset A is called *E-small*. An entourage E is *symmetric* if $E = E^\sharp$.

The first axiom states that each point of X is E-close to itself, for each entourage E. The fourth axiom states that for each entourage E there is an entourage F that is 'twice smaller', at least.

(b) Now, a mapping $f\colon X \to Y$ is said to be *uniformly continuous* if:

(i) for every entourage F of Y there is some entourage E of X such that $(f\times f)(E) \subset F$,

where $f\times f\colon X\times X \to Y\times Y$ takes a point (x,x') to $(f(x),f(x'))$.

(c) Every metric space X has an *associated uniform structure*, where an entourage is any subset $E \subset X\times X$ containing some subset

$$E_\varepsilon = \{(x,y) \in X\times X \mid d(x,y) < \varepsilon\} \qquad (\varepsilon > 0). \tag{4.55}$$

Every uniform space X has an *associated topological structure*, where a subset N is a neighbourhood of a if and only if there is some entourage E such that

$$N \supset E(a) = \{x \in X \mid (a,x) \in E\}. \tag{4.56}$$

A uniform space is said to be Hausdorff (or connected, or compact) if the associated topological space is. If we start from a metric space, the topology associated to the uniform structure associated to the metric is the same as the topology (directly) associated to the metric.

A uniformly continuous mapping $f\colon X \to Y$ between uniform spaces is continuous for the associated topologies. A mapping $f\colon X \to Y$ between metric spaces is uniformly continuous for the metrics if and only if it is for the associated uniform structures.

(d) In a uniform space X, a filter $\mathcal{F}$ (see 4.3.5) is said to be a *Cauchy filter* if, for every entourage E of X, there exists an element $F \in \mathcal{F}$ that is E-*small*, i.e. $F\times F \subset E$. A uniform space is *complete* if every Cauchy filter in it has a limit point.

There is a *Hausdorff-completion theorem*, according to which every uniform space X has a universal uniformly continuous mapping $\eta\colon X \to \hat{X}$ with values in a Hausdorff complete uniform space; η is injective if and only if the uniform space X is Hausdorff. All this extends the completion of metric spaces.

(e) On a compact Hausdorff topological space K there is a unique uniform structure whose associated topology is the given one: its entourages are the neighbourhoods of the diagonal Δ_K in the topological space $K\times K$. Moreover, K is always complete with respect to its uniform structure.

(f) A continuous mapping $f\colon K \to X$ defined on a compact Hausdorff space, with values in a uniform space, is always uniformly continuous with respect to the uniform structure of K.

4.6 From topological groups to topological vector spaces

Algebra and Topology have two main interactions: Topological Algebra and Algebraic Topology. The former studies algebraic structures equipped with a consistent topology, as briefly examined here. We already said something about the latter in the Introduction to Chapter 3.

We end this chapter with some hints at the foundation of Mathematics, in 4.6.7.

Literature. Topological Algebra splits in two main branches: topological groups (see [Bou2], Chapter III) and topological vector spaces, also called topological linear spaces [SW, Da]. The latter open the way to Banach spaces, Hilbert spaces, and Functional Analysis.

4.6.1 Topological groups

A topological group G is a group equipped with a topology consistent with the structure of G as an equational algebra. This means that (in multiplicative notation):

(TG.1) the multiplication $m\colon G\times G \to G$ is a continuous mapping, with respect to the product topology on the domain,

(TG.2) the inversion $i\colon G \to G$, $x \mapsto x^{-1}$ is a continuous mapping.

The unit e is also called the *origin* of G; of course the mapping $e\colon \{*\} \to G$ is automatically continuous. The inversion i is an involutive homeomorphism (inverse to itself).

The group G acts on itself (transitively and freely) by left translations

$$\tau\colon G \to \mathrm{Aut}(|G|), \qquad \tau(g)\colon x \mapsto gx, \tag{4.57}$$

which are homeomorphisms. G is thus a homogeneous topological space (see 3.2.8(c)): all points 'look alike', from the topological point of view (not from the algebraic one, of course).

As a consequence, the set $\mathcal{N}(e)$ of neighbourhoods of the origin determines the set of neighbourhoods of any point g

$$\mathcal{N}(g) = g\mathcal{N}(e) = \{gN \mid N \in \mathcal{N}(e)\}, \tag{4.58}$$

and therefore the whole topology of G. The same holds for open neighbourhoods, fundamental systems of neighbourhoods and local bases.

A consistent topology on a group G can be assigned giving a filter $\mathcal{N}(e)$ of G satisfying some axioms: see [Bou2], III.1.2, Proposition 1.

The natural arrows $f\colon G \to G'$ between topological groups are the *continuous homomorphisms*, which amount to the homomorphism of groups continuous at the origin. They are stable under composition and contain all the identities of topological groups.

An *isomorphism* $f\colon G \to G'$ of topological groups is, at the same time, a homeomorphism of spaces and a homomorphism of groups (actually on isomorphism of groups).

Remarks and complements. (a) Each group G has two obvious consistent topologies, the discrete and the indiscrete topology.

(b) The additive group $\mathbb{R}$ and the multiplicative group $\mathbb{R}^*$ are topological groups, with the euclidean topology (by 3.3.6). The additive group $\mathbb{Z}$ of integers is usually viewed as a discrete group (with the euclidean topology, again).

(c) The inversion $i\colon G \to G$ of a topological group is a homeomorphism; it is an isomorphism of topological groups if and only if G is commutative.

(d) A left translation $\tau(g)\colon G \to G$ is a homeomorphism; it is an isomorphism of topological groups if and only if $g = e$ (and then $\tau(g) = \mathrm{id}\, G$).

(e) (*Hausdorff groups*) A topological group is Hausdorff if and only if the singleton $\{e\}$ (or any singleton of G) is closed. Then it is simply called a *Hausdorff group*.

In fact, the diagonal $\Delta \subset G\times G$ is the preimage of $\{e\}$ with respect to the continuous mapping $G\times G \to G$, $(x, y) \mapsto xy^{-1}$, and is closed if $\{e\}$ is. Conversely, any Hausdorff space is T_1.

(f) (*Action of a topological group*) An *action* of a topological group G on a topological space X is defined as an action of the group $|G|$ on the set $|X|$, such that the mapping

$$\varphi\colon G\times X \to X, \qquad \varphi(g, x) = gx, \tag{4.59}$$

is continuous, with respect to the product topology of $G\times X$.

If G is just a *group*, an action of G on the space X (as defined in 3.5.8) is the same as an action of the discrete group G_d on the space X. (It is a straightforward consequence of Exercise 3.5.4(d).)

4.6.2 ***Exercises and complements*** (Substructures, etc.)

(a) A *topological subgroup* H of a topological group G is a subgroup of G, equipped with the topological structure of a subspace, that is automatically consistent. The inclusion $H \to G$ is a continuous homomorphism.

(b) A *cartesian product* $G = \prod G_i$ of topological groups is equipped with the product of both structures, the algebraic and the topological one: these structures are consistent.

The projections $p_i\colon G \to G_i$ are thus continuous homomorphisms, and satisfy the universal property of the product with respect to continuous homomorphisms.

(c) A *quotient* G/H of a topological group modulo a normal subgroup is the quotient group with the quotient topology (modulo the congruence associated to H): these structures are consistent. The canonical projection $G \to G/H$ is a continuous homomorphism.

Moreover G/H is Hausdorff if and only if H is closed in G.

4.6.3 Topological rings and fields

A *topological ring* R is a ring equipped with a topology consistent with the structure of R as an equational algebra. This means that, after the axioms (TG.1, 2) on the underlying additive structure (in 4.6.1), we also have:

(i) the multiplication $m\colon R \times R \to R$ is a continuous mapping, with respect to the product topology on the domain.

If R is a unital ring, then the opposite $-x = (-1).x$ is determined by the multiplication, and axiom (TG.2) is redundant.

A *topological field* K is a field equipped with a topology consistent with the structure of K as an equational algebra. This means that, after the axioms (TG.1) and (i) we also have:

(ii) the multiplicative inversion $j\colon K^* \to K^*$, $x \mapsto x^{-1}$ is a continuous mapping.

Substructures of topological rings and fields, cartesian products and quotients of topological rings (with respect to a bilateral ideal) are dealt with as in 4.6.2.

Exercises and complements. (a) The real field $\mathbb{R}$ is a topological field, with the euclidean topology. The ring $\mathbb{Z}$ is a discrete topological subring of $\mathbb{R}$.

(b) The complex field $\mathbb{C}$ is a topological field, with the euclidean topology of $\mathbb{R}^2$.

(c) The inclusions $\mathbb{Z} \subset \mathbb{Q} \subset \mathbb{R} \subset \mathbb{C}$ are inclusions of topological subrings (or subfields).

4.6.4 Topological vector spaces

Let K be a topological field. A *topological vector space* X on K is a vector space equipped with a topology consistent with the algebraic structure, in the sense that:

(TVS.1) the sum $s\colon X \times X \to X$ is a continuous mapping, with respect to the product topology on the domain,

(TVS.2) the scalar multiplication $m\colon K \times X \to X$ is a continuous mapping, with respect to the product topology on the domain.

(Also here the continuity of the additive inversion is a consequence.) Substructures, cartesian products and quotients of topological vector spaces are dealt with as in 4.6.2.

4.6.5 Normed vector spaces

Let K be the field $\mathbb{R}$ or $\mathbb{C}$ (we need the modulus of a scalar).

A *normed vector space* on K is a vector space X equipped with a *norm*

$$||-||: X \to \mathbb{R}, \tag{4.60}$$

satisfying the following axioms (for $x, y \in X$ and $\lambda \in K$):

(N.1) $||x+y|| \leqslant ||x|| + ||y||$ (*subadditive property*),

(N.2) $||\lambda x|| = |\lambda|.||x||$ (*absolute homogeneity*),

(N.3) if $||x|| = 0$ then $x = 0$ (*separation*).

Note that (N.2) implies $||0|| = 0$. Dropping the last axiom we have a *seminormed vector space*, equipped with a *seminorm*.

Exercises and complements. (a) Let X be a seminormed vector space on K. Then X is a topological vector space, with the topology determined by the associated semimetric

$$d(x,y) = ||x-y||, \tag{4.61}$$

and $||x|| = d(x, 0)$.

(b) In the previous case, X is normed if and only if d is a metric, if and only if the space X is Hausdorff.

(c) A *Banach space* on K is a normed K-vector space, that is complete as a metric space.

Banach spaces are named after Stefan Banach, a Polish mathematician of the first half of the 20th century. The structural aspects of this beautiful, important theory can be appreciated in [Se].

(d) $\mathbb{R}^n$ is a Banach space on $\mathbb{R}$, with the euclidean norm, also called the l_2-norm (see (3.3)).

*(e) More generally, $\mathbb{R}^n$ is a Banach space with respect to any l_p-norm $||x||_p = d_p(x, 0)$, for $p \in [1, \infty]$ (see 4.5.3(d)). We already know that all the associated metrics are Lipschitz-equivalent.

*(f) Many important Banach spaces come from vector spaces of real valued functions, or complex valued functions, or continuous linear mappings between topological vector spaces.

4.6.6 *Uniform structures on topological 'algebras'

Continuing our hints at uniform spaces, in 4.5.9, we extract now some results from [Bou2], mostly without proof.

(a) A topological abelian group X has an *associated uniform structure* $\mathcal{E}$ (see 4.5.9(a)), formed of all the subsets $E \subset X \times X$ that contain some set of the form

$$E_N = \{(x,y) \in X \times X \mid x - y \in N\} \qquad (\text{for } N \in \mathcal{N}(e)). \tag{4.62}$$

This uniform structure gives back the original topology of X (by the procedure of (4.56)), because the entourage E_N yields back the neighbourhood N of the origin

$$E_N(0) = \{x \in X \mid (x, 0) \in E_N\} = N.$$

A topological abelian group X is said to be *complete* if it is with respect to this uniform structure. Every Hausdorff abelian group has a universal embedding in a Hausdorff complete abelian group, its *group-completion*: see [Bou2], III.3.5, Theorem 2.

On a general topological group, in multiplicative notation, there is a *left uniform structure* $\mathcal{E}'$ generated by the subsets $\{(x, y) \mid x^{-1}y \in N\}$, and – symmetrically – a right uniform structure. Completeness is required for both: see [Bou2].

(b) A topological ring R is always equipped with the uniform structure associated to the underlying topological abelian group, and is said to be *complete* if this uniform structure is.

A Hausdorff topological ring R has a universal embedding in a Hausdorff complete topological ring $\hat{R}$, called its *ring-completion*. If R is unital, or commutative, so is its completion: see [Bou2], III.6.5, Proposition 6.

(c) A Hausdorff topological field K is canonically embedded in its *ring-completion*, the Hausdorff complete topological ring $\hat{K}$.

The latter need not be a field: it is if and only if the inversion mapping $j\colon K^* \to K^*$ of K can be continuously extended to the space $\hat{K}^* = \hat{K} \setminus \{0\}$.

By [Bou2], III.6.8, Proposition 7, this is equivalent to the condition:

(*) for every Cauchy filter $\mathcal{F}$ of K such that $0 \notin \overline{\mathcal{F}}$, the image $j(\mathcal{F})$ is also a Cauchy filter,

where $\overline{\mathcal{F}}$ is the intersection of all $\overline{A}$, for $A \in \mathcal{F}$.

(d) The *ring-completion* of the topological field $\mathbb{Q}$ is the topological field $\mathbb{R}$.

In [Bou2], Section IV.1.3, the euclidean line $\mathbb{R}$ is *constructed* as the group-completion of the topological abelian group $(\mathbb{Q}, +)$.

4.6.7 *A digression on foundations

We began Chapter 1 exploring some basic properties of real numbers, concerned with their main operations and the natural order; we began Chapter 3 exploring basic features of continuity, for real functions. From these starting points we moved on, to investigate algebraic structures and topological structures.

While developing these topics, we have remarked in several instances how the integral, rational, real and complex numbers can be constructed, using the theory which we were developing. *In a foundational setting*, all this should be reorganised in a linear, consequential way, avoiding circular arguments.

Giving a foundation of Mathematics is (obviously) far from the aims of this book, but at this point we can sketch a way of organising things. Such an organisation would necessarily be long and initially quite abstract – of little interest and utility for a beginner.

(a) We start with Set Theory. This includes the construction of the set $\mathbb{N}$ of natural numbers, with the operations of sum and multiplication, derived from disjoint union and cartesian product of sets.

(b) We go on with Algebra. After the first chapters on semigroups, groups, abelian groups, rings and fields, we are able to say that $\mathbb{N}$ is a semiring, and to construct the ring of integers $\mathbb{Z}$ as the ring of formal differences of natural numbers (as in 2.2.7).

Then we construct the rational field $\mathbb{Q}$ as the field of fractions on the ring $\mathbb{Z}$ (as in 2.2.5(c)).

(c) We can now construct the real field $\mathbb{R}$, for instance by the procedure sketched in 4.5.7 and based on a ring of Cauchy sequences of $\mathbb{Q}$.

We already remarked that this procedure, even though inspired by the metric completion theorem, can *not* use the theory of metric spaces – which is not available before introducing $\mathbb{R}$.

(d) We define the order relation of $\mathbb{R}$ and verify the axioms (A.1–16) of Section 1.1; let us recall that they determine $\mathbb{R}$, up to a unique isomorphism of totally ordered fields (see 2.2.8).

(e) The complex field $\mathbb{C}$ can now be constructed from $\mathbb{R}$, as in the first part of Section 2.7.

(f) Topology can now be studied. After developing some parts of Calculus, we can also express complex numbers and complex roots in polar form.

(g) As an alternative to (c), one can proceed with Topology, after (b), *without using the euclidean line*. Then the latter is constructed, as the ring-completion of the topological field $\mathbb{Q}$, and we have (at last!) at our disposal the euclidean spaces $\mathbb{R}^n$, with all the important topological spaces derived from them.

We have seen in 4.6.6 that such a program is followed by Bourbaki.

A reader interested in the history and construction of real numbers and related matters (including complex numbers and quaternions) is referred to a recent book by T.W. Körner [Ko2].

5
Categories and functors

Many aspects of the previous chapters can be organised and unified using the Theory of Categories, as foreshadowed in 1.2.7.

This discipline is now present in most of Mathematics, large parts of theoretical Computer Science and parts of theoretical Physics, often at a deep level. Its unifying power brings together different branches, and can lead to a better understanding of their theoretical roots.

We cover, here and in the next chapter, its basic aspects, starting from examples and exercises taken from the previous chapters, then developing the theory together with new exercises and applications.

Of course, one should not forget the specificity of each theory, that can even be highlighted when the basic parts are unified in a common language.

For further study of category theory there are excellent texts, like Mac Lane [M2], Borceux [Bo1, Bo2, Bo3], Adámek, Herrlich and Strecker [AHS]. At a more elementary level, the author's [G4] is a textbook for beginners, also devoted to applications in Algebra, Topology and Algebraic Topology.

5.1 An overview of Category Theory

5.1.1 The origins

Category Theory originated in an article of Samuel Eilenberg and Saunders Mac Lane [EiM], published in 1945. It has now developed into a branch of mathematics, with its own internal dynamics.

In the previous chapters we have studied structured sets of many kinds. Each kind has its own privileged mappings, that preserve the structure in some sense, like homomorphisms of groups, or order-preserving mappings, or continuous mappings.

Looking at each of these collections – of objects and arrows – *as a whole*, we have the category Gp of groups and their homomorphisms, the category

Ord of ordered sets and monotone mappings, the category Top of topological spaces and continuous mappings, and so on. More elementarily, we have the category Set of sets and their mappings.

In all these instances, the privileged mappings are called morphisms or arrows of the category; an arrow from the object X to the object Y is written as $f \colon X \to Y$.

Two consecutive morphisms, say $f \colon X \to Y$ and $g \colon Y \to Z$, can be composed giving a morphism $gf \colon X \to Z$. This partial composition law is 'as regular as it can be', which means that it is associative (when legitimate) and every object X has an *identity*, written as $\mathrm{id}\, X \colon X \to X$ or 1_X, which acts as a unit for legitimate compositions.

'Concrete categories' are often associated with mathematical structures, in this way; but categories are not limited to these instances, by far.

5.1.2 Universal properties

In these categories (and many 'similar' ones) we have cartesian products, consisting of the cartesian product of the underlying sets equipped with the 'natural' structure of the kind we are considering, be it of algebraic character, or an ordering, or a topology, or something else.

As we have already highlighted, these procedures can be unified: we have a family $(X_i)_{i \in I}$ of objects of a category, indexed by a set I, and we want to find an object X equipped with a family of morphisms $p_i \colon X \to X_i$ $(i \in I)$, called *cartesian projections*, that satisfies the following *universal property*:

- for every object Y and every family of morphisms $f_i \colon Y \to X_i$ $(i \in I)$ in the given category

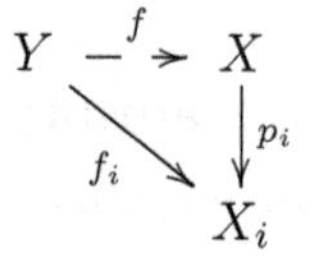

$$\begin{array}{ccc} Y & \xrightarrow{f} & X \\ & \searrow^{f_i} & \downarrow p_i \\ & & X_i \end{array} \tag{5.1}$$

there exists precisely one morphism $f \colon Y \to X$ such that $p_i f = f_i$, for all $i \in I$.

This property determines the solution *up to isomorphism*, i.e. an invertible morphism of the given category. The general proof is quite easy: given two solutions $(X, (p_i))$, $(Y, (q_i))$, we can determine two morphisms $f \colon X \to Y$ and $g \colon Y \to X$ and prove that their composites coincide with the identities of X and Y.

These facts bring to light a crucial aspect: a categorical definition (as the previous one) *is based on morphisms and their composition*, while the objects only step in as domains and codomains of arrows.

If we want to understand what unifies the product of a family (X_i) of sets, or groups, or ordered sets, or topological spaces we should forget the nature of the objects, and think of the family of cartesian projections $p_i\colon X \to X_i$, together with the previous property. Then – in each category we are interested in – we come back to the objects in order to prove that a solution exists.

From a structural point of view, a category only 'knows' its objects by their morphisms and composition, in the same way as a group structurally 'knows' its elements by their composition: any internal structure they may have will only appear at a second order level, e.g. in a proof.

This is even more evident if we think of another procedure, which in category theory is called a 'sum', or 'coproduct'. We start again from a family $(X_i)_{i\in I}$ of objects; its (categorical) *sum* is an object X equipped with a family of morphisms $u_i\colon X_i \to X$ $(i \in I)$, called *injections*, that satisfies the following universal property:

- for every object Y and every family of morphisms $f_i\colon X_i \to Y$ $(i \in I)$ in the given category

$$\begin{array}{ccc} X & \overset{f}{\dashrightarrow} & Y \\ {\scriptstyle u_i}\uparrow & \nearrow_{f_i} & \\ X_i & & \end{array} \qquad fu_i = f_i \ \text{(for } i \in I). \tag{5.2}$$

there exists precisely one morphism $f\colon X \to Y$ such that $fu_i = f_i$, for all $i \in I$.

Again, the solution is determined up to isomorphism; its existence depends on the category.

As we have seen, the sum of a family of objects is easy to construct in the category Set of sets, by a disjoint union. We have similar solutions in Ord and Top. But in Gp the categorical sum of a family of groups is called the *free product* of the family; its construction is rather complex, and its underlying set is not the disjoint union of the sets underlying the groups (see 6.1.6(h)). The categorical approach highlights the fact that we are 'solving the same problem' and – finally – makes clear what we are doing.

It is also important to note that the categorical definitions of product and coproduct are *dual* to each other: each of them is obtained from the other by reversing all arrows and compositions. Structurally, this only makes sense within category theory, because the dual of a given category, formed by reversing its arrows and partial composition law, is a formal construction: the dual of a category of structured sets and mappings is *not* a category of the same kind (even though, *in several cases*, it may essentially be, as a result of a duality theorem).

We have already considered, in 1.2.1, diagrams and commutative diagrams of structured sets. All this makes sense in any category; a formal definition of a diagram will be given in 5.3.4.

5.1.3 Functors

It becomes now possible to view on the same level, so to say, mathematical theories of different branches, and formalise their links. A structure-preserving mapping between categories is called a *functor* (to be defined in Section 5.3).

Among the simplest examples there are the *forgetful functors*, that forget the structure (or part of it), like:

$$\mathsf{Gp} \to \mathsf{Set}, \qquad \mathsf{Ord} \to \mathsf{Set}, \qquad \mathsf{Top} \to \mathsf{Set}, \qquad (5.3)$$

For instance, the forgetful functor $U\colon \mathsf{Gp} \to \mathsf{Set}$ takes a group G to its underlying set $U(G)$, and a homomorphism $f\colon G \to G'$ to its underlying mapping $U(f)\colon U(G) \to U(G')$. The whole procedure is well-behaved, in the sense that it preserves composition and identities.

These functors are so obvious that they are often overlooked, in mathematics; but here it will be important to keep trace of them. In particular, we will see that they often determine other functors 'backwards', which are much less obvious: like the *free-group* functor $F\colon \mathsf{Set} \to \mathsf{Gp}$, 'left adjoint' to $U\colon \mathsf{Gp} \to \mathsf{Set}$ (of which more will be said below).

In the same way, the (obvious) embedding $\mathsf{Ab} \to \mathsf{Gp}$ of the category of abelian groups in the category of groups has a left adjoint, the abelianisation functor $\mathsf{Gp} \to \mathsf{Ab}$ that will be considered in 6.3.6(a).

*The core of Algebraic Topology is constructing *functors from a category of topological spaces to a category of algebraic structures*, and using them to reduce topological problems to simpler algebraic ones. We have thus the sequence of *singular homology functors* $H_n\colon \mathsf{Top} \to \mathsf{Ab}$ $(n \geqslant 0)$, and the fundamental group functor $\pi_1\colon \mathsf{Top}_\bullet \to \mathsf{Gp}$, defined on the category of pointed topological spaces and pointed continuous mappings.*

5.1.4 Natural transformations and adjunctions

The third basic element of category theory is a *natural transformation* $\varphi\colon F \to G$ between two functors $F, G\colon \mathsf{C} \to \mathsf{D}$ with the same domain C and the same codomain D. We also write $\varphi\colon F \to G\colon \mathsf{C} \to \mathsf{D}$.

This simply amounts to a family of morphisms of D, indexed by the objects X of C

$$\varphi_X \colon F(X) \to G(X), \tag{5.4}$$

under a condition of 'naturality' that will be made explicit below.

Here we just give an example, based on sets, groups, and free groups (but the reader can replace groups with semigroups, or abelian groups, or R-modules, or any kind of equational algebras). We have mentioned in the previous point the forgetful functor $U \colon \mathsf{Gp} \to \mathsf{Set}$ and the free-group functor $F \colon \mathsf{Set} \to \mathsf{Gp}$, that takes a set X to the free group generated by X (constructed in 1.6.3).

The insertion of X in $F(X)$, as its basis, is a canonical mapping, in Set

$$\eta_X \colon X \to U(F(X)). \tag{5.5}$$

All of them give a natural transformation

$$\eta \colon \mathrm{id}\,\mathsf{Set} \to UF \colon \mathsf{Set} \to \mathsf{Set}, \tag{5.6}$$

where $\mathrm{id}\,\mathsf{Set}$ is the identity functor of the category Set (taking objects and arrows to themselves) and $UF \colon \mathsf{Set} \to \mathsf{Set}$ is the composed functor (taking each set to the underlying set of the free group on it). This natural transformation links the two functors, making $F \colon \mathsf{Set} \to \mathsf{Gp}$ *left adjoint to* $U \colon \mathsf{Gp} \to \mathsf{Set}$.

The link is represented by the *universal property* of the insertion of the basis, namely the fact that every mapping $f \colon X \to U(G)$ with values in a group (more precisely in its underlying set) can be uniquely extended to a homomorphism $g \colon F(X) \to G$. Formally:

- for every morphism $f \colon X \to U(G)$ in Set, there is precisely one morphism $g \colon F(X) \to G$ in Gp such that $U(g).\eta_X = f$.

We will see many constructions of free algebraic structures, or more generally of left (or right) adjoints of given functors. Many of them are 'real constructions', which give a good idea of the backward procedure; others are so complicated that one can doubt of their constructive character.

In such cases, one can prefer to prove the existence of the adjoint, by the Adjoint Functor Theorem (see 6.3.5) or some other general statement: then the result is determined up to isomorphism, and its universal property allows its use. Much in the same way as we can define a real function as the solution of a certain differential equation with initial data, as soon as we know that the solution exists and is unique.

5.1.5 *Classes of categories*

It is a common feature of Mathematics to look for the natural framework where certain properties should be studied: for instance, the basic properties of polynomials with real coefficients can be examined in general polynomial rings, or in more general algebras. Besides yielding more general results, the natural framework gives a deeper comprehension of what we are studying.

Category theory makes a further step in this sense. For instance, categories of modules are certainly important, but – since Buchsbaum's *Appendix* [Bu] and Grothendieck's paper [Gt] in the 1950's – a consistent part of Homological Algebra finds its natural framework in *abelian categories* and their generalisations (see Section 6.6). Similarly, the categories of structured sets can be viewed as particular *concrete categories* (defined in 5.3.3), and the categories of equational algebras as particular *monadic categories* (see 6.6.3, 6.6.4).

In other words, when studying certain 'classes' of categories, we may (or perhaps had better) look for a *structural definition* including this class, rather than some general way of constructing the important examples we want to study.

This leads to considering categories of (small) categories, and to higher dimensional category theory; an elementary introduction to this domain can be found in [G4], Chapter 7.

5.2 Categories

Categories have two parents and a precise birthday: they were introduced by Eilenberg and Mac Lane [EiM] in 1945, together with the other basic terms of category theory.

5.2.1 *Some examples*

Loosely speaking, before giving a precise definition, a category C consists of *objects* and *morphisms* together with a (partial) *composition law*: given two consecutive morphisms $f\colon X \to Y$ and $g\colon Y \to Z$ we have a composed morphism $gf\colon X \to Z$. This partial operation is associative (whenever composition is legitimate) and every object X has an *identity*, written as $\mathrm{id}\,X\colon X \to X$ or 1_X, that acts as a unit for legitimate compositions.

The prime example is the category Set *of sets* (and mappings), where:

- an object is a set,

- the morphisms $f\colon X \to Y$ between two given sets X and Y are the (set-theoretical) mappings from X to Y,

- the composition law is the usual composition of mappings, where $(gf)(x) = g(f(x))$.

The following categories of structured sets and structure-preserving mappings (with the usual composition) will often be used and analysed:

- the category Gp *of groups* (and their homomorphisms),

- the category Ab *of abelian groups* (and homomorphisms),

- the category Mon *of monoids* (and homomorphisms),

- the category Abm *of abelian monoids* (and homomorphisms),

- the category Rng *of unital rings* (and homomorphisms),

- the category Rng' *of rings* (and homomorphisms),

- the category CRng *of commutative unital rings* (and homomorphisms),

- the category $R\,\mathsf{Mod}$ of left modules on a fixed unital ring R (and homomorphisms),

- the category $K\,\mathsf{Vct}$ $(= K\,\mathsf{Mod})$ of vector spaces on a field K (and homomorphisms),

- the category $R\mathsf{Alg}$ of unital algebras on a fixed commutative unital ring R (and homomorphisms),

- the category $R\mathsf{CAlg}$ of commutative unital R-algebras (and homomorphisms),

- the category Ord of *ordered sets* (and monotone mappings),

- the category pOrd of *preordered sets* (and monotone mappings),

- the category $\mathsf{Set}_\bullet$ *of pointed sets* (and pointed mappings), see 1.2.9(e),

- the category Top *of topological spaces* (and continuous mappings),

- the category Hsd *of Hausdorff spaces* (and continuous mappings),

- the category $\mathsf{Top}_\bullet$ *of pointed topological spaces* (and pointed continuous mappings),

- the category Mtr of *metric spaces and Lipschitz mappings*, see 4.5.1(iii),

- the category Mtr_1 of *metric spaces and weak contractions*, see 4.5.1(iv).

Let us recall that a homomorphism of monoids or unital rings is assumed to preserve the unit. The relationship between Rng and Rng' will be briefly examined in 6.3.9(e).

For $\mathsf{Set}_\bullet$ we recall that a *pointed set* is a pair (X, x_0) consisting of a set X and a base-element $x_0 \in X$, while a pointed mapping $f: (X, x_0) \to (Y, y_0)$ is a mapping $f: X \to Y$ such that $f(x_0) = y_0$.

Similarly, a *pointed topological space* (X, x_0) is a space with a base-point, and a pointed map $f: (X, x_0) \to (Y, y_0)$ is a continuous mapping from X to Y such that $f(x_0) = y_0$.

When a category is named after its objects alone (e.g. the 'category of groups'), this means that the morphisms are understood to be the obvious ones (in this case the homomorphisms of groups), with the obvious composition law. Different categories with the same objects are given different names, like Mtr and Mtr_1 above.

The category $\mathsf{Top}_\bullet$ is important in Algebraic Topology: for instance, the fundamental group $\pi_1(X, x_0)$ is defined for a pointed topological space.

5.2.2 Main definitions

A *category* C consists of the following data:

(a) a set $\mathrm{Ob}\,\mathsf{C}$, whose elements are called *objects* of C,

(b) for every pair X, Y of objects, a set $\mathsf{C}(X, Y)$ (called a *hom-set*) whose elements are called *morphisms* (or *maps*, or *arrows*) of C from X to Y and denoted as $f: X \to Y$,

(c) for every triple X, Y, Z of objects of C, a mapping of *composition*

$$\mathsf{C}(X, Y) \times \mathsf{C}(Y, Z) \to \mathsf{C}(X, Z), \qquad (f, g) \mapsto gf,$$

where gf is also written as $g.f$.

These data must satisfy the following axioms.

(i) (*Associativity*) Given three consecutive arrows, $f: X \to Y$, $g: Y \to Z$ and $h: Z \to W$, one has: $h(gf) = (hg)f$.

(ii) (*Identities*) Given an object X, there exists an *endo*map $e: X \to X$ which acts as an identity whenever composition makes sense; in other words if $f: X' \to X$ and $g: X \to X''$, one has: $ef = f$ and $ge = g$. One shows, in the usual way, that e is determined by X; it is called the *identity* of X and written as 1_X or $\mathrm{id}\,X$.

We generally assume that the following condition is also satisfied:

(iii) (*Separation*) for X, X', Y, Y' objects of C, if $\mathsf{C}(X, Y) \cap \mathsf{C}(X', Y') \neq \emptyset$ then $X = X'$ and $Y = Y'$.

In other words, every map $f: X \to Y$ has a well-determined *domain* $\mathrm{Dom}\, f = X$ and *codomain* $\mathrm{Cod}\, f = Y$.

Concretely, when constructing a category, one can forget about condition (iii), as one can always satisfy it by *redefining* a morphism $\hat{f}\colon X \to Y$ as a triple $(X, Y; f)$ where f is a morphism from X to Y in the original sense (possibly not satisfying the Separation axiom).

$\mathrm{Mor}\,\mathsf{C}$ denotes the set of all the morphisms of C, which is the disjoint union of all hom-sets. Two morphisms f, g are said to be *parallel* when they have the same domain and the same codomain.

If C is a category, the *opposite* (or *dual*) category, written as C^{op}, has the same objects as C, reversed arrows and reversed composition $g * f$

$$\mathsf{C}^{\mathrm{op}}(X, Y) = \mathsf{C}(Y, X), \qquad g * f = fg, \qquad \mathrm{id}^{\,\mathrm{op}} X = \mathrm{id}\, X. \tag{5.7}$$

Every topic of category theory has a dual instance, which comes from the opposite category (or categories). A dual notion is generally distinguished by the prefix 'co-'.

A set X can be viewed as a *discrete* category: its objects are the elements of X, and the only arrows are their (formal) identities; here $X^{\mathrm{op}} = X$.

As usual in category theory, the term *graph* will be used to denote a simplified structure, with objects (or vertices) and morphisms (or arrows) $f\colon x \to y$, but no assigned composition nor identities. (This is called a *directed multigraph* in graph theory). A *morphism of graphs* preserves objects, arrows, domain and codomain.

Every category has an *underlying graph*. Every graph freely generates a category (Exercise 5.4.3(b)).

5.2.3 Small and large categories

Again, we do not insist on set-theoretical foundations. Yet some attention is necessary, to avoid involving 'the set of all sets', or requiring of a category properties of completeness that are 'too large for its size' (as we will see in 6.2.3).

We assume the existence of a (Grothendieck) *universe* $\mathcal{U}$, which is fixed throughout. Its axioms, listed in 5.2.9, say that we can perform inside it the usual operations of set theory. Its elements are called *small* sets, or $\mathcal{U}$-small sets, if useful.

A category is understood to have objects and arrows belonging to this universe, and is said to be *small* if its set of morphisms belongs to $\mathcal{U}$, *large* if it does not (and is just a subset of $\mathcal{U}$). As a consequence, in a small category the set of objects (which is in bijective correspondence with the set of identities) also belongs to $\mathcal{U}$.

More generally, a category C is said to be *locally small*, or *to have small hom-sets*, if all its sets $\mathsf{C}(X, Y)$ are small; in this case C is small if and only if its set of objects is.

The 'usual' categories of structured sets are large $\mathcal{U}$-categories with small hom-sets, like the category Set of small sets, and all the examples listed in 5.2.1. In these cases we speak – as usual – of the 'category of sets', and so on, *leaving understood the term 'small' as referred to these objects.*

*While developing the theory, one often needs a *hierarchy* of universes. For instance, Cat will denote the category of small categories and functors, introduced in 5.3.1. In order to view the (large) categories Set, Top, Ab, etc. as objects of a similar structure we should assume the existence of a second universe $\mathcal{V}$, with $\mathcal{U} \in \mathcal{V}$, and use the category $\mathsf{Cat}_{\mathcal{V}}$ of $\mathcal{V}$-small categories. In a more complex situation one may need a longer chain of universes. Most of the time *these points will be left as understood.**

5.2.4 Isomorphisms and groupoids

In a category C a morphism $f\colon X \to Y$ is said to be *invertible*, or an *isomorphism*, if it has an inverse, i.e. a morphism $g\colon Y \to X$ such that $gf = 1_X$ and $fg = 1_Y$. The latter is uniquely determined, by the usual argument; it is called the *inverse* of f and written as f^{-1}.

(If also $g'\colon Y \to X$ is inverse to f, then $g' = 1_X.g' = gfg' = g$.)

In the categories listed in 5.2.1 this definition gives the usual isomorphisms of the various structures. In some cases there are specific names, as homeomorphisms in Top and bijective isometries in Mtr_1.

The *isomorphism relation* $X \cong Y$ between objects of C (meaning that there exists an isomorphism $X \to Y$) is an equivalence relation.

A *groupoid* is a category where every map is invertible; this structure was introduced before categories, by H. Brandt in 1927 [Bra].

The fundamental groupoid of a topological space X contains all the fundamental groups $\pi_1(X, x)$ for $x \in X$ [Bro1, Bro2].

5.2.5 Subcategories, quotients and products of categories

(a) Let C be a category. A *subcategory* D is defined by assigning:

- a subset $\mathrm{Ob}\,\mathsf{D} \subset \mathrm{Ob}\,\mathsf{C}$, whose elements are called *objects of* D,

- for every pair of objects X, Y of D, a subset $\mathsf{D}(X, Y) \subset \mathsf{C}(X, Y)$, whose elements are called *morphisms of* D from X to Y,

so that the following conditions hold:

(i) for every pair of consecutive morphisms of D, their composite in C belongs to D,

(ii) for every object of D, its identity in C belongs to D.

Then D, equipped with the induced composition law, is a category.

One says that D is a *full* subcategory of C if, for every pair of objects X, Y of D, we have $\mathsf{D}(X,Y) = \mathsf{C}(X,Y)$, so that D is determined by assigning the subset of its objects. One says that D is a *wide* subcategory of C if it has the same objects, so that D is determined by assigning the subset of its morphisms (closed under composition and containing all the identities).

For instance, Ab is a full subcategory of Gp, while Mtr_1 is a wide subcategory of Mtr. The only full and wide subcategory of a category is the total one.

(b) A *congruence* $R = (R_{XY})$ in a category C consists of a family of equivalence relations R_{XY} in each set of morphisms $\mathsf{C}(X,Y)$, that is consistent with composition:

(iii) if $f\, R_{XY}\, f'$ and $g\, R_{YZ}\, g'$, then $gf\, R_{XZ}\, g'f'$.

The *quotient category* $\mathsf{D} = \mathsf{C}/R$ has the same objects of C and $\mathsf{D}(X,Y) = \mathsf{C}(X,Y)/R_{XY}$; in other words, a morphism $[f]\colon X \to Y$ in D is an equivalence class of morphisms $X \to Y$ in C. The composition is induced by that of C, which is legitimate because of condition (iii): $[g].[f] = [gf]$.

For instance, the homotopy relation $f \simeq f'$ in Top is a congruence of categories (see Exercise (a), below). The quotient category $\mathsf{hoTop} = \mathsf{Top}/\simeq$ is called the *homotopy category of topological spaces*, and is important in Algebraic Topology.

*The *weak homotopy category* HoTop comes from a more complex procedure, a 'category of fractions' of Top where the continuous mappings that induce isomorphisms in all homotopy sets and groups are made invertible [GaZ, Bo1]. It can also be obtained as the full subcategory of hoTop whose objects are the spaces homotopically equivalent to the geometric realisation of simplicial sets.*

(c) If C and D are categories, one defines the *product category* $\mathsf{C}\times\mathsf{D}$. An object is a pair (X,Y) where X is in C and Y in D. A morphism is a pair of morphisms

$$(f,g)\colon (X,Y) \to (X',Y'), \qquad f \in \mathsf{C}(X,X'),\;\; g \in \mathsf{D}(Y,Y'), \qquad (5.8)$$

and its composition with $(f',g')\colon (X',Y') \to (X'',Y'')$ is component-wise: $(f',g').(f,g) = (f'f, g'g)$.

Similarly one defines the product $\prod \mathsf{C}_i$ of a family of categories indexed by a small set.

Exercises and complements. (a) The homotopy relation $f \simeq f'$ in Top is defined by the existence of a continuous mapping $\varphi\colon X\times\mathbb{I} \to Y$ which coincides with f and f' on the bases of the *cylinder* $X\times\mathbb{I}$ (with the product topology)

$$\varphi(x,0) = f(x), \qquad \varphi(x,1) = f'(x) \qquad (\text{for } x \in X). \qquad (5.9)$$

Prove that this relation is a congruence of categories.

(b) A continuous mapping $f\colon X \to Y$ is a *homotopy equivalence* if there exists a continuous mapping $g\colon Y \to X$ such that $gf \simeq \mathrm{id}\, X$ and $fg \simeq \mathrm{id}\, Y$.

This is equivalent to saying that its homotopy class $[f]$ is an isomorphism of hoTop.

5.2.6 Monoids and categories

Monoids and their homomorphisms form the category Mon. But we deal here with a different interplay of these notions.

A monoid M can (and will often) be viewed as a category with a single formal object, say $*$. The morphisms $x\colon * \to *$ are the elements of M, composed by the multiplication of the monoid, with identity $\mathrm{id}\,(*) = 1$, the unit of the monoid. If M is a group, the associated category is a groupoid.

On the other hand, in every category C, the *endomorphisms* $X \to X$ of any object form a (possibly large) monoid, under the composition law

$$\mathrm{End}(X) = \mathsf{C}(X, X), \tag{5.10}$$

and the invertible ones form the group $\mathrm{Aut}(X)$ of *automorphisms* of X.

In this way a monoid is essentially the same as a category on a single object, while a category can be viewed as a 'multi-object generalisation' of a monoid. Groups and groupoids have a similar relationship – the restriction of the previous one.

The theory of regular, orthodox and inverse semigroups (see [ClP, Ho, Law]) has a strong interplay with the categories of relations and their applications in Homological Algebra, which is investigated in [G2].

5.2.7 Preorders and categories

Preordered sets and monotone mappings form a category, which we are writing as pOrd. Again, we are also interested in a different interplay of these notions.

A preordered set X will often be viewed as a category, where the objects are the elements of X and the set $X(x, x')$ contains a unique (formal) arrow if $x \prec x'$, written as $(x, x')\colon x \to x'$, and no arrows otherwise. Composition and units are (necessarily) as follows

$$(x', x'').(x, x') = (x, x''), \qquad \mathrm{id}\, x = (x, x).$$

In this way a preordered set is essentially the same as a category where each hom-set has at most one element. All diagrams in these categories commute. Two elements x, x' are isomorphic objects if and only if $x \sim x'$, in the equivalence relation associated to the preorder.

In particular, each ordinal defines a category, written as $\mathbf{0}, \mathbf{1}, \mathbf{2}, \dots$. Thus, $\mathbf{0}$ is the empty category; $\mathbf{1}$ is the *singleton category*, i.e. the discrete category on one object; $\mathbf{2}$ is the *arrow category*, with two objects (0 and 1), and one non-identity arrow, $0 \to 1$.

Exercises and complements. (a) Every small category S has an associated preordered set $\mathrm{po}(\mathsf{S})$, whose elements are the objects of S, preordered by the relation

$$x \prec y \quad \text{if there exists an arrow } x \to y \text{ in } \mathsf{S}. \tag{5.11}$$

Describe the preordered set $\mathrm{po}(\mathsf{S})$ as a quotient of the category S.

(b) Every hom-set $\mathsf{pOrd}(X, Y)$ is canonically ordered by the *pointwise preorder* relation, defined as follows for $f, g \colon X \to Y$

$$f \prec g \quad \text{if for all } x \in X \text{ we have } f(x) \prec g(x) \text{ in } Y. \tag{5.12}$$

This relation is preserved by composition. For ordered sets, every hom-set $\mathsf{Ord}(X, Y)$ has a canonical order $f \leqslant g$.

(c) Describe the following categories, not to be confused with $\mathbf{2}$:

- the category associated to the cardinal set $2 = \{0, 1\}$,
- the category associated to the additive group $(\mathbb{Z}/2, +)$,
- the category associated to the multiplicative monoid $(\mathbb{Z}/2, .)$.

(Something more on their difference will be seen in 5.3.4.)

(d) Similarly, the ordinal category $\boldsymbol{\omega}$, associated to the ordered set $(\mathbb{N}, \leqslant)$, should not be confused with other categories associated to the set $\mathbb{N}$, equipped with other structures. (Again, see 5.3.4.)

5.2.8 Mathematical structures and categories

When studying a mathematical structure with the help of category theory, it is crucial to choose the 'right' kind of structure and the 'right' kind of morphisms, so that the result is sufficiently general and 'natural' to have good properties with respect to the goals of our study – even if we are interested in more particular situations.

(a) A first point to be kept in mind is that the isomorphisms of the category (i.e. its invertible arrows) should indeed preserve the structure we are interested in, or we risk of studying something different from our purpose.

As a trivial example, the category T of topological spaces and *all* mappings between them has little to do with Topology: an isomorphism of T is any bijection between topological spaces. Indeed T is *equivalent to the category of sets* (as we shall see in 5.3.6), and is a modified way of looking at the latter.

Less trivially, we have already seen in 4.5.2 how the category M of metric spaces and continuous mappings misses crucial properties of metric spaces, since its invertible morphisms need not preserve completeness. In fact, M is equivalent to the category of *metrisable topological spaces* and continuous mappings (by 5.3.6, again), and should be replaced with the latter.

On the other hand, the categories Mtr and Mtr_1 do respect metric properties, like being complete or bounded (see 4.5.2); both are important.

(b) Other points will become clearer below. For instance, the category Top of topological spaces and continuous mappings is a classical framework for studying topology. Hausdorff spaces are certainly important, but it is 'often' better to view them *in* Top, as their category Hsd is less well behaved.

*Among the good properties of Top there is the fact that all categorical limits and colimits (extending products and sums) exist, and are computed *as in* Set, then equipped with a suitable topology determined by the structural maps.

(As we will see, this is a consequence of the fact that the forgetful functor Top $\to$ Set has a left and a right adjoint, corresponding to discrete and indiscrete topologies.)

In Hsd limits and colimits exist, but the latter are not computed as in Set, and the simplest way to compute them – generally – is to take the colimit in Top and 'make it Hausdorff' (see 6.5.2(b)).*

*(c) Yet, even Top presents problems, related to the fact that it is not a cartesian closed category, and the category of 'compactly generated spaces' can be preferred (see [M2], Section VII.8).

We will also foreshadow, in 6.5.6, another approach influenced by category theory, and called 'pointless topology'.

(d) Artificial exclusions 'most of the time' give categories of poor properties, like the category of *non-empty semigroups* (already discussed in 1.2.9(b)), or *non-abelian groups*.

A category is a whole, and taking something out can destroy its 'shape' and properties, like taking out an element of a group, or a point of a sphere.

Other comments in this sense can be found in 6.6.4(e).

(e) Finally, the solution of a universal problem, like categorical products and sums, completely depends on the category where we are considering it, as the following examples – and many others – make evident.

The categorical sum $\mathbb{Z}+\mathbb{Z}$ in Ab is the direct sum $\mathbb{Z}\oplus\mathbb{Z}$, i.e. the free abelian group on two generators. In Gp we get the free product $\mathbb{Z}*\mathbb{Z}$, that is the free group on two generators.

The categorical product $\{*\}\times\{*\}$ in Set is the singleton. In the category of sets and partial mappings, this product has three elements (see 5.3.8).

5.2.9 *Grothendieck universes

For an interested reader, we recall the definition of a Grothendieck universe, as given in Mac Lane's [M2], Section I.6. It is named after Alexander Grothendieck, who played a central role in Algebraic Geometry and influenced a large part of Mathematics in the second half of the last century.

A *universe* is a set $\mathcal{U}$ satisfying the following (redundant) properties:

(i) $x \in u \in \mathcal{U}$ implies $x \in \mathcal{U}$,

(ii) $u, v \in \mathcal{U}$ implies that the sets $\{u, v\}$, (u, v), $u \times v$ belong to $\mathcal{U}$,

(iii) $x \in \mathcal{U}$ implies that $\mathcal{P}x$ and $\bigcup x$ belong to $\mathcal{U}$,

(iv) the set $\mathbb{N}$ of finite ordinals belongs to $\mathcal{U}$,

(v) if $f \colon x \to y$ is a surjective mapping, $x \in \mathcal{U}$ and $y \subset \mathcal{U}$, then $y \in \mathcal{U}$.

As usual, $\mathcal{P}x$ is the power set of x and $\bigcup x = \{y \mid y \in z \text{ for some } z \in x\}$.

5.3 Functors and natural transformations

Well-behaved mappings $F \colon \mathsf{C} \to \mathsf{D}$ between categories are called 'functors'. Given two *parallel* functors $F, G \colon \mathsf{C} \to \mathsf{D}$ (between the same categories) there can be 'second-order arrows' $\varphi \colon F \to G$, called 'natural transformations' (and functorial isomorphisms when they are invertible).

'Category', 'functor' and 'natural transformation' are the three basic terms of category theory, since the very beginning of the theory in [EiM].

It is interesting to note that only the last term is taken from the common language: one can say that Eilenberg and Mac Lane introduced categories and functors because they wanted to formalise the natural transformations that they were encountering in algebra and algebraic topology (as remarked in [M2], at the end of Section I.6). Much in the same way as a general theory of *continuity* (a familiar term for a familiar notion) requires the introduction of *topological spaces* (a theoretical term for a more abstract notion).

A difficulty of category theory, for a beginner, is the multi-layered structure of the topic: three layers for the theory itself, to which we have to add, inside each category, two levels for objects and arrows. Working on exercises – as always – is the best way to master this feature.

*A reader acquainted with basic homotopy theory can take advantage of a formal parallelism, where *spaces* correspond to *categories*, *continuous mappings* to *functors*, *homotopies* of mappings to *invertible natural transformations*, and *homotopy equivalence* of spaces to *equivalence* of categories.

This analogy is even deeper in the domain of Directed Algebraic Topology, where *directed homotopies* need not be reversible and correspond to *natural transformations*: see [G1].*

5.3.1 Functors

A (covariant) *functor* $F\colon \mathsf{C} \to \mathsf{D}$ consists of the following data:

(a) a mapping $F_0\colon \mathrm{Ob}\,\mathsf{C} \to \mathrm{Ob}\,\mathsf{D}$, whose action is generally written as $X \mapsto F(X)$,

(b) for every pair of objects X, X' in C, a mapping

$$F_{XX'}\colon \mathsf{C}(X, X') \to \mathsf{D}(F(X), F(X')), \qquad f \mapsto F(f).$$

Composition and identities must be preserved. In other words:

(i) if f, g are consecutive maps of C then $F(gf) = F(g).F(f)$,

(ii) if X is an object of C then $F(\mathrm{id}\,X) = \mathrm{id}\,(F(X))$.

Given a second functor $G\colon \mathsf{D} \to \mathsf{E}$, one defines in the obvious way the *composed* functor $GF\colon \mathsf{C} \to \mathsf{E}$. This composition is associative and has identities: the *identity functor* of each category, that takes each object and each morphism to itself

$$\mathrm{id}\,\mathsf{C}\colon \mathsf{C} \to \mathsf{C}, \qquad X \mapsto X, \quad f \mapsto f. \tag{5.13}$$

An *isomorphism of categories* is a functor $F\colon \mathsf{C} \to \mathsf{D}$ which is invertible; this means that it admits a (unique) *inverse* $G\colon \mathsf{D} \to \mathsf{C}$, namely a functor such that $GF = \mathrm{id}\,\mathsf{C}$ and $FG = \mathrm{id}\,\mathsf{D}$. Obviously, the functor F is an isomorphism if and only if all the mappings F_0 and $F_{XX'}$ considered above are bijective. Being isomorphic categories is an equivalence relation, written as $\mathsf{C} \cong \mathsf{D}$.

Categories linked by an obvious isomorphism are often perceived as 'the same thing'. For instance, Ab is isomorphic to the category $\mathbb{Z}\,\mathsf{Mod}$ of modules on the ring of integers (as we have already seen in Section 2.3); the various equivalent ways of defining a topological space give rise to isomorphic categories that are nearly never distinguished.

A functor between two monoids, viewed as categories (see 5.2.6), is the same as a homomorphism of monoids. A functor between two preordered sets, viewed as categories (see 5.2.7), is the same as a monotone mapping.

A *functor in two variables* is an ordinary functor $F\colon \mathsf{C} \times \mathsf{D} \to \mathsf{E}$ defined on the product of two categories. Fixing an object X_0 in C we have a functor $F(X_0, -)\colon \mathsf{D} \to \mathsf{E}$; and symmetrically.

A *contravariant functor* $F\colon \mathsf{C} \dashrightarrow \mathsf{D}$ can be defined as a covariant functor $\mathsf{C}^{\mathrm{op}} \to \mathsf{D}$. Here, every morphism $f\colon X \to Y$ of C is sent to a morphism

$F(f): F(Y) \to F(X)$ of D, preserving identities and reversing composition: $F(gf) = F(f).F(g)$.

Cat will denote the category of small categories and their functors. (Its 2-dimensional structure, including the natural transformations, will be briefly examined in 5.3.5.)

5.3.2 Forgetful and structural functors

(a) Forgetting structure, or part of it, yields various examples of functors between categories of structured sets, like the following obvious instances

$$\mathsf{Top} \to \mathsf{Set}, \qquad \mathsf{Rng} \to \mathsf{Ab} \to \mathsf{Set}. \tag{5.14}$$

These are called *forgetful* functors, and often denoted by the letter U, which refers to the *underlying* set, or *underlying* abelian group, and so on.

(b) A subcategory D of C yields an *inclusion* functor $\mathsf{D} \to \mathsf{C}$, which we also write as $\mathsf{D} \subset \mathsf{C}$. For instance, $\mathsf{Ab} \subset \mathsf{Gp} \subset \mathsf{Mon}$ and $\mathsf{Mtr}_1 \subset \mathsf{Mtr}$. These functors *forget properties*, rather than structure.

(c) A congruence R in a category C yields a *projection functor* $P: \mathsf{C} \to \mathsf{C}/R$, which is the identity on objects and sends a morphism f to its equivalence class $[f]$. For instance we have the projection functor $\mathsf{Top} \to \mathsf{hoTop} = \mathsf{Top}/\!\simeq$.

(d) A product category $\mathsf{C} = \prod_{i \in I} \mathsf{C}_i$ has (cartesian) *projection functors* $P_i: \mathsf{C} \to \mathsf{C}_i$ (for $i \in I$).

(e) If C has small hom-sets (see 5.2.3), there is a functor of morphisms

$$\begin{gathered} \mathrm{Mor}: \mathsf{C}^{\mathrm{op}} \times \mathsf{C} \to \mathsf{Set}, \\ (X, Y) \mapsto \mathsf{C}(X, Y), \qquad (f, g) \mapsto g.-.f, \end{gathered} \tag{5.15}$$

where, for $f: X' \to X$ and $g: Y \to Y'$ in C

$$g.-.f: \mathsf{C}(X, Y) \to \mathsf{C}(X', Y'), \qquad u \mapsto guf,$$

$$\begin{array}{ccc} X & \xrightarrow{u} & Y \\ {\scriptstyle f}\uparrow & & \downarrow{\scriptstyle g} \\ X' & \underset{guf}{\dashrightarrow} & Y' \end{array}$$

This functor is described as being 'contravariant in the first variable and covariant in the second' (with respect to C, in both cases).

Fixing one of these variables we get a 'representable functor', covariant or contravariant on C (see Section 5.4).

5.3.3 Faithful and full functors

For a functor $F\colon \mathsf{C} \to \mathsf{D}$ let us consider again the mappings (of sets, possibly large):

$$F_{XX'}\colon \mathsf{C}(X, X') \to \mathsf{D}(F(X), F(X')), \qquad f \mapsto F(f). \tag{5.16}$$

F is said to be *faithful* if all these mappings are injective (for X, X' in C). F is said to be *full* if all of them are surjective.

An isomorphism of categories is always full and faithful (and bijective on objects). The inclusion functor $\mathsf{D} \to \mathsf{C}$ of a subcategory is always faithful; it is full if and only if D is a full subcategory of C. A projection functor $P\colon \mathsf{C} \to \mathsf{C}/R$ is always full (and bijective on objects).

As a formal alternative to the notion of 'category of structured sets' (either vague or formalised in different, complicated ways) a *concrete category* is defined as a category A *equipped* with a faithful functor $U\colon \mathsf{A} \to \mathsf{Set}$, called its *forgetful functor*. Concrete categories are extensively studied in [AHS].

It is not easy to find categories that can *not* be made concrete: even $\mathsf{Set}^{\mathrm{op}}$ can be, by the *contravariant functor of subsets*

$$\begin{gathered} \mathcal{P}^*\colon \mathsf{Set}^{\mathrm{op}} \to \mathsf{Set}, \\ X \mapsto \mathcal{P}X, \qquad (f\colon X \to Y) \mapsto (f^*\colon \mathcal{P}Y \to \mathcal{P}X), \end{gathered} \tag{5.17}$$

where $f^*(B) = f^{-1}(B)$ is the preimage of $B \subset Y$. (This functor is easily seen to be faithful.)

As an interesting case, the homotopy category hoTop (see 5.2.5(b)) *cannot* be made concrete, as proved by Peter Freyd [Fr2, Fr3].

More generally, a category A equipped with a faithful functor $U\colon \mathsf{A} \to \mathsf{C}$ is said to be *concrete over* C.

Exercises and complements. There are some obvious preservation and reflection properties of functors, whose verification is left to the reader.

(a) Every functor preserves commutative diagrams and isomorphisms.

(b) A faithful functor $F\colon \mathsf{C} \to \mathsf{D}$ *reflects* commutative diagrams: if a diagram of C has a commutative image in D (by F), then it is commutative.

(c) A full and faithful functor $F\colon \mathsf{C} \to \mathsf{D}$ *reflects* isomorphisms: if f is morphism of C and $F(f)$ is invertible in D, so is f in C.

5.3.4 Diagrams as functors

Diagrams in a category of structured sets have already been presented in 1.2.1. We can now give a precise, more general notion.

Let S be a small category. A *diagram* in the category C, *of type* S (or *based on* S) will be any functor $X\colon \mathsf{S} \to \mathsf{C}$.

It can be written in *index notation*, with indices $i \in S = \mathrm{Ob}\,\mathsf{S}$ and $a\colon i \to j$ in Mor S:

$$X\colon \mathsf{S} \to \mathsf{C}, \qquad i \mapsto X_i, \qquad a \mapsto (X_a\colon X_i \to X_j), \tag{5.18}$$

or as a system $((X_i), (u_a))$, with $u_a = X_a$.

The diagram $X\colon \mathsf{S} \to \mathsf{C}$ is said to be *commutative* if it factorises through the canonical projection $\mathsf{S} \to \mathrm{po}(\mathsf{S})$ with values in the preordered set po(S), defined in 5.2.7(a). This means that:

- X coincides on every pair $a, b\colon i \to j$ of parallel morphisms of S.

Every functor defined on a preordered set S is automatically a commutative diagram.

Examples and complements. (a) A functor $\mathbf{1} \to \mathsf{C}$ amounts to an object of C, while a functor $\mathbf{2} \to \mathsf{C}$ amounts to a morphism of C.

(b) The product $\mathbf{2}\times\mathbf{2}$ has four objects and five non-identity arrows, drawn below at the left

$$\begin{array}{ccc} 00 & \longrightarrow & 10 \\ \downarrow & \searrow & \downarrow \\ 01 & \longrightarrow & 11 \end{array} \qquad\qquad \begin{array}{ccc} X_{00} & \longrightarrow & X_{10} \\ \downarrow & = & \downarrow \\ X_{01} & \longrightarrow & X_{11} \end{array} \tag{5.19}$$

A diagram $X\colon \mathbf{2}\times\mathbf{2} \to \mathsf{C}$ amounts to a commutative square in C, as in the right diagram above.

(c) Let $E = \{1, e\}$ be the idempotent monoid on two elements, with $e^2 = e$ (isomorphic to the multiplicative monoid $(\mathbb{Z}/2, .)$). A functor $F\colon E \to \mathsf{C}$ amounts to an object X of C equipped with an idempotent endomorphism $e\colon X \to X$; or – more simply – to an idempotent endomorphism e (which determines its domain-codomain).

Note that the monoid E is isomorphic to the join-semilattice $(\mathbf{2}, \vee)$, with unit 0 and $1 \vee 1 = 1$. As we already noted in 5.2.7(c), we should not confuse the monoid-category E with the ordinal category $\mathbf{2}$; we have also seen that a functor $X\colon \mathbf{2} \to \mathsf{C}$ amounts to an arbitrary morphism of C.

(d) Let $I = \{1, i\}$ be the group on two elements, with $i^2 = 1$ (isomorphic to the additive group $(\mathbb{Z}/2, +)$). A functor $F\colon I \to \mathsf{C}$ is the same as an object X of C equipped with an involutive automorphism $i\colon X \to X$.

(e) A functor $X\colon \boldsymbol{\omega} \to \mathsf{C}$, defined on the ordinal category $\boldsymbol{\omega}$ (see 5.2.7(d)), amounts to a sequence of consecutive morphisms of C

$$X_0 \to X_1 \to X_2 \to \ldots \to X_n \to \ldots \tag{5.20}$$

On the other hand, a functor $X\colon (\mathbb{N}, +) \to \mathsf{C}$, defined on the additive monoid of natural numbers, amounts to an endomorphism $u_1\colon X_* \to X_*$ (and its powers, $u_n = (u_1)^n$, including $u_0 = \mathrm{id}\, X_*$).

A functor $X\colon (\mathbb{N}, .) \to \mathsf{C}$ is also a different thing.

(f) A *coherent system of isomorphisms* is a functor $X\colon S \to \mathsf{C}$ where S is a set equipped with an equivalence relation $i \sim j$ (i.e. a symmetric preorder). The system amounts thus to a family (X_i) of objects of C together with a family of (iso)morphisms $u_{ij}\colon X_i \to X_j$ for $i \sim j$, such that

$$u_{jk}.u_{ij} = u_{ik}, \qquad u_{ii} = \mathrm{id}\, X_i \qquad (\text{for } i \sim j \sim k \text{ in } S). \tag{5.21}$$

Since the celebrated Coherence Theorem of monoidal categories, by Mac Lane [M1], a 'coherence theorem' in category theory states that, under suitable hypotheses, all the isomorphisms of a certain kind form a coherent system.

*(g) According to a slightly different definition, a diagram in the category C is a morphism of graphs $X : \Gamma \to \mathsf{C}$, where Γ is a small graph.

This is not more general, because every graph generates a free category $\mathsf{C}(\Gamma)$, by finite paths of consecutive arrows, and the morphism X can be extended to a functor $X'\colon \mathsf{C}(\Gamma) \to \mathsf{C}$ (as made explicit in Exercise 5.4.3(b)). Now, the commutativity of X, as expressed in 1.2.1, is equivalent to the commutativity of the functor X', as defined above.

In fact, this second definition is *less expressive*: for instance, commutative squares or idempotent endomorphisms cannot be formalised as *arbitrary* diagrams on a given graph, but as diagrams satisfying certain conditions.

5.3.5 Natural transformations

Given two functors $F, G\colon \mathsf{C} \to \mathsf{D}$ between the same categories, a *natural transformation* $\varphi\colon F \to G$ (or $\varphi\colon F \to G\colon \mathsf{C} \to \mathsf{D}$) consists of the following data:

- for each object X of C a morphism $\varphi X\colon FX \to GX$ in D, called the *component* of φ on X, and also written as φ_X,

so that, for every arrow $f\colon X \to X'$ in C, we have a commutative square in D (the *naturality condition* of φ on f), whose diagonal will be written as $\varphi(f)$, when useful

$$\begin{array}{ccc} FX & \xrightarrow{\varphi X} & GX \\ {\scriptstyle Ff}\downarrow & & \downarrow{\scriptstyle Gf} \\ FX' & \xrightarrow[\varphi X']{} & GX' \end{array} \qquad \varphi X'.F(f) = G(f).\varphi X = \varphi(f). \tag{5.22}$$

In particular, *the identity of a functor* $F\colon \mathsf{C} \to \mathsf{D}$ is the natural transformation $\mathrm{id}\, F\colon F \to F$, with components $(\mathrm{id}\, F)X = \mathrm{id}\,(FX)$.

For instance if $F\colon R\,\mathsf{Mod} \to R\,\mathsf{Mod}$ is the identity functor, every scalar λ *in the centre* of the unital ring R (see 2.1.2(c)) gives a natural transformation $\lambda\colon F \to F$ whose component on the left module A

$$\lambda_A\colon A \to A, \qquad \lambda_A(x) = \lambda x, \tag{5.23}$$

is the multiplication by λ. (See Exercise (a) for the converse.)

A natural transformation $\varphi\colon F \to G$ often comes out of a 'canonical choice' of a family $(\varphi X\colon FX \to GX)_X$ of morphisms, *but these two aspects do not coincide*: there are canonical choices of such families that are not natural, and natural transformations derived from the axiom of choice.

Natural transformations have a *vertical composition*, written as $\psi\varphi$, or $\psi.\varphi$, where $\psi\colon G \to H$ is 'vertically consecutive' to $\varphi\colon F \to G$

$$\mathsf{C} \;\underset{H}{\overset{F}{\substack{\longrightarrow \\ \downarrow\varphi \\ \longrightarrow \\ \downarrow\psi \\ \longrightarrow}}}\; \mathsf{D} \qquad (5.24)$$

$$\psi\varphi\colon F \to H, \qquad (\psi\varphi)(X) = \psi X.\varphi X\colon FX \to GX \to HX.$$

There is also a *whisker composition*, or *reduced horizontal composition*, of natural transformations with functors, written as $K\varphi H$ (or $K{\circ}\varphi{\circ}H$, when useful to distinguish compositions)

$$\mathsf{C}' \xrightarrow{\;H\;} \mathsf{C} \;\underset{G}{\overset{F}{\substack{\longrightarrow \\ \downarrow\varphi \\ \longrightarrow}}}\; \mathsf{D} \xrightarrow{\;K\;} \mathsf{D}' \qquad (5.25)$$

$$K\varphi H\colon KFH \to KGH\colon \mathsf{C}' \to \mathsf{D}', \qquad (K\varphi H)(X') = K(\varphi(HX')).$$

An *isomorphism of functors*, or *natural isomorphism*, or *functorial isomorphism*, is a natural transformation $\varphi\colon F \to G$ which is invertible with respect to vertical composition. It is easy to see that this happens if and only if all the components φX are invertible in D. The inverse is written as $\varphi^{-1}\colon G \to F$.

Isomorphism of (parallel) functors is an equivalence relation, written as $F \cong G$. (The old term 'natural equivalence' for a functorial isomorphism can be confusing and will not be used.)

This 2-dimensional structure of Cat, where natural transformations play the role of 2-dimensional arrows between functors, is the beginning of 2-dimensional and higher dimensional category theory; hints and references for this far-reaching perspective can be found in [G4], Chapter 7.

Replacing the category C with a graph Γ one can consider a *natural transformation* $\varphi\colon F \to G\colon \Gamma \to \mathsf{D}$ between two morphisms of graphs defined on Γ, *with values in a category.*

Exercises and complements. (a) If $F\colon R\,\mathsf{Mod} \to R\,\mathsf{Mod}$ is the identity functor, every natural transformation $\varphi\colon F \to F$ is of the form (5.23), determined by a (unique) scalar λ belonging to $\mathrm{Cnt}(R)$, the centre of the ring.
Hints. Use the component of φ on R, as a left module on itself.

(b) This bijective correspondence shows that the category $R\,\mathsf{Mod}$ *determines* the centre of the ring R, in a structural way, and leads to considering the relation of Morita equivalence of rings (see 5.3.7(e)).

(c) (*Duality*) For a functor $F\colon \mathsf{C} \to \mathsf{D}$ and a natural transformation $\varphi\colon F \to G\colon \mathsf{C} \to \mathsf{D}$, duality acts in a covariant way for functors and a contravariant way for transformations

$$F^{\mathrm{op}}\colon \mathsf{C}^{\mathrm{op}} \to \mathsf{D}^{\mathrm{op}}, \qquad \varphi^{\mathrm{op}}\colon G^{\mathrm{op}} \to F^{\mathrm{op}}\colon \mathsf{C}^{\mathrm{op}} \to \mathsf{D}^{\mathrm{op}}. \tag{5.26}$$

All the compositions we have considered are preserved, taking into account the appropriate variance:

$$\begin{gathered}(GF)^{\mathrm{op}} = G^{\mathrm{op}}F^{\mathrm{op}}, \qquad (\psi\varphi)^{\mathrm{op}} = \varphi^{\mathrm{op}}\psi^{\mathrm{op}},\\ (K\varphi H)^{\mathrm{op}} = K^{\mathrm{op}} \circ \varphi^{\mathrm{op}} \circ H^{\mathrm{op}}.\end{gathered} \tag{5.27}$$

5.3.6 Equivalence of categories

Isomorphisms of categories have been recalled in 5.3.1. More generally, an *equivalence of categories* is a functor $F\colon \mathsf{C} \to \mathsf{D}$ that is invertible up to functorial isomorphism, i.e. there exists a functor $G\colon \mathsf{D} \to \mathsf{C}$ such that $GF \cong \mathrm{id}\,\mathsf{C}$ and $FG \cong \mathrm{id}\,\mathsf{D}$. The functor G can be called a *pseudo inverse* of F.

An *adjoint equivalence of categories* is a coherent version of this notion; namely it is a four-tuple $(F, G, \eta, \varepsilon)$ where

- $F\colon \mathsf{C} \to \mathsf{D}$ and $G\colon \mathsf{D} \to \mathsf{C}$ are functors,
- $\eta\colon \mathrm{id}\,\mathsf{C} \to GF$ and $\varepsilon\colon FG \to \mathrm{id}\,\mathsf{D}$ are natural isomorphisms,
- $F\eta = (\varepsilon F)^{-1}\colon F \to FGF, \qquad \eta G = (G\varepsilon)^{-1}\colon G \to GFG$

(*coherence conditions*).

(The direction of η and ε is written above as in the general case of an adjunction $F \dashv G$, where these transformations need not be invertible and their direction is substantial: see Section 6.3.)

The following conditions on a functor $F\colon \mathsf{C} \to \mathsf{D}$ are equivalent, forming a very useful *characterisation of the equivalence of categories*:

(i) F is an equivalence of categories,

(ii) F can be completed to an adjoint equivalence of categories $(F, G, \eta, \varepsilon)$,

(iii) F is faithful, full and *essentially surjective on objects*.

The last property means that: for every object Y of D there exists some object X in C such that $F(X)$ is isomorphic to Y in D. The proof of the equivalence of these three conditions requires the axiom of choice: see [M2], Section V.4, or [G4], Theorem 1.5.8.

One says that the categories C, D are *equivalent*, written as $\mathsf{C} \simeq \mathsf{D}$, if there exists an equivalence of categories between them (or, equivalently, an *adjoint* equivalence of categories). This is indeed an equivalence relation, as can be easily proved (directly or from the previous characterisation).

5.3.7 *Exercises and complements* (Equivalences and skeletons)

Equivalences of categories play an important role in the theory of categories.

(a) The category of finite sets (and mappings between them) is equivalent to its full subcategory of finite cardinals, which is small and therefore cannot be isomorphic to the former.

(b) A category C is equivalent to the singleton category $\mathbf{1}$ if and only if it is non-empty and *indiscrete*, i.e. each hom-set $\mathsf{C}(X, Y)$ has precisely one element.

(c) Top is a subcategory of a category equivalent to Set.

(d) Extending point (c), prove that any category A concrete over C (by a faithful functor $U \colon \mathsf{A} \to \mathsf{C}$) can be embedded as a subcategory of a suitable category equivalent to C.

*(e) Two rings R, S are said to be *Morita equivalent* [Mo] if their categories of left modules are equivalent. This is an important notion in ring theory, that becomes trivial in the domain of commutative rings: in fact, if two rings are Morita equivalent one can easily prove that their centres are isomorphic (using what we have seen in 5.3.5(a)).

However, quite interestingly, commutative rings can be Morita equivalent to non-commutative ones, like their rings of square matrices. Thus, studying left modules on any matrix ring $M_n(\mathbb{R})$ is equivalent to studying real vector spaces.

(f) A category is said to be *skeletal* if it has no pair of distinct isomorphic objects. Prove that every category has a *skeleton*, i.e. an equivalent skeletal category. The latter can be obtained by *choosing* one and only one object in each class of isomorphic objects.

(g) We have described above a skeleton of the category of finite sets that can be constructed without any choice, even though we do need the axiom of choice to prove that the inclusion of this skeleton has a pseudo inverse.

In a different way, a preordered set X has a natural skeleton *formed by a quotient*, the associated ordered set $X/\!\sim$ (see (1.91)); again we use the axiom of choice to prove that the projection $X \to X/\!\sim$ has a pseudo inverse.

(h) Two categories are equivalent if and only if they have isomorphic skeletons. Loosely speaking, this says that an equivalence of categories amounts to multiplying or reducing isomorphic copies of objects, even though there may be no canonical way of doing this.

5.3.8 Partial mappings

A *partial mapping* (of sets) $f\colon X \nrightarrow Y$ (denoted by a dot-marked arrow) is a mapping $\mathrm{Def}(f) \to Y$ defined on a subset of X, its subset of *definition*.

For a partial mapping $g\colon Y \nrightarrow Z$, the composite $gf\colon X \nrightarrow Z$ is obvious (as in the well-known case of partially defined functions $\mathbb{R} \nrightarrow \mathbb{R}$):

$$\mathrm{Def}(gf) = f^{-1}(\mathrm{Def}(g)), \qquad (gf)(x) = g(f(x)) \ \ (x \in \mathrm{Def}(gf)). \qquad (5.28)$$

We have thus a category $\mathcal{S}$ of sets and partial mappings, that contains Set as a wide subcategory.

The category $\mathcal{S}$ *should not be viewed as essentially different* from the usual categories of structured sets and 'total' mappings, as *it is equivalent* to the category $\mathsf{Set}_\bullet$ of pointed sets, via the functor

$$R\colon \mathsf{Set}_\bullet \to \mathcal{S}, \qquad (X, x_0) \mapsto X \setminus \{x_0\}, \qquad (5.29)$$

that sends a pointed mapping $f\colon (X, x_0) \to (Y, y_0)$ to its restriction

$$R(f)\colon X \setminus \{x_0\} \nrightarrow Y \setminus \{y_0\}, \qquad \mathrm{Def}(R(f)) = f^{-1}(Y \setminus \{y_0\}),$$

as proved in the following exercises.

Exercises and complements. These easy exercises are left to the reader.

(a) Prove the associativity of the composition defined in (5.28).

(b) Every set $\mathcal{S}(X, Y)$ has an order relation, where $f' \leqslant f$ means that f' is a restriction of f to a smaller subset of X. This relation agrees with composition.

(c) The functor R defined above is faithful, full and surjective on objects.

(d) A pseudo inverse functor $S\colon \mathcal{S} \to \mathsf{Set}_\bullet$ can be obtained choosing, for every set X, a pointed set $S(X) = (X \cup \{x_0\}, x_0)$ where $x_0 \notin X$, and defining S on partial mappings in the obvious (and unique) way that gives $RS = 1$. The composite SR is only isomorphic to the identity.

(Set theory can give a canonical – perhaps 'confusing' – choice for S: one can take $x_0 = X$, because $X \notin X$.)

This equivalence will be used in 6.2.5(d) to transfer to $\mathcal{S}$ categorical constructions that are easily computed in $\mathsf{Set}_\bullet$, like categorical products.

5.4 Universal arrows and representable functors

Universal properties have frequently appeared in the previous chapters. They can now be expressed in a general way.

Many topics of the previous chapters can be reconsidered at this light. We begin by revisiting free objects, on sets or 'weaker objects'; other instances, like products and sums, will be reviewed later.

Universal properties are closely related to the representability of functors with values in Set.

5.4.1 Universal arrows

There is a general way of formalising 'universal properties', based on a functor $U\colon \mathsf{A} \to \mathsf{X}$ and an object X of X. It comes in two forms, dual to each other.

First, a *universal arrow from the object X to the functor U* is a pair

$$(A, \eta\colon X \to UA)$$

consisting of an object A of A and an arrow η of X, which is universal, in the sense that every similar pair $(B, f\colon X \to UB)$ factorises uniquely through (A, η).

In other words, there exists a unique map $g\colon A \to B$ in A such that the following triangle commutes in X

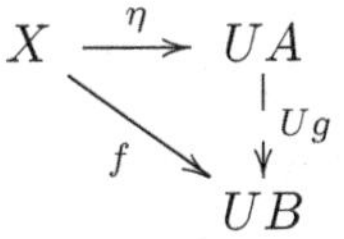

$$\begin{array}{ccc} X & \xrightarrow{\eta} & UA \\ & \searrow^{f} & \downarrow Ug \\ & & UB \end{array} \qquad Ug.\eta = f. \tag{5.30}$$

The pair (A, η) may exist or not, but is determined up to a unique isomorphism: if $(A', \eta'\colon X \to UA')$ is also a solution, the unique map $g\colon A \to A'$ such that $Ug.\eta = \eta'$ is an isomorphism of A.

This is proved extending an argument we have already used in particular cases. There is also a unique map $h\colon A' \to A$ in A such that $Uh.\eta' = \eta$, and the composite $hg\colon A \to A$ coincides with $\mathrm{id}\, A$, because

$$U(hg).\eta = Uh.Ug.\eta = Uh.\eta' = \eta = U(\mathrm{id}\, A).\eta.$$

In the same way $gh = \mathrm{id}\, A'$, so that h is inverse to g in A.

Dually, a *universal arrow from the functor U to the object X* is a pair

$$(A, \varepsilon\colon UA \to X)$$

consisting of an object A of A and an arrow ε of X such that every similar pair $(B, f\colon UB \to X)$ factorises uniquely through (A, ε): there exists a unique $g\colon B \to A$ in A such that the following triangle commutes in X

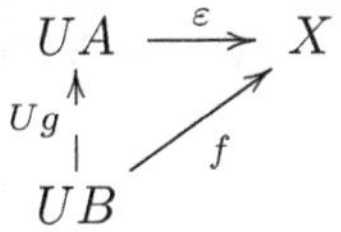

$$\begin{array}{ccc} UA & \xrightarrow{\varepsilon} & X \\ Ug \uparrow & \nearrow_{f} & \\ UB & & \end{array} \qquad \varepsilon.Ug = f. \tag{5.31}$$

Universal arrows compose: for a composed functor $UV\colon \mathsf{B} \to \mathsf{A} \to \mathsf{X}$, given

- a universal arrow $(A, \eta\colon X \to UA)$ from an object X to U,
- a universal arrow $(B, \zeta\colon A \to VB)$ from the previous object A to V,

it is easy to verify that we have a universal arrow from X to UV constructed as follows

$$(B, U\zeta.\eta \colon X \to UA \to UV(B)). \qquad (5.32)$$

This can be better understood at the light of some exercises, below.

Remark. From a formal point of view, *existence* and *uniqueness* of solutions are independent facts: existence means that some solution exists, while uniqueness means that, *if* there are two solutions, *then* they coincide (maybe up to some equivalence relation, as here).

However, mathematics is not written in a formal language – it would be unreadable – and we often express uniqueness in a 'psychologically prudent' way, saying that the solution is unique, 'provided it exists', even though this addition is redundant.

5.4.2 **Free objects on sets**

In the previous definition the functor $U \colon \mathsf{A} \to \mathsf{X}$ is arbitrary.

Taking a concrete category $U \colon \mathsf{A} \to \mathsf{Set}$ (see 5.3.3), we can formalise the notion of a *free* A*-object over the set* X (with respect to the functor U): this is defined as a universal arrow $(A, \eta \colon X \to UA)$ from the set X to the forgetful functor U.

In this way, as we have seen in many istances of the first two chapters (recalled below), the mapping $\eta \colon X \to UA$ plays the role of the insertion of the basis, and the universal property says that every mapping $f \colon X \to UB$ (with values in the underlying set of some object B of A) can be uniquely extended *as a morphism* $g \colon A \to B$ in A.

We will use the same terminology for any functor $U \colon \mathsf{A} \to \mathsf{Set}$ which is viewed, in a given context, as a 'forgetful functor', even if it is not faithful – as in Exercise 5.4.4(d).

Note. If $\eta \colon X \to UA$ is injective, we can view the basis of A as the subset $\eta(X) \subset A$. This is certainly true if there exists some injective mapping $X \to UB$, for some object B of A. But we have seen that this can fail, e.g. for the (trivial) variety of modules on the trivial ring (in 2.3.6(a)).

Examples, exercises and complements. (a) We have already seen various free algebraic structures on a set X, with respect to a forgetful functor $U \colon \mathsf{A} \to \mathsf{Set}$, defined on some variety of equational algebras:

- the free abelian group $\mathbb{Z}X$, in 1.6.1, for $\mathsf{A} = \mathsf{Ab}$,
- the free monoid $M(X) = \sum_{n \in \mathbb{N}} X^n$, in 1.6.2,
- the free semigroup, in 1.6.2(a),
- the free abelian monoid $\mathbb{N}X$, in 1.6.2(b),
- the free group $G(X)$, in 1.6.3,
- the free R-module RX, in 2.3.6.

The free pointed set on the set X, with respect to the forgetful functor $\mathsf{Set}_\bullet \to \mathsf{Set}$ that sends a pointed set (X, x_0) to X, is also obvious: it adds to X a base-point not belonging to it.

(b) For a group G, construct the free G-set on a set X (see 1.6.4). Compare these free objects with the G-sets whose action is fixed-point free.

(c) For a commutative unital ring R, the polynomial ring $R[X]$ in *one* indeterminate is the free commutative unital R-algebra *on the singleton set* $\{X\}$, as we have seen in 2.5.3(g). In particular, the polynomial ring $\mathbb{Z}[X]$ is the free commutative unital ring on the singleton.

(d) Extend the previous point by constructing the free commutative unital R-algebra $R[\mathcal{X}]$ on an arbitrary set $\mathcal{X}$, whose elements $X, Y, ...$ are viewed as 'indeterminates'.

In particular, the polynomial ring $\mathbb{Z}[\mathcal{X}]$ is the free commutative unital ring on the set $\mathcal{X}$. (Taking out the constant non-zero polynomials we have the free commutative ring on the set $\mathcal{X}$ – a less important structure.)

Hints. The forgetful functor of $R\mathsf{CAlg}$ is a composite $R\mathsf{CAlg} \to \mathsf{Abm} \to \mathsf{Set}$ of forgetful functors, and universal arrows compose (as we have seen in 5.4.1).

(e) Dropping commutativity, construct the free unital R-algebra and the free unital ring on a set $\mathcal{X}$.

(f) The *initial object* of a category A is defined as an object $\perp$ such that, for every object A of A, there is a unique A-arrow $\perp \to A$. (It amounts to the empty sum in A, as we will see in Section 6.1.)

This is the same as the free A-object on the empty set, with respect to *any* functor $U\colon \mathsf{A} \to \mathsf{Set}$.

(g) The category Fld of fields (and their homomorphisms) is concrete, by the forgetful functor $U\colon \mathsf{Fld} \to \mathsf{Set}$. It has no free object on the empty set, nor on any set. (Let us recall that Fld is a category of algebraic structures, which are not equational: see 1.2.8.)

(h) The category Fld_0, of fields of characteristic 0 and their homomorphisms, has a free object on the empty set.

It has no free object on any set $X \neq \emptyset$. The category Fld_p, of fields of characteristic p (a prime number), has similar properties.

*(i) Every variety of algebras has a free object on every set: see [G4], Theorem 4.4.2.

5.4.3 Free objects on weaker objects

Extending the previous definition, if $U\colon \mathsf{A} \to \mathsf{X}$ is any functor *viewed*, in a given context, as a 'forgetful functor', a universal arrow $(A, \eta\colon X \to UA)$ from the object X to the functor U can be called the *free* A-*object over the object* X (with respect to the functor U).

Examples, exercises and complements. (a) We have already constructed:

- the field of fractions $Q(R)$ on an integral domain R, in 2.2.4; it gives a universal arrow $\eta\colon R \to Q(R)$ from R to the inclusion functor $\mathsf{C} \to \mathsf{Fld}$ of the category of integral domains and *injective* unital homomorphisms, in the category of fields,

- the free commutative unital R-algebra on a commutative monoid, in the solution of 5.4.2(d),

- the free unital R-algebra on a monoid, in the solution of 5.4.2(e),

- the completion of a metric space, in Theorem 4.5.6.

(b) Construct the free category $\mathsf{C}(\Gamma)$ on a small graph Γ, with respect to the forgetful functor $U\colon \mathsf{Cat} \to \mathsf{Gph}$.

(c) Construct the free commutative semigroup on the semigroup S. The same construction works for monoids and groups.

(d) In particular, the free abelian group G^{ab} on the group G is better known as the *abelianised group* associated to G, and can be constructed using the subgroup of commutators $[G, G]$, introduced in 1.5.5(g). Write its universal property, and observe that, if G is a non-commutative simple group, G^{ab} is trivial.

(e) The universal arrow from a preordered set X to the embedding $\mathsf{Ord} \to \mathsf{pOrd}$ is the canonical projection $\eta\colon X \to X/{\sim}$ onto the associated ordered set (in 1.4.1).

*(f) Every functor $U\colon \mathsf{A} \to \mathsf{X}$ from a variety of algebras to a more general variety, that forgets part of the structure and properties of the domain, has a free object on every object of X: for a precise formulation and proof, see [G4], Theorem 4.4.5.

5.4.4 Other examples

The other way round, examples of universal arrows *from a functor to an object*, based on concrete categories, are less abundant; but we have already seen some of them, and many others will result from limits, in Sections 6.1 and 6.2.

In the examples below, both kinds of universal arrows exist. (The situation will become clearer studying adjoint functors, in Section 6.3.)

(a) If M is a monoid, the inclusion $\varepsilon\colon \mathrm{Inv}(M) \to M$ of the group of its invertible elements gives the universal arrow from the embedding $U\colon \mathsf{Gp} \to \mathsf{Mon}$ to the monoid M, as we have already seen in Exercise 1.5.4(f).

The universal arrow *from* a monoid M *to* U also exists, because of a general result recalled in 5.4.3(f).

The solution can be formally 'constructed' taking the free group $F|M|$ on the underlying set of the monoid, and then its quotient modulo the congruence of groups that forces the embedding $|M| \to F|M|$ to become a homomorphism of monoids.

(b) For an abelian monoid M, the inclusion $\varepsilon\colon \mathrm{Inv}(M) \to M$ gives the universal arrow from the embedding $U\colon \mathsf{Ab} \to \mathsf{Abm}$ to the object M.

Here, the universal arrow *from* an abelian monoid M *to* U can be built in a more constructive way, extending the obvious construction of $\mathbb{Z}$ as a quotient of $\mathbb{N}\times\mathbb{N}$; some attention is required, because an abelian monoid need not satisfy the cancellation law.

Given an abelian monoid M, the abelian group $F(M)$ is the quotient of the monoid $M \times M$ modulo the congruence where $(x, y) \sim (x', y')$ when

$$x + y' + z = x' + y + z, \tag{5.33}$$

for some $z \in M$.

The unit-component $\eta M \colon M \to UF(M)$ sends x to the equivalence class $[x, 0]$. It is injective if and only if M satisfies the cancellation law (and then the previous congruence can be simply expressed as: $x + y' = x' + y$).

(c) (*Discrete and indiscrete topology*) For the forgetful functor $U\colon \mathsf{Top} \to \mathsf{Set}$, it is easy to see that the universal arrow from a set X to U is the identity $\eta\colon X \to UD(X)$, where $D(X)$ is the given set with the discrete topology. On the other hand, the universal arrow from U to the set X is the identity $\varepsilon\colon UC(X) \to X$, where $C(X)$ is the given set with the chaotic, or indiscrete, topology.

(d) (*Discrete and indiscrete categories*) Consider now the (non-faithful) functor $\mathrm{Ob} : \mathsf{Cat} \to \mathsf{Set}$ that takes a small category to its set of objects.

The universal arrow from a set X to the functor Ob is the identity $\eta\colon X \to UD(X)$, where $D(X)$ is the discrete category with objects in X. The universal arrow from Ob to the set X is the identity $\varepsilon\colon UC(X) \to X$, where $C(X)$ is the *chaotic*, or *indiscrete*, category with objects in X, that has a unique arrow $x \to x'$ for each pair of elements of X.

5.4.5 Representable functors

Coming back to studying set-valued functors, as in 5.4.2, we start from an easy example.

The forgetful functor $U\colon \mathsf{Gp} \to \mathsf{Set}$ is isomorphic to the hom-functor $\mathsf{Gp}(\mathbb{Z}, -)$, by a natural isomorphism φ

$$\begin{gathered} \varphi\colon \mathsf{Gp}(\mathbb{Z}, -) \to U\colon \mathsf{Gp} \to \mathsf{Set}, \\ \varphi_G\colon \mathsf{Gp}(\mathbb{Z}, G) \to U(G), \qquad \varphi_G(f) = f(1), \end{gathered} \tag{5.34}$$

which computes a map $f\colon \mathbb{Z} \to G$ at the basis-element $1 \in \mathbb{Z}$. (This expresses the fact that $\mathbb{Z}$ is the free group on the singleton, with universal arrow $1\colon \{*\} \to U(\mathbb{Z})$.)

We say that the functor $U\colon \mathsf{Gp} \to \mathsf{Set}$ is 'represented' by the object $\mathbb{Z}$, via the isomorphism φ, or via the element $1 \in U(\mathbb{Z})$. Similarly, the forgetful functor of any variety of algebras is representable, but the forgetful functor $\mathsf{Fld} \to \mathsf{Set}$ is not.

For a general definition, let A be a category with small hom-sets. A functor $\mathsf{A} \to \mathsf{Set}$ is said to be *representable* if it is isomorphic to a *hom-functor*

$$F = \mathsf{A}(A_0, -) \colon \mathsf{A} \to \mathsf{Set}, \tag{5.35}$$

for some object A_0 in A; this functor is often written as H^{A_0}.

The Yoneda Lemma (in 5.4.6) describes the natural transformations $F \to G$, for every functor $G \colon \mathsf{A} \to \mathsf{Set}$, and characterises the representability of G.

This relatively simple result is at the core of many points in category theory. It is also known for a curious story: it originated in a discussion between Nobuo Yoneda, a Japanese mathematician, and Saunders Mac Lane at Gare du Nord, in Paris, first in a café and then at Yoneda's train, about to leave; likely in August 1954.

Representability can be expressed by a universal arrow, and conversely (see Corollary 5.4.7).

A contravariant functor $\mathsf{A} \dashrightarrow \mathsf{Set}$ is a functor $\mathsf{A}^{\mathrm{op}} \to \mathsf{Set}$; it is representable if it is isomorphic to a functor $H_{A_0} = \mathsf{A}(-, A_0) \colon \mathsf{A}^{\mathrm{op}} \to \mathsf{Set}$.

5.4.6 Yoneda Lemma and Definition

(a) Let A be a category with small hom-sets, and F a hom-functor

$$F = \mathsf{A}(A_0, -) \colon \mathsf{A} \to \mathsf{Set}. \tag{5.36}$$

Then, for every functor $G \colon \mathsf{A} \to \mathsf{Set}$, the Yoneda mapping

$$y \colon \mathrm{Nat}(F, G) \to G(A_0), \qquad y(\varphi) = (\varphi A_0)(\mathrm{id}\, A_0) \in G(A_0), \tag{5.37}$$

is a bijection, between the set of natural transformations $\varphi \colon F \to G$ and the set $G(A_0)$.

(b) The functor G is representable if and only if there is an object A_0 in A and an element $x_0 \in G(A_0)$ such that:

(*) for every A in A and every $x \in G(A)$, there is a unique morphism $f \colon A_0 \to A$ such that $G(f)(x_0) = x$.

Then we say that G *is represented by A_0 via the element $x_0 \in G(A_0)$*, or *via the isomorphism $\varphi \colon \mathsf{A}(A_0, -) \to G$* associated to x_0 (i.e. $x_0 = y(\varphi)$).

Proof (a) The crucial point is that, for every natural transformation $\varphi \colon F \to G$ and every morphism $f \colon A_0 \to A$, the naturality of φ on f

gives (computing everything on $\mathrm{id}\, A_0$)

$$\begin{array}{ccc} \mathsf{A}(A_0, A_0) & \xrightarrow{\varphi A_0} & GA_0 \\ {\scriptstyle f.-}\downarrow & & \downarrow{\scriptstyle Gf} \\ \mathsf{A}(A_0, A) & \xrightarrow[\varphi A]{} & GA \end{array} \qquad (5.38)$$

$$(\varphi A)(f) = (\varphi A)(f.\mathrm{id}\, A_0) = (Gf)(\varphi A_0(\mathrm{id}\, A_0)) = (Gf)(y(\varphi)).$$

We construct the mapping

$$\begin{array}{c} y'\colon G(A_0) \to \mathrm{Nat}(F, G), \\ (y'(x))_A\colon \mathsf{A}(A_0, A) \to GA, \qquad (f\colon A_0 \to A) \mapsto (Gf)(x), \end{array} \qquad (5.39)$$

where the family $\psi = y'(x)$ is indeed a natural transformation $F \to G$. In fact, for $g\colon A \to B$ in A

$$\begin{array}{ccc} \mathsf{A}(A_0, A) & \xrightarrow{\psi A} & GA \\ {\scriptstyle g.-}\downarrow & & \downarrow{\scriptstyle Gg} \\ \mathsf{A}(A_0, B) & \xrightarrow[\psi B]{} & GB \end{array} \qquad (5.40)$$

$$\psi_B(gf) = G(gf)(x) = (Gg)(Gf)(x) = (Gg)(\psi_A f).$$

Now we have to show that the mappings y, y' are inverse to each other.

First, a natural transformation $\varphi\colon F \to G$ is taken to the element $x = y(\varphi) = \varphi_{A_0}(\mathrm{id}\, A_0) \in G(A_0)$, and the latter to the natural transformation $y'(x)\colon F \to G$. By (5.37), the component $(y'(x))_A$ of the latter takes the morphism $f\colon A_0 \to A$ to

$$(Gf)(x) = G(f)(y(\varphi)) = \varphi_A(f),$$

so that $y'(x) = \varphi$.

The other way round, an element $x \in G(A_0)$ is taken to $\varphi = y'(x)\colon F \to G$ and then to

$$y(\varphi) = \varphi_{A_0}(\mathrm{id}\, A_0) = (G(\mathrm{id}\, A_0))(x) = x.$$

(b) We have seen that a natural transformation $\varphi\colon F \to G$ amounts to an element $x_0 \in G(A_0)$, where $x_0 = y(\varphi)$ and $\varphi = y'(x_0)$.

Moreover φ is an isomorphism if and only if each component φA is a bijection of sets; this is equivalent to condition (*), by (5.38). □

5.4.7 Corollary (Universal arrows and representability)

Let A *be a category with small hom-sets.*

(a) (Representable functors by universal arrows) *A functor* $G\colon \mathsf{A} \to \mathsf{Set}$ *is represented by the object* A_0 *of* A*, via* $x_0 \in G(A_0)$*, if and only if:*

(*) *the pair* $(A_0, x_0\colon \{*\} \to G(A_0))$ *is a universal arrow, from the singleton to the functor* G*.*

(b) (Universal arrows by representable functors) *Given a functor* $U\colon \mathsf{A} \to \mathsf{X}$ *and an object* X *in* X *(a category with small hom-sets), a universal arrow* $(A, \eta\colon X \to UA)$ *from* X *to* U *is the same as representing the functor*

$$G\colon \mathsf{A} \to \mathsf{Set}, \qquad G = \mathsf{X}(X, U(-)), \tag{5.41}$$

by the object A *and the element* $\eta \in G(A) = \mathsf{X}(X, UA)$*.*

Proof (a) Letting $F = \mathsf{A}(A_0, -)$, the present condition (*) is equivalent to condition (*) of Lemma 5.4.6.

(b) By definition, a universal arrow $(A, \eta\colon X \to UA)$ gives, for every B in A, a bijection

$$\mathsf{A}(A, B) \to G(B) = \mathsf{X}(X, UB), \qquad g \mapsto U(g)\eta, \tag{5.42}$$

which means that the object A represents the functor G, via the element $\eta \in GA$. □

5.4.8 Functor categories and categories of presheaves

Let S and C be categories. We write as C^{S} the category whose objects are the functors $F\colon \mathsf{S} \to \mathsf{C}$ and whose morphisms are the natural transformations $\varphi\colon F \to G\colon \mathsf{S} \to \mathsf{C}$, with vertical composition; it is called a *functor category*.

This construction is generally used when S is small and C has small hom-sets: then C^{S} is also locally small, because the set of natural transformations $\mathrm{Nat}(F, G)$ can be embedded in the cartesian product $\Pi_i\, \mathsf{C}(Fi, Gi)$ indexed by the small set $\mathrm{Ob}\,\mathsf{S}$. Furthermore, if S and C are small, so is C^{S}.

(One can also consider the category C^{Γ} of graph-morphisms $F\colon \Gamma \to \mathsf{C}$, where Γ is a graph: see 5.3.5.)

For instance, the arrow category $\mathbf{2}$ (with one non-identity arrow $\iota\colon 0 \to 1$) gives the *category of morphisms* $\mathsf{C}^{\mathbf{2}}$ of C, where a map $(f_0, f_1)\colon u \to v$ is a commutative square of C; these are composed *by pasting squares*, as in the right diagram below

$$\begin{array}{ccc} A_0 & \xrightarrow{f_0} & B_0 \\ {\scriptstyle u}\downarrow & & \downarrow{\scriptstyle v} \\ A_1 & \xrightarrow[f_1]{} & B_1 \end{array} \qquad\qquad \begin{array}{ccccc} A_0 & \xrightarrow{f_0} & B_0 & \xrightarrow{g_0} & C_0 \\ {\scriptstyle u}\downarrow & & \downarrow{\scriptstyle v} & & \downarrow{\scriptstyle w} \\ A_1 & \xrightarrow[f_1]{} & B_1 & \xrightarrow[g_1]{} & C_1 \end{array} \qquad (5.43)$$

A functor $\mathsf{S}^{\mathrm{op}} \to \mathsf{C}$, defined on the opposite category S^{op}, is also called a *presheaf* of C on the category S. They form the presheaf category $\mathrm{Psh}(\mathsf{S}, \mathsf{C}) = \mathsf{C}^{\mathsf{S}^{\mathrm{op}}}$, and S is called the *site* of the latter.

We are particularly interested in the category $\mathrm{Psh}(\mathsf{S}, \mathsf{Set}) = \mathsf{Set}^{\mathsf{S}^{\mathrm{op}}}$, of presheaves of sets on S. If the category S has small hom-sets, it is canonically embedded in $\mathrm{Psh}(\mathsf{S}, \mathsf{Set})$, by the *Yoneda embedding*

$$Y \colon \mathsf{S} \to \mathsf{Set}^{\mathsf{S}^{\mathrm{op}}}, \qquad Y(i) = \mathsf{S}(-, i) \colon \mathsf{S}^{\mathrm{op}} \to \mathsf{Set}, \qquad (5.44)$$

which sends every object i in S to the corresponding representable presheaf $Y(i)$.

*Taking as S the small category $\underline{\Delta}$ of finite positive ordinals (and monotone maps), one gets the category $\mathrm{Smp}(\mathsf{C}) = \mathsf{C}^{\underline{\Delta}^{\mathrm{op}}}$ of *simplicial objects* in C, and – in particular – the category $\mathrm{Smp}\mathsf{Set}$ of *simplicial sets*, quite important in Algebraic Topology.

A reader interested in simplicial sets is referred to [May]. For cubical and multiple sets one can see [G4] and references therein; for categories of sheaves [MaM, Bo3].*

Exercises and complements. (a) A natural transformation $\varphi \colon F \to G \colon \mathsf{C} \to \mathsf{D}$ can be viewed as a functor $\mathsf{C}\times\mathbf{2} \to \mathsf{D}$, or equivalently as a functor $\mathsf{C} \to \mathsf{D}^{\mathbf{2}}$. (The first point is related to the notation $\varphi(f)$ in (5.22).)

(b) The category $G\mathsf{Set}$ of G-sets and G-morphisms (introduced in 1.6.4) can be identified with a category of functors

$$G\mathsf{Set} = \mathsf{Cat}(G, \mathsf{Set}), \qquad (5.45)$$

whose domain is the group G, viewed as a category.

Similarly we have the category $G\mathsf{Top} = \mathsf{Cat}(G, \mathsf{Top})$ of *G-spaces*. All this works more generally for monoids.

5.5 Monomorphisms and epimorphisms

In a category, monomorphisms and epimorphisms, or monos and epis for short, are defined by cancellation properties with respect to composition – an intrinsic aspect, independent of any set-valued forgetful functor.

For categories of structured sets, monos and epis represent an 'approximation' to the injective and surjective mappings of the category; many exercises will help a beginner to master the comparison of these aspects.

5.5.1 Main definitions

In a category C a morphism $f\colon X \to Y$ is said to be a *monomorphism*, or *mono*, if it satisfies the following cancellation property: for every pair of maps $u, v\colon X' \to X$ (defined on an arbitrary object X') such that $fu = fv$, one has $u = v$ (see the left diagram below)

$$X' \underset{v}{\overset{u}{\rightrightarrows}} X \xrightarrow{f} Y \qquad\qquad X \xrightarrow{f} Y \underset{v}{\overset{u}{\rightrightarrows}} Y' \tag{5.46}$$

Dually, the morphism $f\colon X \to Y$ is said to be an *epimorphism*, or *epi*, if it satisfies the dual cancellation property: for every pair of maps $u, v\colon Y \to Y'$ such that $uf = vf$, one has $u = v$ (see the right diagram above).

An arrow $\rightarrowtail$ will always denote a monomorphism, while $\twoheadrightarrow$ stands for an epimorphism.

Every isomorphism is mono and epi. A category is said to be *balanced* if the converse holds: every morphism that is mono and epi is invertible.

Suppose now that we have, in a category C, two maps $m\colon A \to X$ and $p\colon X \to A$ such that $pm = \mathrm{id}\, A$. It follows that m is a monomorphism, called a *section*, or a *split monomorphism*. On the other hand, p is an epimorphism, called a *retraction*, or a *split epimorphism*. A is said to be a *retract* of X (a term born in Topology, as we have seen in 3.4.9).

A family of morphisms $f_i\colon X \to Y_i$ $(i \in I)$ with the same domain is said to be *jointly mono* if for every pair of maps $u, v\colon X' \to X$ such that $f_i u = f_i v$ (for all indices i) one has $u = v$. Dually a family $f_i\colon X_i \to Y$ is *jointly epi* if for all $u, v\colon Y \to Y'$ such that $uf_i = vf_i$ (for all i) one has $u = v$.

The general properties of monos, epis and retracts will be examined in 5.5.6. Related notions, like subobjects and quotients, regular monos and epis, will be seen later (in 5.5.9 and Section 6.1).

5.5.2 Comments

In a category of structured sets and structure-preserving mappings, an injective mapping (belonging to the category) is easily seen to be a monomorphism, while a surjective one is an epimorphism. The converse may require a non-trivial proof, or fail. This can only be understood by working out the examples below.

Interestingly, a divergence appears between monos and epis: the theory of categories is self-dual, but our frameworks of structured sets are not! When we classify monos in Set, this tells us everything about the epis of $\mathsf{Set}^{\mathrm{op}}$ but nothing about the epis of Set.

In fact, in all the examples below it will be easy to prove that the monomorphisms coincide with the injective morphisms. Later we will see, in 5.5.8, a condition that ensures this fact, and holds true in 'most' categories of structured sets.

On the other hand, problems can easily occur with epimorphisms: classifying them in many categories of equational algebras leads to difficult problems with no elementary solution (and no real need of it).

5.5.3 Exercises and complements, I

A solution of these exercises is given below, but a beginner should try to give an independent one; this is easy for monos, less easy for epis.

(a) In Set a mono is the same as an injective mapping; an epi is the same as a surjective mapping. The category is balanced.

(b) In Top and Ab monos and epis coincide with the injective and surjective mappings of the category, respectively. Ab is balanced, and Top is not.

(c) In the categories Mon of monoids and Rng of unital rings, the monomorphisms coincide again with the injective homomorphisms. But the inclusion $\mathbb{N} \to \mathbb{Z}$, of additive monoids, is both mono and epi in Mon (which is not balanced). The same holds for the inclusion $\mathbb{Z} \to \mathbb{Q}$ in Rng.

Epimorphisms in Mon and Rng have no elementary characterisation; we will see that the 'regular epimorphisms' (that coincide here with the surjective homomorphisms) are more important.

(d) In a preordered set, viewed as a category, all arrows are mono and epi. The category is balanced under a precise condition on the preordering.

Solutions. (a) If $f\colon X \to Y$ is a monomorphism in Set, let us suppose that $f(x) = f(x')$ for $x, x' \in X$. We consider the mappings u, v

$$\{*\} \underset{v}{\overset{u}{\rightrightarrows}} X \xrightarrow{f} Y \qquad u(*) = x, \ \ v(*) = x'. \tag{5.47}$$

Now we have $fu = fv$, whence $u = v$ and $x = x'$, which shows that f is injective. Note that we are simulating an element of X by a map $\{*\} \to X$.

On the other hand, if $f\colon X \to Y$ is an epimorphism in Set, we define two mappings u, v with values in the set $\{0, 1\}$

$$X \xrightarrow{f} Y \underset{v}{\overset{u}{\rightrightarrows}} \{0, 1\} \tag{5.48}$$

where u is the characteristic function of the subset $f(X) \subset Y$ (with $u(y) = 1$ if and only if $y \in f(X)$) while v is the constant map $v(y) = 1$.

Then $uf = vf$, whence $u = v$ and $f(X) = Y$. (A different proof can be based on the set $Y \times \{0, 1\}$, the disjoint union of two copies of Y; or a quotient of the latter.)

Since the invertible morphisms in Set are the bijective mappings, the category Set is balanced.

(b) For monomorphisms, the proof is similar to the previous one, making use of the singleton in Top and of the group $\mathbb{Z}$ in Ab. Note that the latter allows us to simulate an element $x \in X$ by a homomorphism $u: \mathbb{Z} \to X$, sending the generator 1 to x (because $\mathbb{Z}$ is the free abelian group on the singleton set).

In Top, to prove that an epi is surjective we can proceed as in (5.48), using a two-point indiscrete space (so that all mappings with this codomain are automatically continuous).

In Ab we can use the quotient group $Y/f(X)$ and the homomorphisms

$$X \xrightarrow{f} Y \underset{0}{\overset{p}{\rightrightarrows}} Y/f(X) \tag{5.49}$$

with the canonical projection $p: Y \to Y/f(X)$ and the zero-homomorphism: as the composites coincide, if f is epi we infer that $p = 0$ and $f(X) = Y$.

We are now following a different pattern: constructing arrows is fairly free in Set, somewhat less in Top, much less in categories of algebraic structures.

We conclude here that Ab is balanced, while Top is not: a bijective continuous mapping need not be invertible in Top, i.e. a homeomorphism. We have already remarked that this is an important divide, with respect to algebraic structures.

(c) For monos we proceed again as in (a), making use for Mon of the additive monoid $\mathbb{N}$ (freely generated by 1), and for Rng of the polynomial ring $\mathbb{Z}[X]$ (freely generated by the element X).

The inclusion $\mathbb{N} \to \mathbb{Z}$, of additive monoids, is injective and mono; it is also epi, because a homomorphism $f: \mathbb{Z} \to M$ of monoids (in additive notation) is determined by its values on $\mathbb{N}$: if $n > 0$, then $f(n) + f(-n) = 0_M = f(-n) + f(n)$, whence $f(-n)$ is additively inverse to $f(n)$ in M.

Similarly we prove that the inclusion $\mathbb{Z} \to \mathbb{Q}$ is an epimorphism in Rng: for a homomorphism $f: \mathbb{Q} \to R$ of unital rings, $f(1/k)$ must be inverse to $f(k)$ in R.

(d) The first claim is obvious. Saying that the category X is balanced means here that every arrow is an isomorphism: in other words, the preorder of X is symmetric, i.e. an equivalence relation.

5.5.4 Exercises and complements, II

(a) In pOrd and Ord, monos and epis coincide with the injective and surjective mappings of the category, respectively. These categories are not balanced.

(b) In the category Hsd of Hausdorff spaces, monomorphisms coincide with the injective continuous mappings.

Every map whose image is dense (in its codomain) *is an epimorphism.* *This condition is also necessary: see [G4], Exercise 2.3.6(c).*

*(c) In the category Gp of groups all epimorphisms are surjective: a non-obvious fact, whose proof can be found in [M2], Section I.5, Exercise 5.

(One cannot use the same argument as in Ab, because the image of a homomorphism need not be normal in the codomain.)

(d) Monomorphisms and epimorphisms in Cat are of little interest: the important notions are a faithful functor and a full functor, respectively.

Prove that a functor $F\colon \mathsf{C} \to \mathsf{D}$ is faithful if and only if it satisfies a suitable cancellability property, with respect to whisker composition with natural transformations.

5.5.5 *Exercises and complements* (Retracts)

These exercises are left to the reader. For the sake of simplicity, we replace a monomorphism $m\colon A \rightarrowtail X$, in a category of structured sets, by its image $m(A) \subset X$.

(a) In Set a retract of a set $X \neq \emptyset$ is any non-empty subset.

(b) In R Mod retracts coincide with 'direct summands'. More precisely, a submodule A of an R-module M is a retract if and only if M is the internal direct sum of A and some other submodule $B \subset M$ (see 2.3.4(a)).

(c) In particular, this holds in Ab. The abelian group $\mathbb{Z}$ has no retracts except the obvious ones, the null and the total subgroup.

(d) Retracts in Top have been discussed in 3.4.9.

5.5.6 *General properties of monos and epis*

The following properties, frequently used, can be easily verified.

We have two consecutive maps $f\colon X \to Y$ and $g\colon Y \to Z$ in a category. Property (a) is dual to (a*), and so on; we only need to prove one of them.

(a) If f and g are mono, gf is also; if gf is mono, f is also.

(a*) If f and g are epi, gf is also; if gf is epi, g is also.

(b) If f and g are split mono, gf is also; if gf is a split mono, f is also.

(b*) If f and g are split epi, gf is also; if gf is a split epi, g is also.

(c) If f is a split mono and an epi, then it is invertible.

(c*) If f is a split epi and a mono, then it is invertible.

5.5.7 *Exercises and complements* (Functors and cancellability)

The preservation and reflection properties of functors in 5.3.3 can be completed as follows. Again, the easy verification is left to the reader.

(a) Every functor preserves commutative diagrams and isomorphisms, retracts, split monos and split epis.

(b) A faithful functor $F: \mathsf{C} \to \mathsf{D}$ *reflects* monos, epis and commutative diagrams.

The first statement means: if $f: X \to X'$ belongs to C and $F(f)$ is mono (or epi) in D, so is f in C.

(c) A full and faithful functor reflects isomorphisms, split monos and split epis.

(d) Applying point (a), a common way in Algebraic Topology of proving that two topological spaces X, Y are not homeomorphic is to find a functor $F: \mathsf{Top} \to \mathsf{Ab}$ such that the abelian groups $F(X)$ and $F(Y)$ are not isomorphic.

*The singular homology functor H_n (or any homology functor of spaces) proves in this way that $\mathbb{S}^n$ is not homeomorphic to any other sphere $\mathbb{S}^m$ (for $m \neq n$). This implies the Theorem of Topological Dimension (mentioned in 3.1.3(e)): if the euclidean spaces $\mathbb{R}^m$ and $\mathbb{R}^n$ are homeomorphic, the same is true of their one-point compactifications $\mathbb{S}^m$ and $\mathbb{S}^n$, and then $m = n$.

(We cannot get this result applying – directly – the homology functors to the spaces $\mathbb{R}^n$: the latter are contractible, i.e. homotopy equivalent to a point, and a homotopy-invariant functor gives the same result on all of them.)

The compact surfaces mentioned in 4.4.8 can also be distinguished using homology functors.*

(e) Similarly, to prove that a topological subspace $A \subset X$ is not a retract (in Top), it is sufficient to find a functor $F: \mathsf{Top} \to \mathsf{Ab}$ such that the associated homomorphism $F(A) \to F(X)$ is not a split mono in Ab.

The functor H_n proves thus that the sphere $\mathbb{S}^n$ is not a retract of $\mathbb{R}^{n+1}$.

5.5.8 Representable functors and monomorphisms

(a) A morphism $f: X \to Y$ is mono in C if and only if, for every object X_0 in C, the mapping

$$\mathsf{C}(X_0, f) = f.-: \mathsf{C}(X_0, X) \to \mathsf{C}(X_0, Y), \quad u \mapsto fu,$$

of post-composition by f is injective. In fact, this is a rewriting of the definition of a monomorphism.

Dually, $f: X \to Y$ is epi if and only if, for every object X_0 in C, the mapping $-.f: \mathsf{C}(Y, X_0) \to \mathsf{C}(X, X_0)$ of pre-composition by f is injective.

(b) Thus, a covariant representable functor $F: \mathsf{C} \to \mathsf{Set}$ preserves monomorphisms, while a contravariant one takes epis of C to injective mappings.

(c) We have seen that the representability of a functor $F: \mathsf{C} \to \mathsf{Set}$ amounts to the existence of a universal arrow from the singleton set to F.

In a concrete category $U\colon \mathsf{C} \to \mathsf{Set}$, the free object on a singleton often exists, and *ensures that U preserves monomorphisms* (and reflects them, being faithful). This fact has already been used, in many characterisations of monomorphisms.

(d) The non-faithful functor $\mathrm{Ob} : \mathsf{Cat} \to \mathsf{Set}$ is represented by the singleton category $\mathbf{1}$, and preserves monomorphisms (even though of little interest in Cat, as already remarked). It is easy to see that it does not reflect them.

5.5.9 Subobjects and quotients

Let A be an object of the category C. A subobject of A is defined (below) as an equivalence class of monomorphisms, or better as a *selected* representative of such a class.

(The first definition, as an equivalence class, is common in category theory, but the second is what is normally used in concrete categories of structured sets. Moreover, when working with the restriction of morphisms to subobjects, one is forced to shift to the second setting – even in the abstract theory.)

Explicitly, given two monos m, n *with values in* A, we say that $m \prec n$ if there is a (uniquely determined) morphism u such that $m = nu$

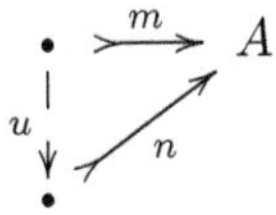

$$\begin{array}{ccc} \bullet & \overset{m}{\rightarrowtail} & A \\ {\scriptstyle u}\downarrow & \nearrow_{n} & \\ \bullet & & \end{array} \tag{5.50}$$

We say that m and n *are equivalent,* or $m \sim n$, if $m \prec n \prec m$, which amounts to the existence of a (unique) isomorphism u such that $m = nu$.

In every class of equivalent monos with codomain A, one is *selected* and called a *subobject* of A; in the class of isomorphisms we always choose 1_A. The subobjects of A in C form the (possibly large) *ordered* set $\mathrm{Sub}A$, with maximum 1_A; here, the induced *order* $m \prec n$ is also written as $m \leqslant n$.

Any category of equational algebras, including Set, has a canonical choice of subobjects, based on the inclusion mappings $m\colon X \subset A$ of substructures. In Top, 'regular subobjects' are more important: see 6.1.4.

Epimorphisms with a fixed domain A are dealt with in a dual way. Their preorder, written as $p \prec q$, means that p factorises through q

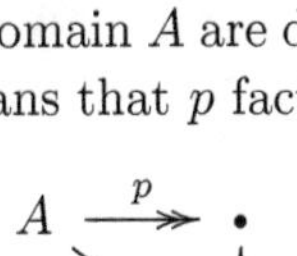

$$\begin{array}{ccc} A & \overset{p}{\twoheadrightarrow} & \bullet \\ & \searrow^{q} & \uparrow{\scriptstyle u} \\ & & \bullet \end{array} \tag{5.51}$$

and the equivalence relation is written as $p \sim q$.

A *quotient* of A is a *selected* representative of an equivalence class of epimorphisms with domain A; they form the ordered set Quo$A$, with maximum 1_A; again the induced order is also written as $p \leqslant q$.

The category C is said to be *well powered* (resp. *well copowered*) if all its sets Sub$A$ (resp. Quo$A$) are small, as is often the case with categories of (small) structured sets.

Duality of categories turns subobjects of C into quotients of C^{op}, preserving their order.

6

Categorical limits and adjunctions

Universal properties lead to crucial developments: categorical limits and adjoint functors, which allow us to revisit various aspects of Algebra and Topology.

In Category Theory, *limits* extend products and equalisers, together with the 'projective limits' of Algebra and Topology (see [Bou1, Bou2]), casting all of them in the same mould. Dually, categorical *colimits* extend and unify sums, coequalisers, and the classical 'inductive limits'.

There is a marginal relationship with limits of real sequences, or functions: see a note in 6.1.2(c).

6.1 Basic limits and colimits

We begin by defining *products* and *equalisers* in a category C, and computing them in various concrete categories. Then we do the same for the dual notions, *sums* and *coequalisers*. We end this section with a brief introduction to biproducts and additive categories.

The general definition of (categorical) limit and colimit is deferred to the next section, where we will see that all limits and colimits can be constructed with the basic ones, dealt with here (Theorem 6.2.4).

Various examples are presented, as exercises. The reader will find useful to complete the missing details, and work out similar results for other categories of structured sets.

Further computations of limits and colimits will be given in the rest of this Chapter.

6.1.1 Products

The simplest case of a limit, in a category C, is the *product* of a family $(X_i)_{i \in I}$ of objects (indexed by a *small set* I).

This is defined as an object X equipped with a family of morphisms $p_i \colon X \to X_i$ ($i \in I$), called (cartesian) *projections*, which satisfies the following universal property (already used many times, after 1.2.3):

$$\begin{array}{ccc} Y & \overset{f}{\dashrightarrow} & X \\ & {\scriptstyle f_i}\searrow & \downarrow {\scriptstyle p_i} \\ & & X_i \end{array} \qquad (6.1)$$

(i) for every object Y and every family of morphisms $f_i \colon Y \to X_i$ there exists a unique morphism $f \colon Y \to X$ such that, for all $i \in I$, $p_i f = f$.

The morphisms f_i are called the *components* of f, and the latter is often written as (f_i) – even though it is not the same as the family of its components, of course.

The product of a family of objects need not exist. It is determined up to a unique *coherent* isomorphism, in the sense that if also Y is a product of the family $(X_i)_{i \in I}$, with projections $q_i \colon Y \to X_i$, then the unique morphism $f \colon X \to Y$ which commutes with all projections (i.e. $q_i f = p_i$, for all indices i) is invertible. This follows easily from the fact that there is also a unique morphism $g \colon Y \to X$ such that $p_i g = q_i$; moreover $gf = \mathrm{id}\, X$ (because $p_i(gf) = p_i(\mathrm{id}\, X)$, for all i) and $fg = \mathrm{id}\, Y$.

Therefore one speaks of *the* product of the family (X_i), denoted as $\prod_i X_i$.

We say that a category C *has products* (resp. *finite products*) if every family of objects indexed by a *small* set (resp. by a finite set) has a product in C.

In particular the product of the empty family of objects $\emptyset \to \mathrm{Ob}\, \mathsf{C}$ means an object X (equipped with no projections) such that for every object Y (equipped with no maps) there is a unique morphism $f \colon Y \to X$ (satisfying no conditions). The solution is called the *terminal* object of C; if it exists, it is determined up to a unique isomorphism and can be written as $\top$.

6.1.2 Examples, exercises and complements (Products)

(a) In the previous chapters we have seen the existence and description of products in many categories of structured sets (constructed on the product of the underlying sets): for Set (in 1.2.3), Ab (in 1.3.4), pOrd and Ord (in 1.4.8), Mon and Gp (in 1.5.6), Rng (in 2.1.4), R Mod (in 2.3.4), RAlg (in 2.4.7(a)), and – informally – for each variety of algebras (in 1.2.8).

Moreover, we have described cartesian products in Top (in 3.5.2), and seen that various full subcategories of Top are stable in the latter, under cartesian products, like Hausdorff spaces, connected spaces, compact spaces (in Theorems 3.6.4, 4.1.4 and 4.2.4).

In all these cases, the terminal object is the singleton, with the (unique) relevant structure.

(b) Products in the categories $\mathsf{Set}_\bullet$ and $\mathsf{Top}_\bullet$ are also obvious

$$\Pi_i\,(X_i, \overline{x}_i) = (\Pi_i\, X_i, (\overline{x}_i)). \tag{6.2}$$

(c) In the category X associated to a preordered set, the categorical product of a family of points $x_i \in X$ amounts to the greatest lower bound $\inf(x_i)$, and the terminal object amounts to $\max X$ (when such elements exist). Note that they are determined up to the equivalence relation associated to our preorder, and uniquely determined in an ordered set.

As a marginal relationship between topological and categorical limits, we can now remark that a decreasing sequence (x_k) in the euclidean line $\mathbb{R}$ has a topological limit x (in $\mathbb{R}$) if and only if it is lower bounded, if and only if $x = \inf(x_k)$ is the product (and categorical limit) of the family (x_k) in the category associated to the ordered set $\mathbb{R}$.

For an increasing sequence in $\mathbb{R}$, the topological limit and categorical colimit are similarly related. Monotone and anti-monotone real functions have similar results.

(d) Products in Cat have been considered in 5.2.5; the terminal object is the singleton category **1**.

(e) A category has finite products if and only if it has binary products $X_1 \times X_2$ and a terminal object. Unary products always exist, trivially.

Hints: a ternary product $X_1 \times X_2 \times X_3$ can be obtained as $(X_1 \times X_2) \times X_3$, by an accurate definition of the ternary projections.

6.1.3 Equalisers and regular monomorphisms

Another basic limit is the *equaliser* of a pair $f, g \colon X \to Y$ of parallel maps of C. This is an object E with a map $m \colon E \to X$ such that $fm = gm$ and the following universal property holds:

$$\begin{array}{ccc} E & \xrightarrow{m} & X \overset{f}{\underset{g}{\rightrightarrows}} Y \\ & {\scriptstyle w}\nwarrow & \uparrow{\scriptstyle h} \\ & & Z \end{array} \qquad m = \mathrm{eq}(f, g) \tag{6.3}$$

(ii) every map $h \colon Z \to X$ such that $fh = gh$ factorises uniquely through m (i.e. there exists a unique map $w \colon Z \to E$ such that $mw = h$).

The equaliser morphism is necessarily a monomorphism, and is called a *regular* monomorphism. It is easy to see that a regular mono which is epi is an isomorphism.

If m is epi in diagram (6.3), the maps f and g must be equal, and we can take $m = \mathrm{id}\,X$ (or any isomorphism with values in X).

Products and equalisers are particular instances of the limit of a functor, that will be introduced in 6.2.0.

6.1.4 Exercises and complements (Equalisers)

(a) In Set the natural solution for the equaliser of a pair of mappings $f, g: X \to Y$ is the subset

$$E = \{x \in X \mid f(x) = g(x)\}, \tag{6.4}$$

with the inclusion $m: E \subset X$.

(b) In Top, Ab, Ord, Set$_\bullet$ and Top$_\bullet$ (and many other concrete categories) we construct the equaliser in the same way, putting on E the structure induced by X as a subspace, or a subgroup, etc. For topological spaces, the equaliser has already been considered in 3.6.7.

(c) In Cat, the equaliser of two functors F, G: C $\to$ D is the subcategory of the objects and arrows of C on which F and G coincide.

(d) In any category, every split monomorphism is regular.

(e) The reader will note that in Top a monomorphism $m: X' \to X$ is any injective map (by 5.5.3(b)), while a regular mono must be the inclusion of a subspace $E \subset X$, up to homeomorphism, that is a *topological embedding* $m: X' \to X$ (as defined in 3.4.7).

The converse is also true: every embedding of a subspace is a regular mono, because every subspace $E \subset X$ is the equaliser of its 'cokernel pair' (as we will see in 6.2.9(g)).

In Top the 'interesting subobjects' are the regular ones.

(f) We have a similar situation in Ord: a monomorphism is any injective monotone mapping, and is a regular mono if and only if it *reflects* the ordering (and induces an isomorphism between the domain and an ordered subset of the codomain).

(g) All subgroups are regular subobjects in Gp (a non-trivial fact, see [AHS], Exercise 7H), while a subsemigroup need not be a regular subobject ([AHS], Exercise 7I).

In the categories of algebraic structures the general subobjects are more important than the regular ones.

6.1.5 Sums

The *sum*, or *coproduct*, of a family $(X_i)_{i \in I}$ of objects of C is dual to its product. Explicitly, as we have already seen in many particular instances, it is an object X equipped with a family of morphisms $u_i \colon X_i \to X$ $(i \in I)$, called *injections*, that satisfy the following universal property:

$$\begin{array}{ccc} X & \overset{f}{\dashrightarrow} & Y \\ {\scriptstyle u_i}\uparrow & \nearrow_{f_i} & \\ X_i & & \end{array} \qquad (6.5)$$

(i*) for every object Y and every family of morphisms $f_i \colon X_i \to Y$, there exists a unique morphism $f \colon X \to Y$ such that, for all $i \in I$, $fu_i = f_i$.

The map f can be written as $[f_i]$, by its *co-components*. If the sum of the family (X_i) exists, it is determined up to a unique coherent isomorphism, and denoted as $\sum_i X_i$, or $X_1 + ... + X_n$ in a finite case. (The symbol $\coprod_i X_i$ is also widely used, when $\sum_i X_i$ might be ambiguous.)

The sum of the empty family is the *initial* object $\bot$: this means that every object X has a unique map $\bot \to X$.

6.1.6 *Examples, exercises and complements* (Sums)

(a) In Set a sum $X = \sum_i X_i$ is realised as a 'disjoint union', as we have seen in (1.34). The initial object is the empty set.

(b) In Top we do the same, equipping the set X with the finest topology that makes all the injections continuous: see 3.5.6. The initial object is the empty space.

Similarly, in Ord we endow the set X with the finest ordering that makes all the injections monotone: $(x, i) \leqslant (y, j)$ if and only if $i = j$ and $x \leqslant y$ in X_i.

(c) In Abm, Ab and $R\,$Mod categorical sums are realised as 'direct sums': see 1.3.8 and 2.3.4. The initial object is the singleton.

(d) In $\mathsf{Set}_\bullet$ or $\mathsf{Top}_\bullet$ a sum $\sum_i (X_i, \overline{x}_i)$ can be constructed as a quotient of the unpointed sum $\sum_i X_i$, by identifying all the base points $\overline{x}_i$.

In $\mathsf{Set}_\bullet$ the sum can also be realised – more simply – as a subset of the cartesian product

$$\{(x_i) \in \textstyle\prod_i X_i \mid x_i = \overline{x}_i \text{ for all indices } i \text{ except one at most}\}, \qquad (6.6)$$

the union of all the axes of the product, pointed again at the 'origin' $(\overline{x}_i)$.

*In $\mathsf{Top}_\bullet$ this second construction works for finite products. For instance, the sum $(\mathbb{R}, 0) + (\mathbb{R}, 0)$ can be realised as the subspace of the euclidean plane formed of the two cartesian axes.

For an infinite product one should put on this subset the topology derived from the first construction (and induced by the box topology of the product, see 3.5.3(i)).*

(e) In the category X associated to a preordered set, the categorical sum of a family of points $x_i \in X$ amounts to its sup, while the initial object amounts to the least element of X.

(f) In Cat a sum $\sum \mathsf{C}_i$ of categories is their obvious disjoint union. The initial object is the empty category $\mathbf{0}$.

(g) In Mon, the sum of a family of monoids (M_i) is called the *free product*, and often written as $\ast_i M_i$. It can be obtained as the free monoid $M(X)$ on the sum $X = \Sigma_i |M_i|$ of the underlying sets, modulo the congruence of monoids that forces all embeddings $|M_i| \to M(X)$ to become homomorphisms of monoids.

Loosely speaking, an element of $\ast_i M_i$ can be written as a finite word $(x_1, ..., x_n)$ in the alphabet X, provided we agree to replace every occurrence $x_h x_{h+1}$ of two elements in the same component M_i with their product, and to take out any unit of an M_i.

(h) If, in the previous construction, all M_i are groups, $\ast_i M_i$ is also and gives the sum in Gp. In Group Theory, this is called a *free product* of groups – a term older than Category Theory.

*(i) The construction in (g) can be extended to any kind of equational algebras; but – for a given kind – there can be clearer ways of describing the result, as we have already seen above.

*(j) In CRng, finite sums are realised as tensor products over $\mathbb{Z}$, and a general sum as a colimit of finite tensor products.

6.1.7 Coequalisers and regular epimorphisms

The *coequaliser* of a pair $f, g\colon X \to Y$ of parallel maps of C is a map $p\colon Y \to C$ such that $pf = pg$ and:

$$\begin{array}{ccccc} X & \overset{f}{\underset{g}{\rightrightarrows}} & Y & \xrightarrow{p} & C \\ & & {\scriptstyle h}\downarrow & \swarrow {\scriptstyle w} & \\ & & Z & & \end{array} \qquad p = \mathrm{coeq}(f, g) \tag{6.7}$$

(ii*) every map $h\colon Y \to Z$ such that $hf = hg$ factorises uniquely through p (i.e. there exists a unique map $w\colon C \to Z$ such that $wp = h$).

A coequaliser morphism is necessarily an epimorphism, and is called a *regular* epimorphism. A regular epi which is mono is an isomorphism. A *regular quotient* is defined as a coequaliser (of some pair of maps); its choice is determined by the choice of quotients (see 5.5.9).

Sums and coequalisers are particular instances of the colimit of a functor, that will be introduced in 6.2.0.

6.1.8 Examples and complements (Coequalisers)

(a) In Set the natural solution of the coequaliser of a pair $f, g\colon X \to Y$ of parallel mappings is the projection $p\colon Y \to Y/R$ on the quotient modulo the equivalence relation spanned by the pairs $(f(x), g(x)) \in Y^2$, for $x \in X$.

(b) In Top we do the same, putting on Y/R the quotient topology, namely the finest that makes the mapping p continuous.

In pOrd we do the same, putting on Y/R the induced preordering (the finest that makes p monotone). In Ord we first compute the preordered set Y/R in pOrd, and then take the associated ordered set. (This fact will be extended to all colimits.)

(c) In Ab we take the quotient Y/H modulo the subgroup

$$H = \{f(x) - g(x) \mid x \in X\},$$

which amounts to the quotient Y/E modulo the congruence of abelian groups generated by the previous equivalence relation R. The reader will adapt this construction to Gp.

(d) Coequalisers in $\mathsf{Set}_\bullet$ and $\mathsf{Top}_\bullet$ are (easy and) left to the reader.

(e) In Top an epimorphism $p\colon Y \to Y'$ is any surjective map, while a regular epi must be the projection on a quotient space $p\colon Y \to Y/R$, up to homeomorphism; this means a *topological projection* (see 3.4.7). The converse is also true: every topological projection is a regular epi, because every quotient space X/R is (obviously) the coequaliser of the partial projections $R \rightrightarrows X$ (see also 6.2.9(g)).

*(f) The coequaliser in Cat of two functors $F, G\colon \mathsf{C} \to \mathsf{D}$ is the quotient of D modulo the *generalised congruence* generated by this pair; the latter, as defined in the paper [BBP], also involves equivalent objects. One can avoid giving a 'construction' of the coequaliser category (necessarily complicated) and just prove its existence by the Adjoint Functor Theorem (see 6.3.5).*

6.1.9 Zero objects, biproducts and additive categories

We end this section by briefly introducing some notions, that will be further analysed in Section 6.6.

First, a *zero object* of a category, often written as 0, is both initial and terminal. This exists in Abm, Ab, $R\,$Mod, Gp, Rng$'$, Set$_\bullet$, Top$_\bullet$,... but not in Set, Top, Rng and Cat, where the initial and terminal objects are distinct.

A category with zero object is said to be *pointed*; then each pair of objects A, B has a *zero-morphism* $0_{AB}\colon A \to B$ (also written as 0), which is given by the composite $A \to 0 \to B$.

As a stronger fact, we have already remarked that, in Abm, Ab and $R\,$Mod, a *finite* product $\prod A_i$ and the corresponding finite sum $\sum A_i$ are realised as the same object $A = \bigoplus A_i$, which satisfies:

- the universal property of the product, by a family of projections $p_i\colon A \to A_i$,

- the universal property of the sum, by a family of injections $u_i\colon A_i \to A$,

- the equations $p_i u_i = \mathrm{id}\, A_i$ and $p_j u_i = 0\colon A_i \to A_j$ (for $i \neq j$).

Such an object, in a pointed category, is called a *biproduct*. The empty biproduct is the zero object.

As a related notion, a *preadditive*, or *$\mathbb{Z}$-linear*, category is a category C where every hom-set C(A, B) is equipped with a structure of abelian group, generally written as $f + g$, so that composition is additive in each variable (or bilinear over $\mathbb{Z}$)

$$(f + g)h = fh + gh, \qquad k(f + g) = kf + kg, \tag{6.8}$$

where $h\colon A' \to A$ and $k\colon B \to B'$. A functor $F\colon \mathsf{C} \to \mathsf{D}$ between preadditive categories is said to be *additive*, or *$\mathbb{Z}$-linear*, if it preserves the sum of parallel morphisms.

A preadditive category on a single object 'is' a unital ring R. An additive functor $R \to S$ between such categories is a homomorphism of unital rings. An additive functor $R \to$ Ab amounts to an R-module.

*An *additive category* C is a preadditive category with finite products, or equivalently with finite sums, which are then biproducts. The equivalence is proved in [M2], Section VIII.2; *in this case*, the sum of parallel maps is determined by the categorical structure. Typical examples are Ab, $R\,$Mod and the category of (real or complex) Banach spaces, with continuous linear mappings.*

6.2 General limits and completeness

This sections contains the main definitions and basic results about limits and colimits in a category C. The main theorem, in 6.2.4, proves that all limits can be constructed from products and equalisers.

Limits can be described as universal arrows, and by representable functors, see 6.2.2. Their relationship with adjoints will be seen in 6.3.4(g) and 6.3.5(c), with terminal objects of comma categories in 6.3.8(b).

6.2.0 Main definitions

Let I be a small category and $X \colon \mathsf{I} \to \mathsf{C}$ a functor, written in index notation (as in (5.18)), for $i \in I = \mathrm{Ob}\,\mathsf{I}$ and $a \colon i \to j$ in I:

$$X \colon \mathsf{I} \to \mathsf{C}, \qquad i \mapsto X_i, \quad a \mapsto (X_a \colon X_i \to X_j). \tag{6.9}$$

A *cone* for X is an object A of C (the *vertex* of the cone) equipped with a family of maps $(f_i \colon A \to X_i)_{i \in I}$ in C such that the following triangles commute (for $a \colon i \to j$ in I)

$$\begin{array}{ccc} A & \xrightarrow{f_i} & X_i \\ & \searrow^{f_j} & \downarrow X_a \\ & & X_j \end{array} \qquad X_a.f_i = f_j. \tag{6.10}$$

The *limit* of $X \colon \mathsf{I} \to \mathsf{C}$ is a universal cone $(L, (u_i \colon L \to X_i)_{i \in I})$.

This means a cone of X such that every cone $(A, (f_i \colon A \to X_i)_{i \in I})$ factorises uniquely through the former; in other words, there is a unique map $f \colon A \to L$ such that, for all $i \in I$

$$\begin{array}{ccc} A & \overset{f}{\dashrightarrow} & L \\ & \searrow^{f_i} & \downarrow u_i \\ & & X_i \end{array} \qquad u_i.f = f_i. \tag{6.11}$$

The solution need not exist; when it does, it is determined up to a unique coherent isomorphism, and the limit object L is denoted as $\mathrm{Lim}(X)$.

The uniqueness part of the universal property amounts to saying that *the family $u_i \colon L \to X_i$ ($i \in I$) is jointly mono* (as defined in 5.5.1). A cone of X that satisfies the existence part of the universal property is called a *weak limit*; of course such a solution is weakly constrained.

Dually, a *cocone* $(A, (f_i \colon X_i \to A)_{i \in I})$ of X satisfies the condition $f_j.X_a = f_i$ for every $a \colon i \to j$ in I. The *colimit* of the functor X is a universal cocone

$(L', (u_i\colon X_i \to L')_{i\in I})$: the universal property says now that for every cocone $(A, (f_i))$ of X there exists a unique map $f\colon L' \to A$ such that $fu_i = f_i$ (for all i).

The colimit object is denoted as $\mathrm{Colim}(X)$. The uniqueness part of the universal property means that the family $u_i\colon X_i \to L'$ is jointly epi. A cocone that satisfies the existence part of the universal property is called a *weak colimit*.

6.2.1 Examples and complements

(a) The product ΠX_i of a family $(X_i)_{i\in I}$ of objects of C is the limit of the corresponding functor $X\colon \mathsf{I} \to \mathsf{C}$, defined on the *discrete* category whose objects are the elements $i \in I$ (and whose morphisms reduce to formal identities of such objects). The sum ΣX_i is the colimit of this functor X.

Let us recall that the projections (p_i) of a product are *jointly mono*. These projections are 'often' epi but *not necessarily*: for instance in Set this fails whenever some of the factors X_i are empty and others are not.

However, one can easily show that, if C has a zero object (defined in 6.1.9), all cartesian projections are split epi.

(b) The equaliser in C of a pair of parallel morphisms $f, g\colon X_0 \to X_1$ is the limit of the obvious functor defined on the category $0 \rightrightarrows 1$ (with two non-identity arrows). The coequaliser is the colimit of this functor.

(c) Pullbacks and pushouts will be defined in 6.2.8.

(d) In Set every non-empty set is weakly terminal; there is only one weakly initial object, namely the initial object $\emptyset$. (This divergence proves that Set is not self-dual.)

In a pointed category, every object is weakly terminal and weakly initial.

*(e) The existence of all limits in Set will follow from the existence of products and equalisers, in Theorem 6.2.4.

However, as an interesting exercise, the reader can prove that the limit of any functor $\mathsf{I} \to \mathsf{Set}$ can be directly constructed as the set of its cones with vertex at the singleton set $\{*\}$, and extend this result to many other categories.

*(f) Weak limits and colimits are important in Homotopy Theory. For instance, a homotopy pullback (resp. pushout) of topological spaces [Mat] gives a weak pullback (resp. pushout) in the homotopy category hoTop.

6.2.2 Limits by universal arrows and representability

Consider the category C^{I} of functors $\mathsf{I} \to \mathsf{C}$ and their natural transformations, for a small category I (see 5.4.8). The *diagonal functor*

$$\Delta\colon \mathsf{C} \to \mathsf{C}^{\mathsf{I}}, \qquad (\Delta A)_i = A, \quad (\Delta A)_a = \mathrm{id}\, A, \tag{6.12}$$

sends an object A to the constant functor at A, and a morphism $f\colon A \to B$ to the natural transformation $\Delta f\colon \Delta A \to \Delta B\colon \mathsf{I} \to \mathsf{C}$ whose components are constant at f.

Let a functor $X\colon \mathsf{I} \to \mathsf{C}$ be given. A natural transformation $f\colon \Delta A \to X$ is the same as a cone of X of vertex A.

The limit of X in C is the same as a universal arrow $(L, \varepsilon\colon \Delta L \to X)$ from the functor Δ to the object X of C^{I} (as defined in 5.4.1).

Dually, the colimit of X in C is the same as a universal arrow $(L', \eta\colon X \to \Delta L')$ from the object X of C^{I} to the functor Δ.

As another characterisation, assuming that C has small hom-sets, consider the contravariant functor G *of sets of cones*

$$G\colon \mathsf{C}^{\mathrm{op}} \to \mathsf{Set}, \qquad G = \mathsf{C}^{\mathsf{I}}(\Delta(-), X). \tag{6.13}$$

Then, by Corollary 5.4.7, X has a limit (L, ε) if and only if the functor G is represented by the object $X_0 = L$, via the element $\varepsilon \in G(L)$. Dually, the existence of a colimit (L', η) of X is equivalent to saying that the covariant functor $G'(A) = \mathsf{C}^{\mathsf{I}}(X, \Delta A)$ of *sets of cocones* is represented by the object L', via the element $\eta \in G'(L')$.

6.2.3 Complete categories and the preservation of limits

A category C is said to be *complete* (resp. *finitely complete*) if it has a limit for every functor $\mathsf{I} \to \mathsf{C}$ defined on a *small* category (resp. a *finite* category).

More precisely one speaks – in the first case – of a *small-complete* category. Of course we cannot expect Set to have products indexed by a large set, or limits for all functors $\mathsf{I} \to \mathsf{Set}$ defined over a large category. In other words, the basis I should be of a 'smaller size' than the category C where we construct limits, or we force C to be just a preordered set (where products are infima, and are not influenced by the size of the set of indices).

The question is settled by a result of P. Freyd: a small category C has the limit of every functor defined on a small category (if and) only if it is a preordered set with all infima (see Exercise 6.2.5(g)). Thus, *in an arbitrary category, completeness only means the existence of small limits, and does not imply the existence of (small) colimits*, as is the case of preordered sets (see 6.2.5(f)).

One says that a functor $F\colon \mathsf{C} \to \mathsf{D}$ *preserves the limit* $(L, (u_i\colon L \to X_i))$ of a functor $X\colon \mathsf{I} \to \mathsf{C}$ if the cone $(FL, (Fu_i\colon FL \to FX_i))$ is the limit of the composed functor $FX\colon \mathsf{I} \to \mathsf{D}$. One says that F *preserves limits* (or *products*, or *equalisers*, etc.) if it preserves all the limits (or products, or equalisers, etc.) which exist in C. Saying that a functor *does not preserve limits* means that it does not preserve *some* of them.

Dually we have *cocomplete* categories and the property of *preservation of colimits.*

We will see that an equivalence of categories preserves all limits and colimits, in 6.3.5(b).

6.2.4 Theorem (Construction and preservation of limits)

A category is complete (resp. finitely complete) if and only if it has equalisers and products (resp. finite products). Moreover, if C is complete (resp. finitely complete), a functor $F\colon \mathsf{C} \to \mathsf{D}$ preserves all limits (resp. all finite limits) if and only if it preserves equalisers and products (resp. finite products).

Dual results hold for colimits. In particular, a category is cocomplete (resp. finitely cocomplete) if and only if it has coequalisers and sums (resp. finite sums).

Proof The reader can complete this outline. (Alternatively, a more detailed argument can be found in [M2], Section V.2, Theorem 1.)

Let the functor $X\colon \mathsf{I} \to \mathsf{C}$ be written in index notation, as in 6.2.0, and consider the (small) products

$$\begin{aligned} &\textstyle\prod_i X_i \quad \text{indexed by the objects } i \in \mathrm{Ob}\,\mathsf{I}, \text{ with projections } p_i,\\ &\textstyle\prod_a X_{j(a)} \text{ indexed by the arrows } a \text{ in } \mathsf{I}, \text{ with projections } q_a, \end{aligned} \tag{6.14}$$

where it is understood that $a\colon i(a) \to j(a)$.

A cone $(A, (f_i))$ of X amounts to a map $f\colon A \to \prod_i X_i$ that equalises the following two maps u, v

$$\begin{gathered} u, v\colon \textstyle\prod_i X_i \to \prod_a X_{j(a)},\\ q_a u = p_{j(a)}, \quad q_a v = X_a.p_{i(a)} \qquad (\text{for } a \text{ in } \mathsf{I}), \end{gathered} \tag{6.15}$$

and gives back the original cone letting $f_i = p_i f\colon A \to X_i$.

Taking the equaliser $m\colon L \to \prod_i X_i$ of the pair u, v, we have a cone $(L, (u_i))$ of X, with $u_i = p_i m\colon L \to X_i$ (for $i \in \mathrm{Ob}\,\mathsf{I}$). This cone is the limit.

The second part of the statement, on the preservation of limits, is an obvious consequence, once we assume that products and equalisers exist in C, so that all its limits can be obtained as above. □

6.2.5 Exercises and complements

(a) Taking into account the computations of Section 6.1 it follows that the categories Set, Top, Ab, Ord, $\mathsf{Set}_\bullet$ and $\mathsf{Top}_\bullet$ are complete and cocomplete (as many others).

The forgetful functors Top $\to$ Set and pOrd $\to$ Set preserve limits and colimits, while Ab $\to$ Set only preserves limits; Ord $\to$ Set preserves limits and sums, but it does not preserve coequalisers (and general colimits, as a consequence).

(b) The reader will note that any limit $L = \mathrm{Lim}X$ in Top is constructed as in Set and equipped with the coarsest topology making all the structural mappings $u_i \colon L \to X_i$ continuous.

Similarly, a colimit $L' = \mathrm{Colim}X$ in Top is constructed as in Set, and equipped with the finest topology making all the structural mappings $u_i \colon X_i \to L'$ continuous.

(c) One can proceed in a similar way in the category pOrd of preordered sets. In Ord this procedure works for limits and sums; for coequalisers and general colimits it must be 'corrected', replacing the colimit in pOrd with the associated ordered set (as in 6.1.8(b)).

(d) The equivalence between $\mathsf{Set}_\bullet$ and $\mathcal{S}$ (in 5.3.8) *proves that also the latter has all limits and colimits.* It is interesting to work out the case of products, showing that the product 1×1 in $\mathcal{S}$ has *three* elements, while 2×3 has *eleven*.

Sums in $\mathcal{S}$ can also be constructed in this way, but they simply amount to disjoint unions of sets.

(e) If the category C is complete (resp. finitely complete), so is the category C^{S} (for any small category S), with limits computed pointwise on each object of S. Similarly for colimits.

(f) In a preordered set X, viewed as a category, equalisers and coequalisers are (obviously) trivial, while we have seen in Section 6.1 that products and sums amount to infima and suprema, respectively.

Therefore X is a complete category if and only if it has all infima, which is equivalent to the existence of all suprema (Exercise 1.4.5(a)): in other words, the category X is complete if and only if it is cocomplete. In the ordered case this means that X is a complete lattice.

Similarly, an ordered set is a bounded lattice if and only if, as a category, it is finitely complete and cocomplete.

*(g) Freyd's result recalled in 6.2.3 can be rewritten in this form: a small category C with all small products is a preordered set, with all infima.

6.2.6 Representable functors preserve limits

Let C be a category with small hom-sets, and $X\colon \mathsf{S} \to \mathsf{C}$ a functor defined on a small category. The universal property of the limit of X can be rewritten in Set, as follows:

- a cone $(L, (u_i\colon L \to X_i))$ of X is a limit in C if and only if, for every A in C, the set $\mathsf{C}(A, L)$ is the limit in Set of the sets $\mathsf{C}(A, X_i)$, with limit-cone $u_i.-\colon \mathsf{C}(A, L) \to \mathsf{C}(A, X_i)$.

This rewriting tells us that:

(i) each representable covariant functor $H^A = \mathsf{C}(A, -)\colon \mathsf{C} \to \mathsf{Set}$ preserves all (the existing) limits,

(ii) the family (H^A), indexed by the objects $A \in \mathrm{Ob}\,\mathsf{C}$, reflects them.

Colimits in C are limits in C^{op}. Thus, a representable contravariant functor $H_A = \mathsf{C}(-, A)\colon \mathsf{C}^{\mathrm{op}} \to \mathsf{Set}$ takes *colimits* of C to *limits* of sets; globally, their family reflects limits of Set into colimits of C.

For instance, for the contravariant functor of subsets $\mathsf{Set}^{\mathrm{op}} \to \mathsf{Set}$ of (5.17), we (plainly) have: $\mathcal{P}(\Sigma_i X_i) \cong \Pi_i \mathcal{P}(X_i)$, because a subset $A \subset \Sigma_i X_i$ is the same as a family of subsets $A_i \subset X_i$.

6.2.7 The creation of limits

There is another interesting property, related to the preservation of limits; again, it does not assume their existence.

One says that a functor $F\colon \mathsf{C} \to \mathsf{D}$ *creates limits* for a functor $X\colon \mathsf{I} \to \mathsf{C}$ if:

(i) for every limit cone $(L', (v_i\colon L' \to FX_i)_{i \in I})$ of the composed functor $FX\colon \mathsf{I} \to \mathsf{D}$, there is precisely one cone $(L, (u_i\colon L \to X_i)_{i \in I})$ of X in C, taken by F to $(L', (v_i))$,

(ii) the latter is a limit of X in C (preserved by F).

For instance the forgetful functor $U\colon \mathsf{Ab} \to \mathsf{Set}$ creates all small limits (which do exist). Indeed, a limit of abelian groups can be constructed by taking the limit $(L', (v_i))$ of the underlying sets and putting on L' the unique structure that makes the mappings v_i into homomorphisms.

*The same holds for all varieties of algebras (but it fails for non-equational algebraic structures like fields). Note that the forgetful functors of such varieties do not create *colimits*: they do not even preserve them, generally.*

On the other hand the forgetful functor Top → Set preserves all limits and colimits but does not create them: for instance the product topology is the coarsest topology on the product set that makes all cartesian projections continuous, and is not the only one with this outcome – generally.

The relationship among creating, preserving and reflecting limits is examined in [G4], Exercises 2.2.9.

6.2.8 Pullbacks and pushouts

We end this section by an interesting limit, and the corresponding colimit.

(a) In a category C, the *pullback* of a pair of morphisms

$$f\colon X_0 \to X \leftarrow X_1 : g$$

with the same codomain is the limit of the corresponding functor, defined on the category $0 \to \iota \leftarrow 1$. (We are only drawing the non-trivial arrows of the latter, of course.)

This amounts to a definition that can be familiar to the reader, in some concrete context: we are looking for an object A equipped with a 'span' $u_i\colon A \to X_i$ $(i = 0, 1)$ which forms a commutative square with f and g, in a universal way:

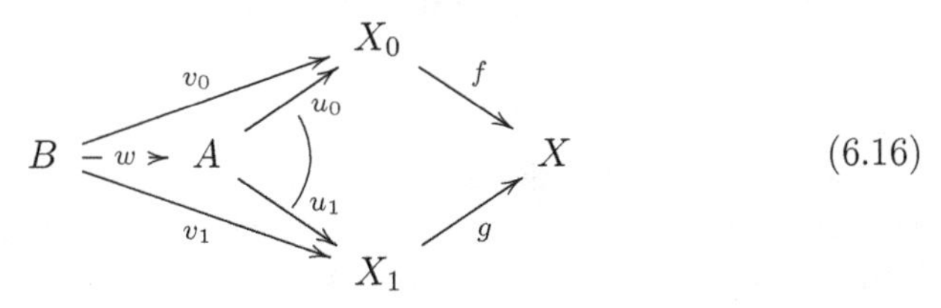

(6.16)

that is, $fu_0 = gu_1$, and for every span (B, v_0, v_1) such that $fv_0 = gv_1$ there is a unique map $w\colon B \to A$ such that $u_0w = v_0$, $u_1w = v_1$. The span (u_0, u_1) is jointly mono. The pullback-object A is also called a *fibred product* over X and written as $X_0 \times_X X_1$. In a diagram, its corner can be marked as above.

In the ordered set X the pullback of the diagram $x_0 \to x \leftarrow x_1$ is the meet $x_0 \wedge x_1$. Saying that X has pullbacks means that every *upper-bounded* pair of elements has a meet (and is weaker than saying that it has binary products).

In Set (resp. Top, Ab) the pullback-object can be realised as a subset (resp. subspace, subgroup) of the product $X_0 \times X_1$:

$$A = \{(x_0, x_1) \in X_0 \times X_1 \mid f(x_0) = g(x_1)\}, \qquad (6.17)$$

a fact that can be extended to any category (see Exercise 6.2.9(a)).

If $f = g$, the pullback of the diagram $X_0 \to X \leftarrow X_0$ is called the *kernel pair* of f. In Set, it can be realised as $\{(x, x') \in X_0 \times X_0 \mid f(x) = f(x')\}$, and amounts to the equivalence relation associated to f. (But note that, in any category, the equaliser of f and f is the identity of its domain: the equaliser is a limit based on a different category, with two objects.)

(b) Dually, the *pushout* of a span $(f, g) = (X_0 \leftarrow X \to X_1)$ is the colimit of the corresponding functor defined on the category $0 \leftarrow \iota \to 1$.

This amounts to an object A equipped with a 'cospan' $u_i \colon X_i \to A$ $(i = 0, 1)$ which forms a commutative square with f and g, in a universal way:

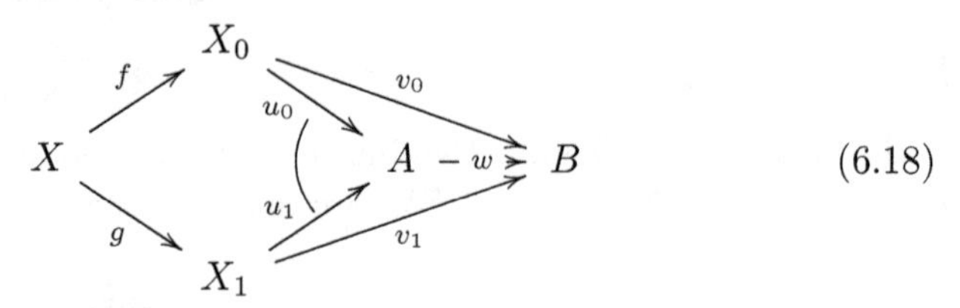

(6.18)

that is, $u_0 f = u_1 g$, and for every triple (B, v_0, v_1) such that $v_0 f = v_1 g$ there is a unique map $w \colon A \to B$ such that $wu_0 = v_0$, $wu_1 = v_1$. 1. If $f = g$, this pushout is called the *cokernel pair* of f.

The pushout-object A is also called a *pasting* over X and written as $X_0 +_X X_1$.

In Set this means a quotient of the sum $X_0 + X_1$ modulo the equivalence relation generated by identifying all pairs $f(x)$, $g(x)$ for $x \in X$. In Top a pasting $X_0 +_X X_1$ is constructed as in Set and equipped with the quotient topology of the sum – a useful way of constructing or describing spaces from more elementary ones (see Exercise 6.2.9(k)).

The well-known Seifert – van Kampen Theorem, one of the main tools to compute the fundamental group of a space, is based on a pushout of groups. Ronald Brown's version generalises this result, using fundamental groupoids [Bro1, Bro2].

6.2.9 Exercises and complements (Pullbacks and pushouts)

The following facts are important and often used; their dual is only written down in a few cases.

(a) Generalising the construction in (6.17), one proves that a category that has binary products and equalisers also has pullbacks. On the other hand, the product $X_0 \times X_1$ amounts to the pullback of X_0 and X_1 over the terminal object, when the latter exists.

Dually, a category that has binary sums and coequalisers also has pushouts. A category that has pushouts and initial object also has finite sums.

(b) (*Characterising monos*) Prove that the following condition on a morphism $f\colon X \to Y$ are equivalent

(i) f is mono,

(ii) the left square below is a pullback (i.e. f has kernel pair $(1_X, 1_X)$),

(iii) if the right square below is a pullback, then $h = k$,

(iv) if the right square below is a pullback, then $h = k$ is invertible

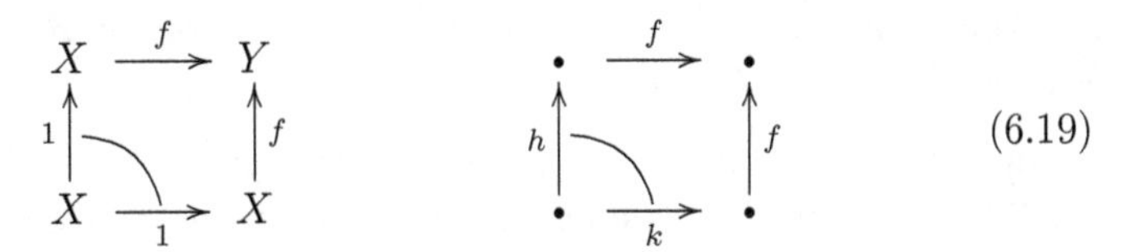

(6.19)

These characterisations do not require the existence of pullbacks in our category.

(c) (*Preimages*) If the following square is a pullback and n is a monomorphism, so is m

$$\begin{array}{ccc} X & \xrightarrow{f} & Y \\ {\scriptstyle m}\uparrow & & \uparrow{\scriptstyle n} \\ \bullet & \longrightarrow & \bullet \end{array} \qquad (6.20)$$

Then m is determined up to equivalence of monomorphisms (see 5.5.9): it is called the *preimage* of n along f and written as $f^*(n)$. Working with subobjects, $f^*(n)$ can be determined by their choice.

(d) (*Symmetry*) If (P, u, v) is a pullback of (f, g), then (P, v, u) is – obviously – *a* pullback of (g, f). However, when a fixed choice of pullbacks is used, *one cannot assume that the choice is strictly symmetric.*

(e) (*Pasting property*) If the two squares below are pullbacks, so is their 'pasting', i.e. the outer rectangle

$$\begin{array}{ccccc} \bullet & \longrightarrow & \bullet & \longrightarrow & \bullet \\ \uparrow & & {\scriptstyle f}\uparrow & & \uparrow \\ \bullet & \longrightarrow & \bullet & \xrightarrow[g]{} & \bullet \end{array} \qquad (6.21)$$

(f) (*Depasting property*) If in the commutative diagram above the outer rectangle is a pullback and the pair (f, g) is jointly mono, then the left square is a pullback.

The hypothesis on the pair (f, g) is automatically satisfied in two cases, frequently used: when the right square is a pullback or one of the morphisms f, g is mono.

(g) In Set (resp. Top, pOrd) every subset (resp. subspace, preordered subset) $E \subset X$ is the equaliser of its cokernel pair. Similarly, every quotient set (resp. quotient space, quotient preordered set) X/R is the coequaliser of its kernel pair $R \rightrightarrows X$.

(h) In Rng, the ideals of a unital ring R are not objects of the category (generally). A quotient $R \to R/I$ has a kernel pair, in the category, the congruence E associated to the ideal. On the other hand, in Rng′ the ideals of a ring R are objects of the category, and a quotient $R \to R/I$ has also a kernel $I \to R$ in the categorical sense (see 6.6.1).

(i) In Set, if $X = X_0 \cup X_1$, then X is the pushout of X_0 and X_1 over $X_0 \cap X_1$. This fact can be extended to Top, under convenient hypotheses on the subspaces $X_i \subset X$.

*(j) A category with pullbacks and terminal object has equalisers, and therefore all finite limits.

*(k) Revisiting the geometry of spheres, as analysed in 4.4.1, the euclidean sphere $\mathbb{S}^2$ can be presented as a pushout in Top, the cokernel pair of the inclusion $\mathbb{S}^1 \to \mathbb{D}^2$ of the circle in the compact disc; concretely, we are pasting two discs along their boundary.

This fact can be extended to any dimension, starting with $\mathbb{S}^1$ as the pasting of two compact intervals over $\mathbb{S}^0$.

The homology groups of all spheres are computed, inductively, on this basis, using the Mayer–Vietoris exact sequence.

6.3 Adjoint functors

Adjunctions, a global way of presenting universal properties and a crucial step in category theory, were introduced in 1958 by Daniel Kan, a Dutch mathematician working in homotopy theory, who made essential contributions to category theory [Ka].

We begin by two well known situations, which help us to introduce adjoint functors. Then we give the definition, in four equivalent forms, and examine properties and many examples.

The proofs of the main results are referred to [M2, G4].

Adjunctions between ordered sets or related to Topology are deferred to the next two sections.

6.3.1 Examples

(a) The forgetful functor $U\colon \mathsf{Top} \to \mathsf{Set}$ has two well known 'best approximations' to an inverse which does not exist, the functors $D, C\colon \mathsf{Set} \to \mathsf{Top}$, where DX (resp. CX) is the set X with the discrete (resp. indiscrete) topology. In fact they provide the set X with the finest (resp. the coarsest) topology.

For every set X and every space Y we can 'identify' the following hom-sets

$$\mathsf{Top}(DX, Y) = \mathsf{Set}(X, UY), \qquad \mathsf{Set}(UY, X) = \mathsf{Top}(Y, CX), \qquad (6.22)$$

because every mapping $X \to Y$ becomes continuous if we put on X the discrete topology, while every mapping $Y \to X$ becomes continuous if we put on X the indiscrete one.

These two facts will tell us that D is left adjoint to U (written as $D \dashv U$) and C is right adjoint to U (i.e. $U \dashv C$), respectively.

(b) In other cases, it may be convenient to follow another approach, based on universal arrows.

To wit, let us start from the forgetful functor $U\colon R\,\mathsf{Mod} \to \mathsf{Set}$. We have already seen, in 5.4.2(a), that, for every set X, there is a universal arrow from X to the functor U

$$F(X) = \textstyle\bigoplus_{x \in X} R, \qquad \eta\colon X \to UF(X), \qquad (6.23)$$

consisting of the free R-module $F(X)$ on the set X, with the insertion of the basis η.

The fact that this universal arrow exists *for every* X *in* Set allows us to construct a 'backward' functor $F\colon \mathsf{Set} \to R\,\mathsf{Mod}$, *left adjoint* to U.

The functor F is already defined on the objects. For a mapping $f\colon X \to Y$ in Set

$$\begin{array}{ccc} X & \xrightarrow{\eta X} & UF(X) \\ {\scriptstyle f}\downarrow & & \downarrow{\scriptstyle U(Ff)} \\ Y & \xrightarrow[\eta Y]{} & UF(Y) \end{array} \qquad\qquad \begin{array}{c} F(X) \\ \downarrow{\scriptstyle Ff} \\ F(Y) \end{array} \qquad (6.24)$$

the universal property of ηX implies that there is precisely one homomorphism $F(f)\colon F(X) \to F(Y)$ such that $U(F(f))$ makes the previous square commute.

(Concretely, $F(f)$ is just the linear extension of f; but the general pattern is better perceived if we proceed in a formal way.)

Further applications of the universal property would show that F preserves composition and identities (but we will see this in a general context). Let us note that we have extended the given object-function $F(X)$ in the only way that makes the family (η_X) into a natural transformation $\eta\colon 1 \to UF\colon \mathsf{Set} \to \mathsf{Set}$, called the *unit* of the adjunction.

(c) The relationship between these functors U and F (assuming now that we have defined both) can be equivalently expressed in a form similar to that used in (a), by a family of bijective mappings

$$\varphi_{XA}\colon R\,\mathsf{Mod}(FX, A) \to \mathsf{Set}(X, UA) \quad (X \text{ in } \mathsf{Set},\ A \text{ in } R\,\mathsf{Mod}), \tag{6.25}$$

which is natural in X and A, as will be made precise in 6.3.2.

Concretely, φ_{XA} takes the R-homomorphism $g\colon FX \to A$ to its restriction $Ug.\eta X\colon X \to UA$ on the basis X of $F(X)$; its bijectivity means that every mapping $f\colon X \to UA$ has a unique extension to a homomorphism $g\colon FX \to A$. The other way round, given the family (φ_{XA}), we define

$$\eta X = \varphi_{X,FX}(\mathrm{id}\,FX)\colon X \to UFX.$$

6.3.2 Main definitions

An *adjunction* $F \dashv G$, with a functor $F\colon \mathsf{C} \to \mathsf{D}$ *left adjoint* to a functor $G\colon \mathsf{D} \to \mathsf{C}$, can be equivalently presented in four main forms.

(i) We assign two functors $F\colon \mathsf{C} \to \mathsf{D}$ and $G\colon \mathsf{D} \to \mathsf{C}$ together with a family of bijections

$$\varphi_{XY}\colon \mathsf{D}(FX, Y) \to \mathsf{C}(X, GY) \qquad (X \text{ in } \mathsf{C}, Y \text{ in } \mathsf{D}),$$

which is natural in X, Y. More formally, the family (φ_{XY}) is a functorial isomorphism

$$\varphi\colon \mathsf{D}(F(-), =) \to \mathsf{C}(-, G(=))\colon \mathsf{C}^{\mathrm{op}} \times \mathsf{D} \to \mathsf{Set}.$$

(ii) We assign a functor $G\colon \mathsf{D} \to \mathsf{C}$ and, for every object X in C, a universal arrow (see 5.4.1)

$$(F_0X, \eta X\colon X \to GF_0X) \qquad \text{from the object } X \text{ to the functor } G.$$

(ii*) We assign a functor $F\colon \mathsf{C} \to \mathsf{D}$ and, for every object Y in D, a universal arrow

$$(G_0Y, \varepsilon Y\colon FG_0Y \to Y) \qquad \text{from the functor } F \text{ to the object } Y.$$

(iii) We assign two functors $F\colon \mathsf{C} \to \mathsf{D}$ and $G\colon \mathsf{D} \to \mathsf{C}$, together with two natural transformations

$$\eta\colon \mathrm{id}\,\mathsf{C} \to GF \quad \text{(the } unit\text{)}, \qquad \varepsilon\colon FG \to \mathrm{id}\,\mathsf{D} \quad \text{(the } counit\text{)},$$

which satisfy the *triangular identities*:

$$\varepsilon F.F\eta = \mathrm{id}\,F, \qquad\qquad G\varepsilon.\eta G = \mathrm{id}\,G \tag{6.26}$$

$$F \xrightarrow{F\eta} FGF \xrightarrow{\varepsilon F} F \qquad G \xrightarrow{\eta G} GFG \xrightarrow{G\varepsilon} G$$

(with the composites $\mathrm{id}\,F\colon F \to F$ and $\mathrm{id}\,G\colon G \to G$ drawn as the lower arrows)

A proof of the equivalence can be found in [M2], Section IV.1, Theorem 2, or in [G4], Theorem 3.1.5. Essentially:

- given (i) one defines

$$\eta X = \varphi_{X,FX}(\mathrm{id}\,FX)\colon X \to GFX,$$
$$\varepsilon Y = (\varphi_{GY,Y})^{-1}(\mathrm{id}\,GY)\colon FGY \to Y,$$

- given (ii) one defines $F(X) = F_0X$, the morphism $F(f\colon X \to X')$ by the universal property of ηX and the morphism $\varphi_{XY}(g\colon FX \to Y)$ as $Gg.\eta X\colon X \to GY$,

- given (iii) one defines the mapping $\varphi_{XY}\colon \mathsf{D}(F(X), Y) \to \mathsf{C}(X, G(Y))$ as above, and also a backward mapping $\psi_{XY}(f\colon X \to GY) = \varepsilon Y.Ff$; then one proves that these mappings are inverse to each other, by the triangular identities.

The previous forms have different features.

Form (i) is the classical definition of an adjunction, and is at the origin of the name, by analogy with adjoint maps of Hilbert spaces.

Form (ii) is used when we start from a given functor G and want to construct its left adjoint (possibly less easy to define). Form (ii*) is used in a dual way.

The 'algebraic' form (iii) is adequate to the formal theory of adjunctions, as exploited below. *In fact, it makes sense in an abstract 2-category.*

6.3.3 The structure of adjunctions

(a) As a consequence of the composition of universal arrows (in 5.4.1), two consecutive adjunctions

$$\begin{aligned} &F\colon \mathsf{C} \rightleftarrows \mathsf{D} : G, \quad \eta\colon 1 \to GF,\ \varepsilon\colon FG \to 1, \\ &H\colon \mathsf{D} \rightleftarrows \mathsf{E} : K, \quad \rho\colon 1 \to KH, \sigma\colon HK \to 1, \end{aligned} \tag{6.27}$$

give a composed adjunction from the first to the third category

$$\begin{aligned} HF\colon \mathsf{C} &\rightleftarrows \mathsf{E}: GK, \\ (G\rho F).\eta\colon 1 &\to GF \to GK.HF, \\ \sigma.(H\varepsilon K)\colon HF.GK &\to HK \to 1. \end{aligned} \tag{6.28}$$

Accordingly, an adjunction will be written as a (dot-marked) arrow, *conventionally directed as the left adjoint*:

$$(F, G, \eta, \varepsilon)\colon \mathsf{C} \dashrightarrow \mathsf{D}.$$

There is a category AdjCat, of small categories and adjunctions, with the previous composition and obvious identities.

We also write $(\eta, \varepsilon)\colon F \dashv G$, but one should not view this as an arrow from F to G in a category: adjunctions do not compose in this way.

(b) Applying duality on categories, functors and transformations (see Exercise 5.3.5(c)), an adjunction $(F, G, \eta, \varepsilon)\colon \mathsf{C} \dashrightarrow \mathsf{D}$ is transformed into an adjunction

$$(G^{\mathrm{op}}, F^{\mathrm{op}}, \varepsilon^{\mathrm{op}}, \eta^{\mathrm{op}})\colon \mathsf{D}^{\mathrm{op}} \dashrightarrow \mathsf{C}^{\mathrm{op}}, \tag{6.29}$$

with left adjoint $G^{\mathrm{op}}\colon \mathsf{D}^{\mathrm{op}} \to \mathsf{C}^{\mathrm{op}}$ and unit $\varepsilon^{\mathrm{op}}\colon \mathrm{id}\,\mathsf{D}^{\mathrm{op}} \to F^{\mathrm{op}}G^{\mathrm{op}}$.

(AdjCat has thus an involutive contravariant endofunctor $(-)^{\mathrm{op}}$ and is self-dual: isomorphic to its dual category.)

(c) Let I be a small category. A functor $F\colon \mathsf{C} \to \mathsf{D}$ has an obvious 'extension' to functor categories on I, composing a functor $X\colon \mathsf{I} \to \mathsf{C}$ with F

$$F^{\mathsf{I}}\colon \mathsf{C}^{\mathsf{I}} \to \mathsf{D}^{\mathsf{I}}, \qquad F^{\mathsf{I}}(X) = FX\colon \mathsf{I} \to \mathsf{D}. \tag{6.30}$$

A natural transformation $\varphi\colon F \to G\colon \mathsf{C} \to \mathsf{D}$ can be similarly extended

$$\varphi^{\mathsf{I}}\colon F^{\mathsf{I}} \to G^{\mathsf{I}}\colon \mathsf{C}^{\mathsf{I}} \to \mathsf{D}^{\mathsf{I}}, \qquad \varphi^{\mathsf{I}}(X) = \varphi X\colon FX \to GX\colon \mathsf{I} \to \mathsf{D}, \tag{6.31}$$

and these extensions preserve all compositions, of functors and natural transformations.

(d) (*Functor categories and adjunctions*) It follows that an adjunction $(\eta, \varepsilon)\colon F \dashv G$ has a canonical extension to an adjunction

$$(\eta^{\mathsf{I}}, \varepsilon^{\mathsf{I}})\colon F^{\mathsf{I}} \dashv G^{\mathsf{I}}\colon \mathsf{C}^{\mathsf{I}} \dashrightarrow \mathsf{D}^{\mathsf{I}}. \tag{6.32}$$

6.3.4 Exercises and complements

(a) An adjoint equivalence $(F, G, \eta, \varepsilon)$, defined in 5.3.6, amounts to an adjunction $(\eta, \varepsilon)\colon F \dashv G$ where the unit and counit are invertible, so that $F \dashv G \dashv F$.

(b) For a 'forgetful functor' $U\colon \mathsf{A} \to \mathsf{X}$ the existence of the left adjoint $F \dashv U$ means that every object X of X has a free object $(FX, \eta\colon X \to UFX)$ in A (as defined in 5.4.3).

The exercises of 5.4.2 and 5.4.3 give an explicit construction of the left adjoint to the following forgetful functors between varieties of algebras:

$$\begin{array}{ccc} \mathsf{Ab} \to \mathsf{Set}, & \mathsf{Mon} \to \mathsf{Set}, & \mathsf{Abm} \to \mathsf{Set}, \\ \mathsf{Gp} \to \mathsf{Set}, & R\,\mathsf{Mod} \to \mathsf{Set}, & \mathsf{Set}_\bullet \to \mathsf{Set}, \\ R\mathsf{CAlg} \to \mathsf{Set}, & \mathsf{CRng} \to \mathsf{Set}, & R\mathsf{Alg} \to \mathsf{Set}, \\ \mathsf{Rng} \to \mathsf{Set}, & R\mathsf{CAlg} \to \mathsf{Abm}, & R\mathsf{Alg} \to \mathsf{Mon}, \\ \mathsf{Cat} \to \mathsf{Gph}, & \mathsf{Abm} \to \mathsf{Mon}, & \mathsf{Ab} \to \mathsf{Gp}. \end{array} \qquad (6.33)$$

(Actually $\mathsf{Cat} \to \mathsf{Gph}$ is not of this type. But Set is a variety of algebras: see 1.2.9(d).) Many other similar left adjoints can be of interest.

In fact, every functor $U\colon \mathsf{A} \to \mathsf{X}$ from a variety of algebras to a 'weaker' variety, that forgets part of the structure and properties of the domain, has a left adjoint: a result already referred to in 5.4.3(f).

(c) The forgetful functor $U\colon \mathsf{Top} \to \mathsf{Set}$ has both adjoints $D \dashv U \dashv C$, as already remarked above.

(d) The forgetful functor $U\colon \mathsf{pOrd} \to \mathsf{Set}$ of the category of preordered sets has a chain of adjunctions $\pi_0 \dashv D \dashv U \dashv C$.

(e) The (non-faithful) functor $\mathrm{Ob}\colon \mathsf{Cat} \to \mathsf{Set}$ has a similar chain $\pi_0 \dashv D \dashv \mathrm{Ob} \dashv C$, that extends the previous one.

A category C is said to be *connected* if $\pi_0(\mathsf{C})$ is the singleton, a property that can be easily characterised.

Note that this definition – the usual one – excludes the empty category. Every category is the sum of its connected components (i.e. its maximal connected subcategories).

(f) The embedding $U\colon \mathsf{pOrd} \to \mathsf{Cat}$ has a left adjoint.

(g) An important property: a left adjoint preserves (the existing) colimits; dually, a right adjoint preserves limits.

(h) Prove that the chain $D \dashv U \dashv C$ in (c) cannot be extended.

(i) Prove that all the Set-valued functors in the list (6.33) have no right adjoint. On the other hand, the embeddings $\mathsf{Gp} \to \mathsf{Mon}$ and $\mathsf{Ab} \to \mathsf{Abm}$ (which only forget properties) also have a right adjoint.

(j) The embedding $\mathsf{Rng} \to \mathsf{Rng}'$ has a left adjoint, which universally adds a unit to a ring, and no right adjoint.

6.3.5 Main properties of adjunctions

Some properties of adjunctions have already been examined in 6.3.3: composition, duality and extension to functor categories.

(a) (*Uniqueness and existence*) Given a functor, its left (or right) adjoint is uniquely determined, up to isomorphism: this follows from the uniqueness property of universal arrows, in 5.4.1.

An effective way of proving the existence, under suitable hypothesis, is the Adjoint Functor Theorem of P. Freyd: see [M2], Section V.6, Theorem 2, or [G4], Section 3.5.

(b) (*Preserving limits and colimits*) We have already seen that a left adjoint preserves colimits, while a right adjoint preserves limits (Exercise 6.3.4(g)).

We also know that an equivalence $F\colon \mathsf{C} \to \mathsf{D}$ can always be completed to an adjoint equivalence, with $G \dashv F \dashv G$; therefore F preserves limits and colimits.

(c) (*Limit functor as a right adjoint*) The description of the limit of a functor $X\colon \mathsf{I} \to \mathsf{C}$ as a universal arrow $(LX, \varepsilon_X\colon \Delta(LX) \to X)$ from the diagonal functor $\Delta\colon \mathsf{C} \to \mathsf{C}^{\mathsf{I}}$ to the object X of C^{I} (in 6.2.2) shows that the existence of all I-limits in C amounts to the existence of a right adjoint to Δ, the *limit functor* $L\colon \mathsf{C}^{\mathsf{I}} \to \mathsf{C}$, with counit $\varepsilon\colon \Delta(LX) \to \mathrm{id}\,\mathsf{C}$.

Dually, a *colimit functor* $L'\colon \mathsf{C}^{\mathsf{I}} \to \mathsf{C}$ is left adjoint to Δ, and the unit $\eta\colon \mathrm{id}\,\mathsf{C} \to \Delta L'$ gives the universal cocone $\eta_X\colon X \to \Delta(L'X)$ of X.

*(d) (*Limits in functor categories*) We can now give a ‘synthetic’ proof of a result already stated in 6.2.5(e): if C has all I-limits, the same holds in any functor category C^{S}, with limits computed pointwise on S. The same applies to colimits.

In fact, we are assuming that the diagonal functor $\Delta\colon \mathsf{C} \to \mathsf{C}^{\mathsf{I}}$ has a (say) right adjoint $L\colon \mathsf{C}^{\mathsf{I}} \to \mathsf{C}$, with counit $\varepsilon\colon \Delta L \to 1$. By 6.3.3(d), the extension $\Delta^{\mathsf{S}}\colon \mathsf{C}^{\mathsf{S}} \to (\mathsf{C}^{\mathsf{I}})^{\mathsf{S}}$ has right adjoint $L^{\mathsf{S}}\colon (\mathsf{C}^{\mathsf{I}})^{\mathsf{S}} \to \mathsf{C}^{\mathsf{S}}$, with counit ε^{S}, and the canonical isomorphisms

$$(\mathsf{C}^{\mathsf{I}})^{\mathsf{S}} = \mathsf{C}^{I\times S} = (\mathsf{C}^{\mathsf{S}})^{\mathsf{I}}$$

show that L^{S} is right adjoint to the diagonal $\mathsf{C}^{\mathsf{S}} \to (\mathsf{C}^{\mathsf{S}})^{\mathsf{I}}$, with counit ε^{S}.

(e) (*Faithful and full adjoints*) Suppose we have an adjunction $(\eta, \varepsilon)\colon F \dashv G$. As proved in [M2], Section IV.3, Theorem 1, or [G4], Theorem 3.2.5:

(i) G is faithful if and only if all the components εY are epi,
(ii) G is full if and only if all the components εY are split mono,
(iii) G is full and faithful if and only if the counit ε is invertible,

(i*) F is faithful if and only if all the components ηX are mono,
(ii*) F is full if and only if all the components ηX are split epi,
(iii*) F is full and faithful if and only if the unit η is invertible.

6.3.6 Reflective and coreflective subcategories

A subcategory $\mathsf{D} \subset \mathsf{C}$ is said to be *reflective* if the inclusion functor $U\colon \mathsf{D} \to \mathsf{C}$ has a left adjoint, and *coreflective* if U has a right adjoint.

For a *full reflective subcategory* the counit ε is invertible (by 6.3.5(e)). One can always choose the reflector $F \dashv U$ so that $FU = \mathrm{id}\,\mathsf{D}$ and $\varepsilon = \mathrm{id}$.

(In fact, one can constrain the unit $\eta\colon 1 \to UF$ so that $\eta U = \mathrm{id}$, and construct the reflector F from the constrained unit. Then $U\varepsilon = \mathrm{id}$ and $\varepsilon = \mathrm{id}$.)

Examples. (a) Ab is reflective in Gp: the reflector $(-)^{\mathrm{ab}}\colon \mathsf{Gp} \to \mathsf{Ab}$ sends a group G to the abelianised group $G^{\mathrm{ab}} = G/[G, G]$.

The unit $\eta G\colon G \to G^{\mathrm{ab}}$ is given by the canonical projection, and the counit $\varepsilon A\colon A \to A/[A, A]$ is invertible. On the other hand, Ab is not coreflective in Gp, because the embedding $\mathsf{Ab} \to \mathsf{Gp}$ does not preserve binary sums.

(b) Ord is reflective in pOrd, with unit the canonical projection $X \to X/{\sim}$ of 5.4.3(e). It is not coreflective, as the embedding $\mathsf{Ord} \to \mathsf{pOrd}$ does not preserve coequalisers.

(c) In Ab the full subcategory tAb formed by all torsion abelian groups (see 1.3.9(h)) is coreflective: the counit $\varepsilon A\colon tA \to A$ is the embedding of the torsion subgroup of an abelian group.

The full subcategory tfAb formed by all torsion-free abelian groups is reflective in Ab, with unit the canonical projection $\eta A\colon A \to A/tA$.

6.3.7 Comma categories and slice categories

This important construction is due to F. William Lawvere.

For functors $F\colon \mathsf{X} \to \mathsf{Z}$ and $G\colon \mathsf{Y} \to \mathsf{Z}$ with the same codomain Z one constructs a *comma category* $F{\downarrow}G$, equipped with functors P, Q and a natural transformation $\pi\colon FP \to GQ\colon (F{\downarrow}G) \to \mathsf{Z}$

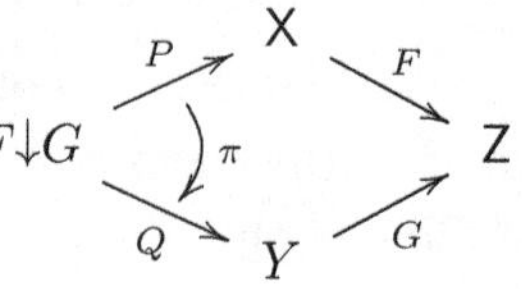

$$\begin{array}{ccccc} & & \mathsf{X} & & \\ & {\scriptstyle P}\nearrow & & \searrow{\scriptstyle F} & \\ F{\downarrow}G & & \Downarrow\pi & & \mathsf{Z} \\ & {\scriptstyle Q}\searrow & & \nearrow{\scriptstyle G} & \\ & & \mathsf{Y} & & \end{array} \tag{6.34}$$

(The original notation was (F, G), whence the name.) The objects of $F{\downarrow}G$ are the triples

$$(X, Y, z\colon FX \to GY)$$

formed of an object of X, an object of Y and a morphism of Z. A morphism

$$(f, g)\colon (X, Y, z\colon FX \to GY) \to (X', Y', z'\colon FX' \to GY'),$$

comes from a pair of maps $f\colon X \to X'$, $g\colon Y \to Y'$ that form a commutative square in Z

$$\begin{array}{ccc} FX & \xrightarrow{Ff} & FX' \\ {\scriptstyle z}\downarrow & & \downarrow{\scriptstyle z'} \\ GY & \xrightarrow[Gg]{} & GY' \end{array} \qquad z'.Ff = Gg.z. \tag{6.35}$$

Composition and identities come from those of $\mathsf{X}{\times}\mathsf{Y}$

$$(f', g').(f, g) = (f'f, g'g), \qquad \mathrm{id}\,(X, Y, z) = (\mathrm{id}\,X, \mathrm{id}\,Y).$$

There is an obvious universal property, that makes the triple (P, Q, π) a sort of *directed 2-dimensional pullback* of categories: for every category C equipped with similar data

$$P'\colon \mathsf{C} \to \mathsf{X}, \qquad Q'\colon \mathsf{C} \to \mathsf{Y}, \qquad \pi'\colon FP' \to GQ',$$

there is precisely one functor $W\colon \mathsf{C} \to (F{\downarrow}G)$ which commutes with the structural data:

$$\begin{array}{c} W(C) = (P'C, Q'C, \pi'C), \qquad W(c\colon C \to C') = (P'c, Q'c), \\ PW = P', \qquad QW = Q', \qquad \pi W = \pi'. \end{array} \tag{6.36}$$

As a matter of notation, one writes: $F{\downarrow}\mathrm{id}\,\mathsf{Z}$ as $F{\downarrow}\mathsf{Z}$, and $\mathrm{id}\,\mathsf{Z}{\downarrow}G$ as $\mathsf{Z}{\downarrow}G$.

Moreover on object Z_0 of Z can be viewed as a functor $Z_0\colon \mathbf{1} \to \mathsf{Z}$; therefore the comma category $F{\downarrow}Z_0$ has objects $(X, z\colon FX \to Z_0)$, while $Z_0{\downarrow}G$ has objects $(Y, z\colon Z_0 \to GY)$.

In particular we have the *slice categories*

$$Z_0{\downarrow}\mathsf{Z} = \mathsf{Z}\backslash Z_0, \qquad \mathsf{Z}{\downarrow}Z_0 = \mathsf{Z}/Z_0, \tag{6.37}$$

of *objects* $(Z, z\colon Z_0 \to Z)$ *below* Z_0 and *objects* $(Z, z\colon Z \to Z_0)$ *above* Z_0, respectively.

The categories $\mathsf{Set}_\bullet$ and $\mathsf{Top}_\bullet$ can be identified with the slice categories $\mathsf{Set}\backslash\{*\}$ and $\mathsf{Top}\backslash\{*\}$, of objects under the terminal object. Rings and unital rings present a dually similar situation, see Exercise 6.3.9(e).

6.3.8 Comma categories, adjunctions and limits

(a) The *graph* of an adjunction $(F, G, \eta, \varepsilon)\colon \mathsf{X} \nrightarrow \mathsf{Y}$ will be the comma category

$$\mathsf{G}(F, G) = F{\downarrow}\mathsf{Y}, \tag{6.38}$$

with objects $(X, Y, c\colon FX \to Y)$ and morphisms

$$(f, g)\colon (X, Y, c\colon FX \to Y) \to (X', Y', c'\colon FX' \to Y'), \qquad c'.Ff = g.c.$$

The adjunction $F \dashv G$ has a *graph factorisation*, for the composition defined in 6.3.3(a):

$$\mathsf{X} \underset{G'}{\overset{F'}{\rightleftarrows}} \mathsf{G}(F, G) \underset{G''}{\overset{F''}{\rightleftarrows}} Y \tag{6.39}$$

$$F'(X) = (X, FX, 1_{FX}), \quad G'(X, Y, c) = X, \quad \eta' X = 1_X,$$
$$F''(X, Y, c) = Y, \quad G''(Y) = (GY, Y, \varepsilon Y\colon FG(Y) \to Y), \quad \varepsilon'' Y = 1_Y.$$

*For small categories, this forms a *natural weak factorisation system* in AdjCat (as defined in [GT]); the factorisation is *mono-epi*, because $G'F'$ and $F''G''$ are identities.*

One can replace $\mathsf{G}(F, G)$ with the comma category $\mathsf{G}'(F, G) = \mathsf{X}{\downarrow}G$, which is isomorphic to the former because of the adjunction.

(b) Limits and colimits in the category C can be viewed as terminal or initial objects in comma categories of the diagonal functor $\Delta\colon \mathsf{C} \to \mathsf{C}^{\mathsf{I}}$ (defined in 6.2.2).

In fact a cone $(A, (f_i\colon A \to X_i))$ of $X\colon \mathsf{I} \to \mathsf{C}$ is the same as an object of the comma category $\Delta{\downarrow}X$ and the limit of X is the same as the terminal object of $\Delta{\downarrow}X$. Dually a cocone (or the colimit) $(A, (f_i\colon X_i \to A))$ of X is an object (or the initial object) of the comma category $X{\downarrow}\Delta$.

6.3.9 Exercises and complements

(a) As we have seen in Exercise 6.2.1(e), the limit of a functor $X\colon \mathsf{I} \to \mathsf{Set}$ defined on a small category can be computed as the set of its cones with vertices at the singleton. This can now be quickly proved, using an adjunction.

(b) Study the adjoints of the diagonal functor $\Delta\colon \mathsf{C} \to \mathsf{C}^2$ of a category C into $\mathsf{C}\times\mathsf{C}$.

(c) In particular, for $\mathsf{C} = R\,\mathsf{Mod}$, we have a periodic chain of adjunctions

$$... \;\; B \dashv \Delta \dashv B \dashv \Delta \;\; ...,$$

where $B(X, Y) = X \oplus Y$ is the direct sum, or biproduct (see 6.1.9).

(d) Prove that, for a given $n \in \mathbb{N}$, the endofunctor $F(X) = X^n$ of $R\,\mathsf{Mod}$ is adjoint to itself.

Prove that, for any small set I, the endofunctor $G(Y) = Y^I$ of $R\,\mathsf{Mod}$ has a left adjoint; extend this fact to a category C, under suitable hypotheses.

(e) To better understand the relationship between the categories Rng, of unital rings, and Rng', it is good to note (and prove) that Rng' is equivalent to the slice category $\mathsf{Rng}/\mathbb{Z}$ of *copointed unital rings*, or unital rings over the initial object $\mathbb{Z}$. *Hints:* use Exercise 6.3.4(j).

6.4 Adjunctions between ordered sets

An adjunction between ordered sets is called a (covariant) Galois connection.

It is a particularly simple form of adjunction, because in Ord a natural transformation $h \to k\colon X \to Y$ between two functors (i.e. two monotone mappings) simply amounts to the relation $h \leqslant k$, and is determined by h and k.

6.4.1 Galois connections

Given a pair X, Y of ordered sets, a (covariant) Galois connection between them is an adjunction between the associated categories.

It can be presented in the following equivalent ways, corresponding to the general definition of an adjunction in 6.3.2. (See Exercise (a).)

(i) We assign two increasing mappings $f\colon X \to Y$ and $g\colon Y \to X$ such that:

$$f(x) \leqslant y \text{ in } Y \quad \Leftrightarrow \quad x \leqslant g(y) \text{ in } X \qquad (\text{for } x \in X \text{ and } y \in Y).$$

(ii) We assign an increasing mapping $g\colon Y \to X$ such that for every $x \in X$ there exists in Y:

$$f(x) = \min\{y \in Y \mid x \leqslant g(y)\}.$$

(ii*) We assign an increasing mapping $f\colon X \to Y$ such that for every $y \in Y$ there exists in X:

$$g(y) = \max\{x \in X \mid f(x) \leqslant y\}.$$

(iii) We assign two increasing mappings $f\colon X \to Y$, $g\colon Y \to X$ such that

$$\mathrm{id}\,X \leqslant gf \quad \text{(the unit)}, \qquad fg \leqslant \mathrm{id}\,Y \quad \text{(the counit)}. \tag{6.40}$$

By these formulas g strictly determines f (called its *left adjoint*) and f strictly determines g (its *right adjoint*). Also here, we write $f \dashv g$. The relations (6.40) imply that

$$f = fgf, \qquad g = gfg, \tag{6.41}$$

which will have peculiar consequences. (In a general adjunction $F \dashv G$ between categories, the functor F need not be isomorphic to FGF, as most left adjoints of the list (6.33) show.)

Of course an isomorphism of ordered sets is, at the same time, left and right adjoint to its inverse. More generally, an increasing mapping *may* have one or both adjoints, which can be viewed as 'best approximations' to an inverse, of different kinds.

Exercises. (a) Prove (directly) the equivalence of conditions (i)–(iii), above.

(b) The embedding of ordered sets $i\colon \mathbb{Z} \to \mathbb{R}$ has both adjoints: the subcategory $\mathbb{Z} \subset \mathbb{R}$ is reflective and coreflective. (The right adjoint is well known.)

(c) The embedding $\mathbb{Q} \to \mathbb{R}$ of the rational numbers has neither a left nor a right adjoint: the subcategory $\mathbb{Q} \subset \mathbb{R}$ is neither reflective nor coreflective.

6.4.2 Properties

Let us come back to a general Galois connection $f \dashv g$ between ordered sets X, Y.

The mapping f *preserves all the existing joins*, while g *preserves all the existing meets*. It is a consequence of 6.3.4(g), that is easy to verify directly.

In fact, if $x = \vee x_i$ in X then $f(x_i) \leqslant f(x)$ (for all indices i). Supposing that $f(x_i) \leqslant y$ in Y (for all i), it follows that $x_i \leqslant g(y)$ (for all i); but then $x \leqslant g(y)$ and $f(x) \leqslant y$.

From the relations $f = fgf$ and $g = gfg$ it follows that:

(a) $gf = \mathrm{id}$ $\Leftrightarrow$ f is injective $\Leftrightarrow$ f is a split mono $\Leftrightarrow$ g is surjective $\Leftrightarrow$ g is a split epi $\Leftrightarrow$ f reflects the order relation,

(a*) $fg = \mathrm{id}$ $\Leftrightarrow$ f is surjective $\Leftrightarrow$ f is a split epi $\Leftrightarrow$ g is injective $\Leftrightarrow$ g is a split mono $\Leftrightarrow$ g reflects the order relation.

Moreover the connection restricts to an isomorphism (of ordered sets) between the sets of *closed elements* of X and Y, defined as follows

$$\begin{aligned} \mathrm{cl}(X) = g(Y) = \{x \in X \mid x = gf(x)\}, \\ \mathrm{cl}(Y) = f(X) = \{y \in Y \mid y = fg(y)\}. \end{aligned} \tag{6.42}$$

Again, an adjunction $f \dashv g$ will be written as a dot-marked arrow $(f, g) \colon X \dashrightarrow Y$, conventionally directed as the left adjoint $f \colon X \to Y$. Such arrows have an obvious composition

$$(f', g').(f, g) = (f'f, gg') \qquad (\text{for } (f', g') \colon Y \dashrightarrow Z), \tag{6.43}$$

and form the category AdjOrd *of ordered sets and Galois connections*; it is a self-dual category (as in 6.3.3(b)).

Each hom-set AdjOrd(X, Y) is canonically ordered: for two adjunctions $(f, g), (f', g') \colon X \dashrightarrow Y$ we let $(f, g) \leqslant (f', g')$ if the following equivalent conditions hold

$$f \leqslant f', \qquad g' \leqslant g, \tag{6.44}$$

as $f \leqslant f'$ gives $g' \leqslant gfg' \leqslant gf'g' \leqslant g$.

The relationship between AdjOrd and the usual category Ord (on the same objects) can be made clearer by amalgamating them in a 'double category' with horizontal arrows in Ord, vertical arrows in AdjOrd, and suitable double cells (see [G4], Section 7.2).

6.4.3 Direct and inverse images of subsets

The transfer of subobjects along morphisms is an important feature, that we examine here in Set; its use in various frameworks related to Homological Algebra is dealt with in [G2, G3].

As we have seen in 1.2.2, every set X has an ordered set $\mathrm{Sub}X = \mathcal{P}X$ of subsets, which actually is a complete boolean algebra. A mapping $f \colon X \to Y$ gives two increasing mappings, of *direct* and *inverse image*

$$\begin{aligned} &f_* \colon \mathrm{Sub}X \to \mathrm{Sub}Y, \qquad f^* \colon \mathrm{Sub}Y \to \mathrm{Sub}Y, \\ &f_*(A) = f(A), \qquad f^*(B) = f^{-1}(B) \qquad (A \subset X,\ B \subset Y). \end{aligned} \tag{6.45}$$

These mappings form a Galois connection $f_* \dashv f^*$

$$A \subset f^* f_*(A), \qquad f_* f^*(B) = B \cap f(X) \subset B. \tag{6.46}$$

(In fact f^* has also a right adjoint, not used here; this would not be the case in Ab.)

All this defines a *transfer functor* for subobjects of Set

$$\text{Sub}\colon \mathsf{Set} \to \mathsf{AdjOrd}, \qquad \text{Sub}(f) = (f_*, f^*)\colon \text{Sub}X \twoheadrightarrow \text{Sub}Y, \qquad (6.47)$$

with values in the category of ordered sets and Galois connections defined above.

This functor 'includes' the *covariant functor of subsets* $\mathcal{P}\colon \mathsf{Set} \to \mathsf{Set}$, that takes the mapping $f\colon X \to Y$ to $f_*\colon \mathcal{P}X \to \mathcal{P}Y$, together with the contravariant functor of subsets $\mathcal{P}^*\colon \mathsf{Set} \dashrightarrow \mathsf{Set}$, that takes f to $f^*\colon \mathcal{P}Y \to \mathcal{P}X$ (already considered in (5.17)).

6.4.4 The graph of a Galois connection

As a particular case of 6.3.8, the *graph* of the adjunction $(f, g)\colon X \twoheadrightarrow Y$ will be the following ordered set

$$\begin{aligned} G(f,g) &= \{(x,y) \in X \times Y \mid f(x) \leqslant y\} \\ &= \{(x,y) \in X \times Y \mid x \leqslant g(y)\}, \end{aligned} \qquad (6.48)$$

with the order induced by the cartesian product $X \times Y$ (a categorical product *in* Ord, not in AdjOrd).

The given adjunction has a canonical factorisation in two adjunctions

$$X \underset{g'}{\overset{f'}{\rightleftarrows}} G(f,g) \underset{g''}{\overset{f''}{\rightleftarrows}} Y \qquad (6.49)$$

$$\begin{aligned} f'(x) &= (x, f(x)), & g'(x,y) &= x, \\ f''(x,y) &= y, & g''(y) &= (g(y), y). \end{aligned}$$

*Again, this is a *natural weak factorisation system* in AdjOrd, in the sense of [GT].*

6.5 Adjoints and categorical limits in Topology

We already know that the forgetful functor $U\colon \mathsf{Top} \to \mathsf{Set}$ has both adjoints $D \dashv U \dashv C$, where DX (resp. CX) is the set X with the discrete (resp. indiscrete) topology (see 6.3.1).

Here we examine various adjunctions of interest in Topology. Some of them are used to compute colimits. Many other similar facts can be found in the literature, under the heading of 'Categorical Topology'. Adjunctions related to homotopy theory are investigated in [G4], Section 5.2, at an elementary level.

6.5.1 Proposition (The reflector of Hausdorff spaces)

The full subcategory Hsd ⊂ Top *of Hausdorff spaces is reflective in* Top, *with a unit whose general component* $p_X \colon X \to X/R$ *is a projection on a quotient.*

The universal property says that X/R *is Hausdorff, and every continuous mapping* $f \colon X \to Y$ *with values in a Hausdorff space factorises as* $f = gp_X$, *for a unique continuous mapping* $g \colon X/R \to Y$.

Proof In the space X, the equivalence relation xRx' is defined as follows:

- for every $f \in \mathsf{Top}(X, Y)$ with values in a Hausdorff space, $f(x) = f(x')$.

Let $H(X) = X/R$ be the quotient space and $p \colon X \to X/R$ the projection.

Plainly, every map $f \colon X \to Y$, with values in a Hausdorff space, factorises as $gp \colon X \to X/R \to Y$, by a unique map g. It is thus sufficient to prove that the quotient space X/R is Hausdorff.

For two equivalence classes $[x] \neq [x']$ there exist some map $f \colon X \to Y$, with values in a Hausdorff space, such that $f(x) \neq f(x')$. If V, V' are disjoint open neighbourhoods of these two points in Y, their preimages U, U' are disjoint open neighbourhoods of x, x' in X.

Moreover the sets U, U' are saturated for the equivalence relation R_f associated to f, whence they also are for the (finer) equivalence relation R. It follows that $p(U)$ and $p(U')$ are disjoint open neighbourhoods of $[x], [x']$ in X/R. □

6.5.2 Exercises and complements (Reflectors for separation axioms)

(a) The 'constructive' character of the previous definition $H(X) = X/R$ is limited, because it may be difficult to determine the equivalence relation R of a given space.

However, let us consider – for a given space X – the equivalence relation R' generated by all pairs of points x, x' that do not have a pair of disjoint neighbourhoods in X.

Then $R' \subset R$. *If it happens that* the quotient X/R' is Hausdorff, these equivalence relations coincide and our goal is reached. If it is not the case, one can reiterate the same procedure on $X' = X/R'$; and so on.

(b) The functor $U \colon \mathsf{Hsd} \subset \mathsf{Top}$ creates all limits and all sums, which thus exist in Hsd. It does not create general colimits, but it is easy to prove that they exist: for a functor $\mathsf{I} \to \mathsf{Hsd}$, one can compute the colimit in Top and apply the left adjoint $H \colon \mathsf{Top} \to \mathsf{Hsd}$ (following the same pattern used for colimits of ordered sets, in 6.2.5(c)).

(c) These results can be easily adapted to many other separation axioms (see 3.6.8).

The construction is quite easy for axiom T_0.

- The full embedding $U: T_0\mathsf{Top} \subset \mathsf{Top}$ of T_0-spaces has a left adjoint $T_0: \mathsf{Top} \to T_0\mathsf{Top}$, where $T_0(X)$ is the quotient of X modulo the equivalence relation $\mathcal{N}(x) = \mathcal{N}(x')$, that identifies pairs of points that are not distinguished by the topology.

- This functor U creates all limits and all sums. To compute the colimit of a functor $\mathsf{I} \to T_0\,\mathsf{Top}$, one can compute the colimit in Top and then apply the left adjoint T_0.

6.5.3 Topological spaces and preorder relations

The relationship between topology and preorder studied in 3.2.7 gives an adjunction

$$A\colon \mathsf{pOrd} \rightleftarrows \mathsf{Top} : S, \qquad A \dashv S. \tag{6.50}$$

Let us recall that, for a space X, the preordered set $S(X)$ has the same underlying set and the specialisation preorder $x \prec y$, meaning that $x \in \overline{y}$ (or $\overline{x} \subset \overline{y}$, equivalently). A continuous mapping $f\colon X \to Y$ becomes a monotone mapping $f\colon S(X) \to S(Y)$.

The other way round, for a preordered set X, $A(X)$ is the same set equipped with the Alexandrov topology, whose closed subsets $C \subset X$ are the downward closed ones. A monotone mapping $f\colon X \to Y$ becomes a continuous mapping $f\colon A(X) \to A(Y)$.

For a preordered set X, the unit $\eta_X\colon X \to SA(X)$ is the identity (because the closure of $x \in X$ in the space $A(X)$ is $\downarrow\! x$). For a space X, the counit is the continuous mapping

$$\varepsilon_X\colon AS(X) \to X, \qquad |\varepsilon_X| = 1_X, \tag{6.51}$$

where $AS(X)$ has a finer topology than X: every closed subset C of X is downward closed for the specialisation preorder. The triangular identities are trivially satisfied.

Therefore A embeds pOrd in Top, as the full coreflective subcategory of spaces with Alexandrov topology.

6.5.4 Exercises and complements (Topology and order)

(a) The previous adjunction can be restricted to an adjunction

$$A\colon \mathsf{Ord} \rightleftarrows T_0\mathsf{Top} : S, \qquad A \dashv S, \tag{6.52}$$

between ordered sets and T_0-spaces. Now A embeds Ord in $T_0\mathsf{Top}$, as the full coreflective subcategory of T_0-spaces with Alexandrov topology.

(b) This adjunction and the previous one (in 6.5.3) can be inserted in a commutative square of adjunctions (i.e. a commutative square in AdjCat, leaving apart questions of size)

$$\begin{array}{ccc} \mathsf{Ord} & \underset{S}{\overset{A}{\rightleftarrows}} & T_0\mathsf{Top} \\ F \uparrow \downarrow U & & T_0 \uparrow \downarrow U \\ \mathsf{pOrd} & \underset{S}{\overset{A}{\rightleftarrows}} & \mathsf{Top} \end{array} \qquad (6.53)$$

where the right adjoints are dashed. The adjunction $F \dashv U$ between ordered and preordered sets can be found in 6.3.6(b), while the adjunction $T_0 \dashv U$ between T_0-spaces and spaces is in 6.5.2(c).

(c) Consider the specialisation order of a T_1-space.

(d) Describe the Alexandrov topology of the space $A(\mathbb{R}, \leqslant)$ associated to the ordered real line.

6.5.5 *Examples and complements* (Metric spaces)

Metric spaces are defined in 4.5.0, by axioms (M.0–3). We have considered the categories

- Mtr *of metric spaces and Lipschitz mappings,*
- Mtr_1 *of metric spaces and weak contractions,*

with obvious forgetful functors

$$\mathsf{Mtr}_1 \subset \mathsf{Mtr} \to \mathsf{Hsd} \subset \mathsf{Top}. \qquad (6.54)$$

(a) The full subcategory $\mathsf{CplMtr} \subset \mathsf{Mtr}$ of complete metric spaces is reflective: the unit is the embedding $\eta X \colon X \to \hat{X}$ constructed in the Completion Theorem 4.5.6.

The same holds for the full subcategory $\mathsf{CplMtr}_1 \subset \mathsf{Mtr}_1$.

(b) The classical notion of a metric has been generalised in various way. For a *pseudometric* (as in Bourbaki [Bou3]), the condition (M.0) is replaced by the following more general one and the separation axiom (M.3) is dropped (while (M.1–2) are kept as in 4.5.0)

(M.0′) $\quad d(x, y) \in [0, +\infty]$ and $d(x, x) = 0$.

The reader can see that the categories

- psMtr *of pseudometric spaces and Lipschitz mappings,*

- psMtr_1 *of pseudometric spaces and weak contractions,*

have better categorical properties than the classical ones: for instance, they have arbitrary products. (Of course we read here the Lipschitz condition letting $+\infty.L = +\infty$, for any real number $L > 0$.)

Removing also the symmetry axiom (M.2) gives a further extension, which is important in category theory [Lw] and Directed Algebraic Topology [G1].

6.5.6 *Frames, locales and pointless topology

Letting Lth be the category of lattices and their homomorphisms, there is a natural functor

$$\mathcal{O}\colon \mathsf{Top} \to \mathsf{Lth}^{\mathrm{op}}, \tag{6.55}$$

that assigns to a space X the lattice $\mathcal{O}(X)$ of its open subsets, and to a continuous mapping $f\colon X \to Y$ the preimage homomorphism $f^*\colon \mathcal{O}(Y) \to \mathcal{O}(X)$.

In fact, $\mathcal{O}(X)$ is a *frame*, i.e. a complete lattice where binary meets distribute over arbitrary joins

$$(\vee_i U_i) \wedge V = \vee_i (U_i \wedge V),$$

as these operations are unions and intersections in $\mathcal{P}X$. By the same argument, $f^*\colon \mathcal{O}(Y) \to \mathcal{O}(X)$ is a *homomorphism of frames*, which means that it preserves arbitrary joins and binary meets.

The *category of locales* Loc, defined as the formal opposite of the category of frames and their homomorphisms, can thus be viewed as a surrogate of Top, and the functor (6.55) can be reinterpreted as

$$\mathcal{O}\colon \mathsf{Top} \to \mathsf{Loc}. \tag{6.56}$$

Most topological notions can be studied in Loc. This is the domain of 'pointless topology', which is important in Topos Theory. The interested reader can see [Bo3].

6.6 *Complements

We end with some hints at abelian categories, monads and 'algebraic categories'.

6.6.1 Kernels and cokernels

The beginning of category theory, in the 1950's, was focused on the study of homological algebra in abelian categories, which provided an extension of the categories of modules with two crucial advantages: the theory is self-dual and includes all categories of sheaves of modules on a given ring [Bu, Gt].

The influential books by Barry Mitchell and Peter Freyd [Mit, Fr1] proved that every small abelian category has a full exact embedding in a category of modules (on an appropriate ring), so that many results are automatically extended from categories of modules to all abelian categories.

Here we briefly review, without proofs, the definition of kernels and cokernels, abelian categories and exact functors. Further information and references can be found in [M2] and [G4].

Let E be a pointed category. As already said in 6.1.9, this means that there is a zero object 0, that is both initial and terminal in E. Given two objects A, B in E, the composite $A \to 0 \to B$ is called the zero morphism from A to B, and written as $0_{AB}\colon A \to B$, or also as 0 when the context identifies it.

In the pointed category E, the *kernel* of a morphism $f\colon A \to B$ is defined as a limit, namely the equaliser of f and $0_{AB}\colon A \to B$. This means a (mono)morphism $\ker f\colon \operatorname{Ker} f \to A$ such that:

(i) $f.(\ker f) = 0$, and for every map h such that $fh = 0$ there is a unique morphism u such that

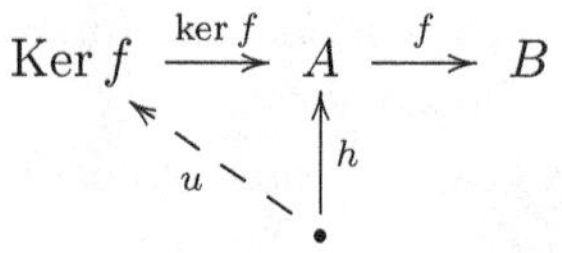

$$\begin{array}{ccccc} \operatorname{Ker} f & \xrightarrow{\ker f} & A & \xrightarrow{f} & B \\ & \nwarrow^{u} & \uparrow{\scriptstyle h} & & \\ & & \bullet & & \end{array} \qquad h = (\ker f)u. \tag{6.57}$$

A *normal monomorphism* is any kernel of a morphism, and is always a regular monomorphism. (The existence of kernels does not require the existence of all equalisers, as many examples in [G4] can show.)

Dually, the *cokernel* of f is the coequaliser of f and the zero morphism $A \to B$, namely the quotient $\operatorname{cok} f\colon B \to \operatorname{Cok} f$ such that

(i*) $(\operatorname{cok} f).f = 0$, and for every map h such that $hf = 0$ there is a unique morphism u such that

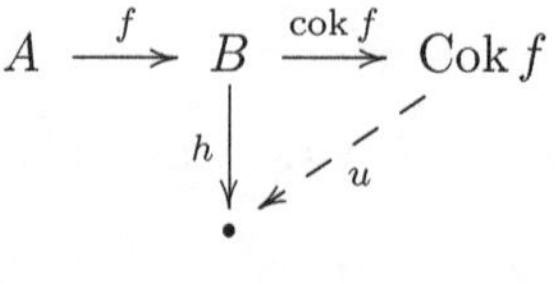

$$\begin{array}{ccccc} A & \xrightarrow{f} & B & \xrightarrow{\operatorname{cok} f} & \operatorname{Cok} f \\ & & {\scriptstyle h}\downarrow & \swarrow_{u} & \\ & & \bullet & & \end{array} \qquad h = u(\operatorname{cok} f). \tag{6.58}$$

A *normal epimorphism* is any cokernel of a morphism, and is always a regular epimorphism.

In $R\,\mathsf{Mod}$, $\ker f$ is the inclusion of the (usual) submodule $\mathrm{Ker}\, f$; every subobject $H \subset A$ is normal in A, as the kernel of the projection $A \to A/H$. Moreover $\mathrm{cok}\, f$ is the projection $B \to \mathrm{Cok}\, f = B/f(A)$; every quotient $B \to B/K$ is normal, as the cokernel of the inclusion $K \to B$.

(The interested reader can examine kernels and cokernels in other pointed categories, like Gp, Rng', $\mathsf{Set}_\bullet$, $\mathsf{Top}_\bullet$, where not all subobjects or not all quotients are normal.)

6.6.2 Abelian categories

An *abelian category* A is a category satisfying the following three (redundant) self-dual axioms:

(ab.1) A is pointed and every morphism has a kernel and a cokernel,

(ab.2) in the *canonical factorisation* of a morphism $f\colon A \to B$ through its *coimage* and its *image*

$$\begin{array}{ccccccc} \mathrm{Ker}\, f & \xrightarrow{\ker f} & A & \xrightarrow{f} & B & \xrightarrow{\mathrm{cok}\, f} & \mathrm{Cok}\, f \\ & & {\scriptstyle q}\downarrow & & \uparrow{\scriptstyle n} & & \\ & & \mathrm{Coim}\, f & \xrightarrow[g]{} & \mathrm{Im}\, f & & \end{array} \tag{6.59}$$

$$\mathrm{Coim}\, f = \mathrm{Cok}\,(\ker f), \qquad \mathrm{Im}\, f = \mathrm{Ker}\,(\mathrm{cok}\, f),$$

$$q = \mathrm{coim}\, f = \mathrm{cok}\,(\ker f), \qquad n = \mathrm{im}\, f = \ker\,(\mathrm{cok}\, f),$$

the unique morphism g such that $f = ngq$ is an *isomorphism*.

(ab.3) A has finite products and finite sums.

An abelian category has finite limits and colimits, and a (unique) additive structure [M2]; finite products and finite sums are biproducts; every subobject is normal and every quotient is normal.

Every category $R\,\mathsf{Mod}$, and every category of sheaves of R-modules on a topological space, is abelian. More generally, this holds for sheaves with values in any abelian category [Bo3].

A functor between abelian categories is said to be *left exact* (resp. *right exact*) if it preserves kernels (resp. cokernels), in the usual sense of preserving limits or colimits. It is *exact* if it is left and right exact.

A right adjoint functor between abelian categories is (obviously) left exact. A left exact functor preserves all finite limits and the additive structure.

Let us note that the previous definition of image and coimage is appropriate in the presence of axiom (ab.2). In Gp this condition does not hold, and the subgroup Ker (cok f) is the 'normal image' of f, i.e. the least normal subobject of the codomain containing $f(A)$.

6.6.3 Monads

Monads and their algebras give a wide formalisation of the 'algebraic character' of a category over another, typically used over the category of sets but also of interest in many other cases.

A classical exposition can be found in Mac Lane [M2], Chapter VI. Monads are also called 'triples' by other authors [Bc, Du], or 'dual standard constructions'.

A *monad* on the category X is a triple (T, η, μ) where $T\colon \mathsf{X} \to \mathsf{X}$ is an endofunctor, while $\eta\colon 1 \to T$ and $\mu\colon T^2 \to T$ are natural transformations (called the *unit* and *multiplication* of the monad) that make the following diagrams commute

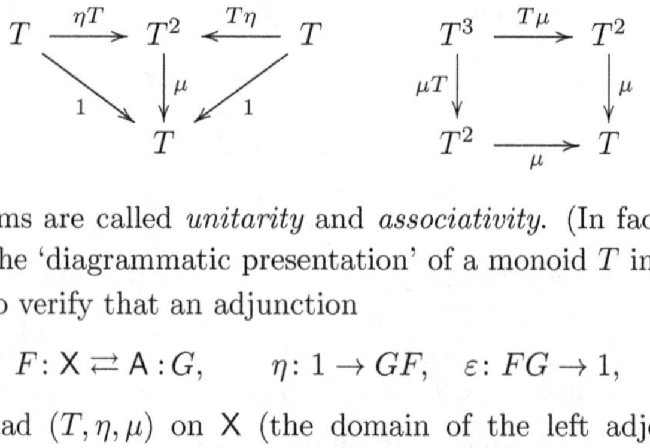

$$\begin{array}{ccccc} T & \xrightarrow{\eta T} & T^2 & \xleftarrow{T\eta} & T \\ & {\scriptstyle 1}\searrow & \downarrow{\scriptstyle \mu} & \swarrow{\scriptstyle 1} & \\ & & T & & \end{array} \qquad \begin{array}{ccc} T^3 & \xrightarrow{T\mu} & T^2 \\ {\scriptstyle \mu T}\downarrow & & \downarrow{\scriptstyle \mu} \\ T^2 & \xrightarrow[\mu]{} & T \end{array} \qquad (6.60)$$

These axioms are called *unitarity* and *associativity*. (In fact they are a rewriting of the 'diagrammatic presentation' of a monoid T in Set.)

It is easy to verify that an adjunction

$$F\colon \mathsf{X} \rightleftarrows \mathsf{A} : G, \qquad \eta\colon 1 \to GF, \quad \varepsilon\colon FG \to 1, \qquad (6.61)$$

yields a monad (T, η, μ) on X (the domain of the left adjoint), where $T = GF\colon \mathsf{X} \to \mathsf{X}$, $\eta\colon 1 \to T$ is the unit of the adjunction and $\mu = G\varepsilon F\colon GF.GF \to GF$.

The other way round, given a monad (T, η, μ) on the category X one defines the category X^T of *T-algebras*, or *Eilenberg–Moore algebras* for T: these are pairs $(X, a\colon TX \to X)$ consisting of an object X of X and a map a (called the *algebraic structure*), satisfying two coherence axioms:

$$a.\eta X = 1_X, \qquad\qquad a.Ta = a.\mu X. \qquad (6.62)$$

Now a functor $G\colon \mathsf{A} \to \mathsf{X}$ is said to be *monadic* (or *algebraic*), or to make A *monadic over* X, if it has a left adjoint $F\colon \mathsf{X} \to \mathsf{A}$ and a certain comparison functor $K\colon \mathsf{A} \to \mathsf{X}^T$ is an *isomorphism* of categories.

There are various 'monadicity theorems' (also called 'tripleability theorems') that give sufficient (or necessary and sufficient) conditions for a functor to be monadic: the interested reader can see [M2], Section VI.8, or [Bo2], Section 4.4.

6.6.4 Examples and comments

(a) It is easy to present Set$_\bullet$ or Mon as a category of T-algebras over Set, starting from their forgetful functor and its left adjoint.

(b) More generally, one can prove that the forgetful functor $U \colon \mathsf{A} \to \mathsf{Set}$ of any variety of algebras is monadic (see [M2], Section VI.8).

(c) But the present *formalisation of the algebraic character of a category* is much wider: for instance the category of compact Hausdorff spaces is monadic over Set, as we recall below; this monadicity depends on the closure operator of such spaces, a sort of 'infinitary operation'.

(d) We are following Mac Lane ([M2], Section VI.3) in defining monadicity in a strong sense, *up to isomorphism of categories*, because concrete examples, based on structured sets, generally fall in this case.

On the other hand [Bo2], in Section 4.4, and other authors only ask that the comparison K be an *equivalence* of categories.

The difference can be appreciated at the light of this example. The category Set$_\bullet$ of pointed sets is a variety of algebras, produced by a single zeroary operation (the base point) under no axioms, and is thus monadic over Set, via its forgetful functor U. The equivalent category $\mathcal{S}$ of sets and partial mappings (in 5.3.8) inherits a composed functor $\mathcal{S} \to \mathsf{Set}_\bullet \to \mathsf{Set}$, and the associated comparison is now an equivalence of categories. We prefer to view $\mathcal{S}$ as 'weakly algebraic' over Set, rather than as 'algebraic'.

(e) We also note that the free semigroup on the empty set is empty: the category of 'non-empty semigroups', considered in 5.2.8(d), lacks such a free object and is not monadic over Set.

6.6.5 Compact Hausdorff spaces as algebras

The full subcategory CmpHsd $\subset$ Top of *compact Hausdorff spaces* has peculiar properties among the categories of topological spaces.

To begin with, the category CmpHsd is balanced, as a consequence of Theorem 4.2.7(b): any bijective continuous mapping in this category is a homeomorphism. Moreover, an epimorphism has a dense image (by the same argument used for Hsd, in 5.5.4(b)), and must be surjective (by 4.2.7(a)).

We now review some important properties of CmpHsd, as dealt with in [M2], Chapter V.

The embedding $V\colon \mathrm{CmpHsd} \subset \mathsf{Top}$ has a left adjoint

$$\beta\colon \mathsf{Top} \to \mathrm{CmpHsd}, \tag{6.63}$$

called the *Stone–Čech compactification* (see [M2], Section V.8). The unit $\eta X\colon X \to V\beta(X)$ is a topological embedding if and only if X is a completely regular space. (Classically a 'compactification' was meant to be an embedding, and this procedure was only considered for completely regular spaces.)

The composed adjunction, where D equips a set with the discrete topology

$$\mathsf{Set} \underset{W}{\overset{D}{\rightleftarrows}} \mathsf{Top} \underset{V}{\overset{\beta}{\rightleftarrows}} \mathrm{CmpHsd} \qquad\qquad \beta D \dashv WV, \tag{6.64}$$

gives the left adjoint $\beta D\colon \mathsf{Set} \to \mathrm{CmpHsd}$ of the forgetful functor $U = WV\colon \mathrm{CmpHsd} \to \mathsf{Set}$.

The latter is monadic, so that a compact Hausdorff space can be viewed as an algebra $(X, a\colon TX \to X)$, where $T = U\beta D\colon \mathsf{Set} \to \mathsf{Set}$ is the monad associated to the adjunction (6.64) (see [M2], Section VI.9).

The existence of the left adjoint β can be proved by the Special Adjoint Functor Theorem (as in [M2]) or by some 'constructive' procedure which can be found in many books on General Topology (for instance in [Dg, Ke]).

Here we only sketch, without proofs, one of the most frequently used – based on the compact interval $[0,1]$ of the euclidean line.

We consider the set $C = \mathsf{Top}(X, [0,1])$ and the set $[0,1]^C = \mathsf{Set}(C, [0,1])$; the latter is given the product topology (of the euclidean interval), which is compact Hausdorff, by Tychonoff's theorem.

Then one takes the mapping of *evaluation*

$$\begin{gathered} \mathrm{ev}_X\colon X \to [0,1]^C, \\ \mathrm{ev}_X(x)\colon C \to [0,1], \qquad \mathrm{ev}_X(x)(f) = f(x), \end{gathered} \tag{6.65}$$

that is continuous, because all its components are:

$$p_f.\mathrm{ev}_X = f\colon X \to [0,1].$$

Now we let βX be the closure of the subset $\mathrm{ev}_X(X)$ in the space $[0,1]^C$, which is a compact Hausdorff space, and take the codomain-restriction of the mapping ev_X

$$\eta_X\colon X \to V\beta X, \qquad \eta_X(x) = \mathrm{ev}_X(x). \tag{6.66}$$

The crucial point is proving the universal property of this arrow, from the space X to the functor V.

The interval $[0, 1]$ works in this construction because it is a *cogenerator* of the category CmpHsd: this means that if A and B are compact Hausdorff spaces, and $f, g\colon A \to B$ are distinct maps, there is a map $h\colon B \to [0, 1]$ that distinguishes them: $hf \neq hg$.

7
Solutions and hints

Easy exercises and exercises marked with * are often left to the reader.

7.1 Exercises of Chapter 1

7.1.1 Exercises of Section 1.1

Solutions of 1.1.4. (a) In the ring $\mathbb{Z}$ the empty subset satisfies (trivially) the conditions (i), (iii), (iv), while the subset $\mathbb{N}$ of natural numbers satisfies (i), (ii), (iv).

(It is easy to see that one can replace the first three conditions with: R' is non-empty and stable in R under difference.)

(b) The subset $2\mathbb{Z}$ plainly satisfies the conditions (i)–(iv). More generally, the same is true of the subset $n\mathbb{Z}$ of integral multiples of any natural number n (i.e. the integers of the form nh, for $h \in \mathbb{Z}$)

$$nh + nk = n(h+k), \qquad 0 = n0,$$
$$-nh = n(-h), \qquad nh.nk = n(nhk).$$

This family contains the null subring $0\mathbb{Z} = \{0\}$ and the total subring $1\mathbb{Z} = \mathbb{Z}$. We will see, in Exercise 2.1.6(a), that each subring of $\mathbb{Z}$ is of the form $n\mathbb{Z}$, for a unique $n \geqslant 0$.

Finally, if R is a unital subring of $\mathbb{Z}$, it also contains any positive integer n, as the latter can be obtained as a sum $1 + 1 + ... + 1$ of n addends. (An interested reader can write a precise proof of this obvious fact, by induction.) Therefore R also contains the opposite $-n$, and all integers.

(c) Any unital subring of $\mathbb{Q}$ contains $\mathbb{Z}$, with the same argument as above. Therefore a subfield of $\mathbb{Q}$ contains all quotients h/k with $h \in \mathbb{Z}$ and $k \in \mathbb{Z}^*$, and is total. In the same way, any subfield of $\mathbb{R}$ contains $\mathbb{Q}$.

Solutions of 1.1.7. (a), (c) The verification of the axioms (VS.1–8) is straightforward, using the axioms of fields for $\mathbb{R}$, or K.

(d) An n-tuple $x = (x_1, ..., x_n) \in K^n$ can (and will often) be viewed as a mapping $x\colon \{1, ..., n\} \to K$, and we can identify K^n with the set $F(\{1, ..., n\}, K)$, consistently with the operations of sum and scalar multiplication in these two structures.

(e), (f) As in similar exercises of 1.1.3. For $0.x = \underline{0}$, we can observe that $0.x + x = (0 + 1).x = x = \underline{0} + x$ and apply the additive cancellation law.

Solutions of 1.1.9. (a) The relation $x < y$ satisfies the following properties, in $\mathbb{R}$:

(i) for every x, y, one and only one of the following relations holds: $x < y$, or $x = y$, or $y < x$,

(ii) for every x, y, z, if $x < y$ and $y < z$ then $x < z$,

(iii) for every x, y, z, if $x < y$ then $x + z < y + z$,

(iv) for every x, y, z, if $x < y$ and $0 < z$ then $xz < yz$.

It is easy to see that, in any field, these conditions imply (A.10–15) for the relation ($x < y$ or $x = y$), while (A.10–15) imply (i)–(iv) for the relation ($x \leqslant y$ and $x \neq y$).

(b) From $x \leqslant y$, adding $-x$ (at the right hand) we get $0 \leqslant y - x$; then, adding $-y$ (at the left) we get $-y \leqslant -x$. (This argument does not depend on the commutativity of the sum, provided that axiom (A.14) is written in a complete form.)

Now, if $x \leqslant y$ and $z \leqslant 0$, we can multiply by $-z \geqslant 0$ getting $-xz \leqslant -yz$, which gives $xz \geqslant yz$.

(c) For every x we have $x \geqslant 0$ or $x \leqslant 0$; multiplying by x we get $x^2 \geqslant 0$, in both cases. Since $0 \neq 1$, it follows that $0 < 1$.

(d) As to the first inequality in (1.13), either $x + y \geqslant 0$ or $x + y < 0$; accordingly, we apply the first or the second inequality below

$$|x + y| = x + y \leqslant |x| + |y|, \qquad |x + y| = -x - y \leqslant |x| + |y|.$$

The rest is obvious, or a straightforward consequence.

(e) The set of lower bounds of $A =]0, +\infty[$ is $]-\infty, 0]$; the greatest element of the latter is 0, which does not belong to A.

(f) The set $A = \{k \in \mathbb{Z} \mid k \leqslant x\}$ is non-empty and upper bounded, and we take $a = \sup A$. The number $a - 1$ is not an upper bound of A, and there is some $k \in A$ with $a - 1 < k \leqslant a$. But then $a < k + 1$, so that $k + 1 \notin A$ and k is the greatest element of A (and coincides with a).

7.1.2 Exercises of Section 1.2

Solutions of 1.2.1. (b) It is a well-known point of combinatorics (besides being a consequence of viewing Y^X as a cartesian power, see (1.30)).

A mapping $f\colon X \to Y$ is obtained by choosing, for every element $x \in X$, its value $f(x) \in Y$. Each of these choices is independent of the others, and has n possible outcomes; there are m choices to be made. Finally there are n^m mappings.

The remaining points are obvious.

Solutions of 1.2.2. (a) By straightforward verifications. Note that two disjoint subsets A, A' of X can have non-disjoint images in Y.

(c) One can easily verify that χ is bijective, or use the inverse mapping

$$\varphi\colon \mathrm{Map}(X, \{0,1\}) \to \mathcal{P}X, \qquad \varphi(f) = f^{-1}\{1\}. \tag{7.1}$$

(d) Using the binomial formula, we have

$$\sharp(\mathcal{P}X) = \Sigma_k \binom{n}{k} = (1+1)^n.$$

(e) X is in bijective correspondence with the set of its singletons.

Solutions of 1.2.3. (a) The inclusion mapping $j\colon X \to X'$ of the set $X = \bigcup_i A_i$ gives a bijection

$$A \to A', \qquad x \mapsto jx, \tag{7.2}$$

by composing a mapping $x\colon I \to X$ which belongs to A, with $j\colon X \to X'$.

(d) All the triangles (1.36) commute if and only if the mapping $f : A \to B$ is defined as $f(x,i) = f_i(x)$, for each $i \in I$ and $x \in A_i$.

As to essential uniqueness, if the pair $(A', (v_i)_{i\in I})$ is also a solution of our property, we have two (well determined) mappings

$$\begin{aligned} f\colon A \to A', &\qquad fu_i = v_i \quad (\text{for } i \in I), \\ g : A' \to A, &\qquad gv_i = u_i \quad (\text{for } i \in I), \end{aligned} \tag{7.3}$$

and they are inverse to each other. This comes out of the fact that $(gf)u_i = fv_i = u_i = (\mathrm{id}\, A)u_i$, for all indices i, so that $gf = \mathrm{id}\, A$; similarly, $fg = \mathrm{id}\, A'$.

Solutions of 1.2.4. (c) Given an equivalence relation R, its equivalence classes form a partition of X. Conversely, given a partition $(A_i)_{i\in I}$ of X, we define xRy if and only if there is an $i \in I$ for which $x, y \in A_i$.

(Note that the partition associated to the equivalence relation R cannot be written as $([x])_{x \in X}$, because its terms must not be repeated.)

(d) To prove the existence of E, it is sufficient to verify that the intersection in $X \times X$ of any family of equivalence relations is an equivalence relation, and define E as the intersection of all the equivalence relations of X that contain R.

More concretely, one can construct E in two steps. First we form a reflexive, symmetric relation $R_1 = R \cup \Delta_X \cup R'$, where Δ_X is the diagonal of $X \times X$ and $R' = \{(x, y) \in X \times X \mid (y, x) \in R\}$. Then we let E be the transitive extension of R_1, where $(x, y) \in E$ if and only if there is a finite sequence $x_0, x_1, ..., x_n \in X$ such that

$$x_0 = x, \quad x_n = y, \qquad x_{i-1} \, R_1 \, x_i \ \text{ for } i = 1, 2, ..., n. \tag{7.4}$$

Solutions of 1.2.5. (a) As to the uniqueness of i, for every $a \in A$ there exists some $x \in X$ such that $p(x) = a$, and we must have $i(a) = p'(x)$. As to its existence, the mapping i is well defined in this way: if we also have $x' \in X$ such that $p(x') = a$, it follows that

$$m'p'(x') = mp(x') = m(a) = mp(x) = m'p'(x),$$

and $p'(x') = p'(x)$, by the injectivity of m'. Moreover this mapping i does give the relations in (1.45).

Solutions of 1.2.6. (a) In fact this is true for $n = 0$. Assuming that it is true for some $n \in \mathbb{N}$, it also holds for $n' = n + 1$, as we have:

$$2s_{n'} = 2(s_n + n + 1) = n(n+1) + 2n + 2 = (n+1)(n+2) = n'(n'+1).$$

The reader will note that this proof is simple, but requires that we already know (or just guess) how to express the sum s_n in a compact formula. If we want to *find* such a formula, we can follow a longer procedure, in two steps, using the associative and commutative properties of the sum of integers.

- If n is odd, $s_n = 0 + ... + n$ has $n + 1$ summands (an even number), and we can reorganise it as the sum of $(n+1)/2$ binary sums, each of them equal to n

$$(0 + n) + (1 + (n-1)) + ... \ = n.(n+1)/2.$$

- If n is even, $s_n = 1 + ... + n$ has n summands (even again), and we can reorganise it as the sum of $n/2$ integers, all equal to $n + 1$

$$(1 + n) + (2 + (n-1)) + ... \ = (n+1).n/2.$$

Once the formula is found (or guessed), one can prefer to prove it by induction, which is simpler.

(b) The proof is similar to that of 'simple' induction: if there is some natural number that is not in A, we can let m be the least of them. Then $m > 0$, by (i), and $\{0, 1, ..., m-1\} \subset A$. Applying (ii′) we get $m \in A$, a contradiction.

In fact, our statement amounts to apply simple induction to the set

$$A' = \{n \in \mathbb{N}^* \mid \{0, 1, ..., n-1\} \subset A\}.$$

Complete induction is thus a reformulation of the simple one, of an easier application in some cases, like the following one.

(c) We proceed by complete induction. Let A be the set of natural numbers $n \geqslant 2$ which are products of prime numbers. Then $2 \in A$, and assuming that any number k with $2 \leqslant k < n$ is in A, we can prove that $n \in A$. In fact, if n is not a prime number, it can be expressed as a product $n = ab$ with $a, b \neq 1$; then $2 \leqslant a, b < n$, whence a, b are products of prime numbers, and n is also.

7.1.3 Exercises of Section 1.3

Solutions of 1.3.0. (e) As to associativity, one verifies that both sets

$$A \,\Delta\, (B \Delta C), \qquad (A \Delta B) \,\Delta\, C,$$

contain all $x \in X$ which belong to precisely one of A, B, C or to all of them.

Commutativity is obvious; the identity element is $\emptyset$ and the opposite of A is A itself.

(f) The operation $A \cap B$ is plainly associative and commutative, and admits the total subset X as unit. As to distributivity, writing $A \cap B \cap C = H$ for short, we have:

$$\begin{aligned}(A \Delta B) \cap C &= ((A \cup B) \setminus (A \cap B)) \cap C \\ &= ((A \cup B) \cap C) \setminus H = ((A \cap C) \cup (B \cap C)) \setminus H \\ &= ((A \cap C) \setminus H) \cup ((B \cap C) \setminus H) = (A \cap C) \,\Delta\, (B \cap C).\end{aligned}$$

The last points are obvious.

(h) One can adapt to abelian groups the presentation of rings as equational algebras, in 1.2.8, taking out all points involving the multiplication.

Solutions of 1.3.2. (a) Working in the abelian group $\mathbb{Z}$ of the integers:

- the subset $\{-1, 0, 1\}$ only satisfies (ii) and (iii),
- the empty subset only satisfies (i) and (iii),
- the subset $\mathbb{N}$ only satisfies (i) and (ii).

(b), (c) Straightforward verifications.

(d) If $\mathrm{Ker}\, f = \{0\}$ and $f(x) = f(y)$, we have $f(x - y) = f(x) - f(y) = 0$ and $x = y$; the converse is also obvious, as $f(0_A) = 0_B$.

(e) For $f, g \in \mathrm{Hom}(A, B)$, the pointwise sum $f + g$ is a homomorphism

$$\begin{aligned}(f+g)(x+y) &= f(x+y) + g(x+y) = f(x) + f(y) + g(x) + g(y)\\ &= f(x) + g(x) + f(y) + g(y) = (f+g)(x) + (f+g)(y).\end{aligned}$$

Note that *we have used the commutativity of* B: this construction fails for non-commutative groups. The rest of the proof is straightforward.

(f) Consider the mapping that evaluates every homomorphism at $1 \in \mathbb{Z}$

$$\varphi\colon \mathrm{Hom}(\mathbb{Z}, A) \to A, \qquad \varphi(f) = f(1), \tag{7.5}$$

and prove that it is a bijective homomorphism of abelian groups. (This essentially means that $\mathbb{Z}$ is the free abelian group generated by the element 1; see 1.6.1.)

(g) We already know that composition is an operation in $\mathrm{End}(A)$, that it is associative and has a unit, namely $\mathrm{id}\, A$. We are left with the distributive properties, of composition with respect to pointwise sum:

$$\begin{aligned}(g(f+f'))(x) &= g(f(x) + f'(x)) = g(f(x)) + g(f'(x)) = (gf + gf')(x),\\ ((g+g')f)(x) &= (g+g')(f(x)) = g(f(x)) + g'(f(x)) = (gf + g'f)(x).\end{aligned}$$

(In the monoid $\mathrm{End}(|A|)$ of all endomappings of the set $|A|$, the pointwise sum does not satisfy the first distributive law above, and we do not have a ring.)

Solutions of 1.3.5. (b) This can be verified directly. More quickly, it is sufficient to note that the subset $E \subset A \times A$ is the kernel of the composed homomorphism

$$p(p_1 - p_2)\colon A \times A \to A \to A/E, \qquad (x, y) \mapsto [x - y].$$

Solutions of 1.3.9. (a) Defining $x \equiv_H x'$ as in 1.88 we do get a congruence of abelian groups:

$x - x = 0 \in H$,
$x - x' \in H \Rightarrow x' - x \in H$,
if $x - x' \in H$ and $x' - x'' \in H$, then $x - x'' = x - x' + x' - x'' \in H$,
if $x - x' \in H$ and $y - y' \in H$, then $(x + y) - (x' + y') \in H$.

In the last point one rewrites $(x + y) - (x' + y')$ as $(x - x') + (y - y')$, using the commutativity of A. (Quotients of general groups will be dealt with in 1.5.8.)

(b) In fact, for every $x \in A$, we have a canonical bijection

$$\varphi \colon [0] \rightleftarrows [x] : \psi, \qquad \varphi(h) = x + h, \quad \psi(x') = x' - x.$$

The relation (1.89) follows from the fact that the cosets $[x] = x + H$ form a partition of the set A, and each of them has the same cardinal $\sharp H$.

(c) It is easy to see that the image $f(H)$ of a subgroup $H \subset A$ is a subgroup of B, while the pre-image $f^{-1}(K)$ of a subgroup $K \subset B$ is a subgroup of A.

(d) If A is a cyclic group, generated by its element x, consider the homomorphism

$$f \colon \mathbb{Z} \to A, \qquad f(k) = kx.$$

Then f is surjective, and induces an isomorphism $\mathbb{Z}/\mathrm{Ker}\, f \to A$. But we already know that each quotient of $\mathbb{Z}$ is of the form $\mathbb{Z}/n$, for $n \geqslant 0$.

(e) A quotient A/H of a cyclic group $A = \langle x \rangle$ is plainly generated by the class $[x]$ of the generator of A, as $[kx] = k[x]$ (for $k \in \mathbb{Z}$).

As to subgroups, we only need to consider the subgroups of $A = \mathbb{Z}/n$, for $n \geqslant 0$, by the previous exercise. We already know that any subgroup of $\mathbb{Z}$ is of the form $k\mathbb{Z}$, generated by some $k \geqslant 0$.

Consider now the canonical projection $p \colon \mathbb{Z} \to \mathbb{Z}/n$, for some $n > 0$, and a subgroup $K \subset \mathbb{Z}/n$. Its preimage $H = p^{-1}(K) \subset \mathbb{Z}$ is a cyclic subgroup, and $K = p(H)$ (because p is surjective); thus K is isomorphic to a quotient of H (by 1.3.7), and is cyclic as well.

(f) Let A be a finite group of prime order p, and $x \in A$. Then the subgroup $\langle x \rangle$ generated by the latter has an order which divides p, either 1 or p. It follows that A is generated by any non-zero element, whence cyclic and isomorphic to $\mathbb{Z}/p$.

(g) The abelian group $\mathbb{Z}/2 \oplus \mathbb{Z}/3$ is generated by $x = ([1], [1]')$, denoting by $[-]$ and $[-]'$ the equivalence classes modulo $2\mathbb{Z}$ and $3\mathbb{Z}$. In fact:

$$\begin{array}{lll} 0x = 0, & x = ([1], [1]'), & 2x = (0, [2]'), \\ 3x = ([1], 0), & 4x = (0, [1]'), & 5x = ([1], [2]'). \end{array}$$

The abelian group $\mathbb{Z}/2 \oplus \mathbb{Z}/2$ is not cyclic: in fact each element $x \neq 0$ has $2x = 0$, and generates a subgroup of order 2, namely $\{0, x\}$.

(h) The first part follows from point (d). Then tA is obviously a subgroup of A, and the inclusion $\varepsilon \colon tA \to A$ is universal in this sense: every homomorphism $f \colon T \to A$ defined on a *torsion abelian group* (with $tT = T$) factorises as $f = \varepsilon g$, for a unique homomorphism $g \colon T \to tA$.

*(i) A torsion element of $\mathbb{C}^*$ is a complex number z such that $z^n = 1$ for

some $n > 0$, i.e. a complex root of the unit (see Section 2.7). The group $t(\mathbb{C}^*)$ contains an isomorphic copy of the cyclic group $\mathbb{Z}/n$, for all $n > 0$.

7.1.4 Exercises of Section 1.4

Solutions of 1.4.2. (b) For the ordered line $\mathbb{R}$ we have the involution $r(x) = -x$. For the power set $\mathcal{P}X$, the involution $r(A) = X \setminus A$.

(c) The divisibility relation in the set $\mathbb{N}$ of natural numbers is plainly reflexive, transitive, anti-symmetric and non-total. To show that it is not self-dual, one can note that for any $a \in \mathbb{N}^*$, the set $\downarrow a$ of its divisors in $\mathbb{N}$ is finite, while the set $\uparrow a = a\mathbb{N}$ of its multiples is infinite. The rest is obvious.

(d) In the ring $\mathbb{Z}$ of integers the divisibility relation $h|k$ (defined by $k \in h\mathbb{Z}$) is a preorder. The associated equivalence relation can be written as $h = \pm\, k$, and the ordered quotient set $|\mathbb{Z}|/\!\sim$ is isomorphic to $\mathbb{N}_D$. There are two least elements, 1 and -1, and the atoms are all elements $\pm\, p$, where p is a positive prime number.

(e) The canonical factorisation $f = mgp$ in (1.43) gives a canonical projection $p\colon X \to X/R_f$ with values in a quotient preordered set, then a bijective monotone mapping $g\colon X/R_f \to \mathrm{Im}\, f$ and the inclusion of a preordered subset $m\colon \mathrm{Im}\, f \to Y$.

Note that the mapping $g\colon X/R_f \to \mathrm{Im}\, f$ need not be an isomorphism of preordered sets, as shown by the bijective monotone mapping

$$f\colon (\mathbb{R}, =) \to (\mathbb{R}, \leqslant)$$

of (1.46) (which essentially gives $g = f$).

(f) The preordered quotient on X/R has the chaotic preorder, as $[0] \prec [x] \prec [1] = [0]$ (for every x). The associated ordered set is the singleton.

Solutions of 1.4.3. It is sufficient to prove (d). If A is non-empty and lower bounded in $X = \mathbb{R}^n$, its projection A_i on the i-th axis is non-empty and lower bounded in $\mathbb{R}$; letting $x_i = \inf_{\mathbb{R}} A_i$, we get the point $(x_1, ..., x_n) = \inf_X A$.

Solutions of 1.4.5. (a) If the preordered set X has all infima, the supremum of any $A \subset X$ can be obtained as the infimum of the set of its upper bounds

$$\xi = \inf(U(A)) = \max(LU(A)), \tag{7.6}$$

because this maximum ξ is bigger than any element of $A \subset LU(A)$, and any $\alpha \in U(A)$ is bigger than any element of $LU(A)$, and bigger than ξ.

(b) The first claim is obvious. Every $A \subset X$ is the union of all singletons $\{x\}$, for $x \in A$. For a mapping $f\colon X \to Y$, the required preservation properties are easily checked. But one can easily have $A \cap B = \emptyset$ and $f(A) \cap f(B) \neq \emptyset$; moreover, $f(X) = Y$ only holds for a surjective mapping.

(c) The ordered subset $\mathrm{Sub}(A)$ is obviously stable in the power set $\mathcal{P}(|A|)$, under arbitrary intersections. By (a), it is a complete lattice and the join $H = \vee H_i$ of any family of subgroups is the intersection of all subgroups which contain all H_i.

This procedure applies to all equational algebraic structures: in any case we obtain a complete lattice of substructures, where arbitrary meets are set-theoretical intersections. A similar process was used to prove that any relation in a set generates an equivalence relation, in Exercise 1.2.4(d).

Here, extending what we have seen in 1.3.2(c), the subgroup H can be described as the set $\Sigma_i H_i$ of essentially finite sums $\Sigma_i h_i$ in A, where $h_i \in H_i$ (for all indices i).

(d) The relation $(m) \subset (n)$ is equivalent to $m \in n\mathbb{Z}$, which means that m is a multiple of n, or equivalently that n divides m. This proves that the bijective mapping

$$\varphi\colon \mathbb{N}_D \to \mathrm{Sub}(\mathbb{Z}), \qquad \varphi(n) = (n) = n\mathbb{Z}, \tag{7.7}$$

and its inverse ψ are order-reversing. Therefore the ordered set $\mathbb{N}_D$ is isomorphic to the complete lattice $(\mathrm{Sub}(\mathbb{Z}))^{\mathrm{op}}$, and formulas (1.102) hold true.

Plainly $6 = \mathrm{lcm}(2, 3)$ is a join of atoms (those that divide 6), while the only atom of $\mathbb{N}_D$ smaller than 4 is 2. (The reader can easily characterise the natural numbers which are joins of atoms.) In $\mathrm{Sub}(\mathbb{Z})$, for any subgroup $(n) \neq (0)$ there are strict inclusions $(0) \subset (n^2) \subset (n)$.

(e) For all positive integers m, n we can consider the mapping

$$f\colon \mathbb{Z}/mn \to \mathbb{Z}/m \oplus \mathbb{Z}/n, \qquad f([k]) = ([k]_m, [k]_n), \tag{7.8}$$

which is well defined, because $mn\mathbb{Z} \subset m\mathbb{Z} \cap n\mathbb{Z}$, and is a homomorphism.

If m, n are coprime, then f is injective, because of the previous point

$$([k]_m, [k]_n) = 0 \quad \Leftrightarrow \quad k \in (m) \cap (n) = (mn).$$

It is also surjective, because A and $\mathbb{Z}/mn$ have the same order mn, so that f is an isomorphism.

Conversely, if $A = \mathbb{Z}/m \oplus \mathbb{Z}/n$ is cyclic (of order mn) with generator $x \in A$, then the homomorphism $g\colon \mathbb{Z} \to A$ defined by $g(k) = kx$ is surjective, and induces an isomorphism $\mathbb{Z}/\mathrm{Ker}\, g \to A$; this proves that $\mathrm{Ker}\, g = (mn)$.

But any multiple of m and n annihilates x, whence

$$(mn) = \text{Ker}\, f \supset (m) \cap (n) = (m \vee_D n).$$

Therefore mn divides the least common multiple of m and n, and coincides with the latter, which means that m, n are coprime.

Solutions of 1.4.7. (a) Let X be a totally ordered set, and therefore a lattice. As x and y play the same role in axiom (D), we can suppose that $x \leqslant y$. Then $(x \wedge z) \leqslant (y \wedge z)$ and

$$(x \vee y) \wedge z = y \wedge z = (x \wedge z) \vee (y \wedge z).$$

(b) If x' and x'' are complements of x in X, we have

$$x'' = x'' \wedge 1 = x'' \wedge (x \vee x') = (x'' \wedge x) \vee (x'' \wedge x') = 0 \vee (x'' \wedge x') = x'' \wedge x',$$

whence $x'' \leqslant x'$, and symmetrically $x' \leqslant x''$.

As to the second point, if $x \leqslant y$ then $y^* \leqslant x^*$, as follows from

$$\begin{aligned} y^* &= y^* \wedge 1 = y^* \wedge (x \vee x^*) = (y^* \wedge x) \vee (y^* \wedge x^*) \\ &= (y^* \wedge x \wedge y) \vee (y^* \wedge x^*) = y^* \wedge x^*. \end{aligned}$$

The third point is a consequence.

(With some care, one can deduce the first point from a rewriting of the argument we have given for the second, supposing that x' and y' be any complements of x and y. We have followed a 'more prudent' organisation of the proof, hopefully clearer.)

(c) For distributive lattices we take the algebraic presentation of lattices, in 1.4.6(a), and add the distributivity axiom (D). For modular lattices we add the modularity axiom, in the form (M″).

For boolean algebras we start from the algebraic presentation of bounded lattices, in 1.4.6(b). Then we add a unary operation $(-)^* \colon X \to X$, with the distributivity axiom (D) and the axiom of complements (C).

(d) We know that, in the lattice Sub(A), we have $H \vee K = H + K$ and $H \wedge K = H \cap K$ (Exercise 1.3.2(c)). Now, for H, K, L in Sub(A), with $H \subset L$, we only have to verify condition (M′):

$$(H + K) \cap L \subset H + (K \cap L).$$

We take an element $x = h + k \in L$, with $h \in H$, $k \in K$; then $k = (h + k) - h \in L$ (because $H \subset L$), and x belongs to $H + (K \cap L)$.

(e) The following subgroups of $A = \mathbb{Z} \times \mathbb{Z}$ form a non-distributive triple

$$\begin{aligned} &H = \mathbb{Z} \times \{0\}, \qquad \Delta = \{(k, k) \mid k \in \mathbb{Z}\}, \qquad L = \{0\} \times \mathbb{Z}, \\ &(H + \Delta) \cap L = A \cap L = L, \qquad (H \cap L) + (\Delta \cap L) = \{0\}. \end{aligned} \tag{7.9}$$

One can also remark that H is a complement of Δ in A, in the sense that $H + \Delta = A$ and $H \cap \Delta = \{0\}$; but the same is true of L, whence our lattice cannot be distributive.

(f) To verify that f is a homomorphism is easy. Let p be an atom of X.

First, if $p \leqslant x \vee y$ then $p = (p \wedge x) \vee (p \wedge y)$; if $p \wedge x \neq 0$, the inequalities $0 < p \wedge x \leqslant p$ give $p \wedge x = p$ and $p \in f(x)$; otherwise $p \wedge y \neq 0$ and $p \in f(y)$; we conclude that $p \in f(x) \cup f(y)$. Conversely, if this is true, then $p \leqslant x$ or $p \leqslant y$ and $p \leqslant x \vee y$.

Second, the condition $p \leqslant x \wedge y$ is plainly equivalent to $p \in f(x) \cap f(y)$.

Third, if $p \leqslant x^*$ then $p \notin f(x)$. Conversely, if the atom $p \notin f(x)$, then $p \leqslant x^*$, because $p \wedge x = 0$ and $p = p \wedge (x \vee x^*) = p \wedge x^* \leqslant x^*$.

To prove that f is injective, we prove that every $x \in X$ is the join of its atoms: $x = \vee f(x)$ (without requiring that X be complete). In fact x is an upper bound of $f(x)$; if y is also, suppose that $z = x \wedge y < x$. Then $x \wedge z^* \neq 0$ has some atom $p \in f(x) \setminus f(z)$ which is not an atom of y, a contradiction. Thus $x = x \wedge y \leqslant y$.

We conclude remarking that $f(0) = \emptyset$ and $f(1) = A$.

If X is complete, every set of atoms $B \subset A$ can be obtained as $f(\vee B)$. Conversely, we know that every power set is a complete boolean algebra.

(g) A finite boolean algebra X is obviously complete. It is also atomic: if the element $x > 0$ is not an atom, there is some x' such that $0 < x' < x$; and so on. Since X is finite, the process ends at some atom smaller than x.

*(h) The subset $\mathcal{X}$ is obviously a bounded sublattice of $\mathcal{P}S$, stable under complements. Its atoms are the singletons.

We want now to prove that $\mathcal{X}$ is not a complete boolean algebra (in its own right). Take a subset $A \subset S$ which is neither finite nor cofinite: there must be some, as we can start from an injective mapping $f \colon \mathbb{N} \to S$ and take $A = f(2\mathbb{N})$, which is infinite and has an infinite complement. Now the family of singletons $(\{a\})_{a \in A}$ has no upper bound in $\mathcal{X}$: if B contains all of them, it must be cofinite in X and we can always take out of it some element of $S \setminus A$, obtaining a smaller upper bound of our family.

Solutions of 1.4.9. The existence of E can be proved as in 1.2.4(d), taking the intersection of all the congruences in X that contain R.

Constructivly, one can take $E = \bigcup R_n$, letting $R_0 = R$ and

$$R_{2n+1} = (R_{2n})', \qquad R_{2n+2} = (R_{2n+1})'' \qquad (n \geqslant 0), \tag{7.10}$$

where, for any relation $S \subset X \times X$

$$S' = \{(x \vee y, x' \vee y') \mid xSx',\, ySy'\} \cup \{(x \wedge y, x' \wedge y') \mid xSx',\, ySy'\},$$

and S'' is the equivalence relation generated by S.

7.1.5 Exercises of Section 1.5

Solutions of 1.5.1. (b) If X has at least two distinct elements a, b, we can consider the mapping $f\colon X \to X$ constant at a (i.e. $f(x) = a$, for $x \in X$) and the mapping g constant at b. Then $gf = g$ and $fg = f$.

(c) Associativity is obvious: for the first operation, both $x(yz)$ and $(xy)z$ give x; for the second, we always get z.

(d) Property (1.116) is proved by induction on $n \geqslant 1$. The initial step holds by definition, for every m. If (1.116) holds, then

$$x^m.x^{n+1} = x^m.x^n.x = x^{m+n}.x = x^{m+n+1}.$$

In (1.117), the first property is obvious. The second trivially holds for $n = 1$; if it holds for n, then:

$$(x^m)^{n+1} = (x^m)^n.x^m = x^{mn}.x^m = x^{mn+m} = x^{m(n+1)}.$$

In (1.118), both properties hold for $n = 1$. If the first holds for n (and every m), then

$$x^m.y^{n+1} = x^m.y^n.y = y^n.x^m.y = y^n.y.x^m = y^{n+1}.x^m.$$

Supposing that the second holds for n, we have:

$$x^{n+1}.y^{n+1} = x.(xy)^n.y = xy.(xy)^n = (xy)^{n+1}.$$

(f) For semigroups this is obvious: we have a binary operation, under the equational axiom of associativity. For monoids we add a zeroary operation $e\colon S^0 \to S$ with the axiom $ex = x = xe$ (for all x).

Solutions of 1.5.2. (c) We write $i(x) = x^{-1}$, and recall that $i(xy) = i(y).i(x)$.

The property $x^h.x^k = x^{h+k}$ is already known for positive exponents, and obvious if one of them is 0.

If h, k are both negatives, we have

$$x^h.x^k = i(x^{-h}).i(x^{-k}) = i(x^{-k}.x^{-h}) = i(x^{-k-h}) = x^{h+k}.$$

If $h < 0$ and $k > 0$, we distinguish two cases. If $h + k \geqslant 0$, we have

$$x^{-h}.(x^h.x^k) = x^k = x^{-h}.(x^{h+k})$$

(as $-h > 0$), and it is sufficient to cancel x^{-h}. If $h + k < 0$, we have

$$(x^h.x^k).x^{-k} = x^h = x^{h+k}.x^{-k}.$$

The symmetric case can be turned into the present one, applying the mapping i.

The remaining properties are similarly extended to integral exponents.

(d) Starting from the equational presentation of monoids (in 1.5.1(f)), we further add a unary operation $(-)'\colon S \to S$ and the axiom $x'x = e = xx'$ (for all x).

(e) For instance, we can take the injective homomorphism $f\colon \mathbb{Z} \to \mathbb{Q}^*$, $f(k) = 2^k$.

*(f) The exponential mapping $\exp\colon \mathbb{R} \to \mathbb{R}$ turns addition into multiplication:

$$\exp(x + y) = \exp(x).\exp(y),$$

and is injective. It gives an isomorphism $(\mathbb{R}, +) \to (\mathbb{R}^*_+, .)$, whose inverse is the natural logarithm. We can note that both mappings preserve the natural order, forming an isomorphism of totally ordered abelian groups.

Solutions of 1.5.5. (a) Plainly $\varphi_g(xy) = \varphi_g(x).\varphi_g(y)$, for all $g, x, y \in G$.

(b) If $g^{-1}Hg \subset H$ (for all $g \in G$), we also have $gHg^{-1} \subset H$ and $H = g^{-1}(gHg^{-1}) \subset g^{-1}Hg$. The other implications are similarly proved.

(d) The mapping

$$\varphi\colon G \to \mathrm{Sym}(X), \qquad \varphi(g)(x) = g.x, \tag{7.11}$$

sends an element g to the left multiplication by g, an invertible endomapping of X; φ is a homomorphism, because

$$(\varphi(g)\varphi(g'))(x) = g(g'x) = (gg')(x) = \varphi(gg')(x).$$

It is injective, because $\varphi(g) = \mathrm{id}\, X$ implies $ge = e$. Finally, φ gives an isomorphism from G to the subgroup $\varphi(G) \subset \mathrm{Sym}(X)$.

(e) Consider the mapping $\varphi\colon S \to \mathrm{End}(X)$ that sends an element $a \in S$ to the mapping $\varphi(a)\colon X \to X$, constant at a. Then $\varphi(a).\varphi(b) = \varphi(a) = \varphi(ab)$. As φ is obviously injective, it induces an isomorphism from S to its image $\varphi(S)$, which is the subsemigroup of $\mathrm{End}(X)$ of all constant mappings.

(f) The proof is straightforward, taking into account that an inner automorphism (see (1.125)) preserves powers (as any homomorphism).

(g) In fact, $\varphi_g[x, y] = [\varphi_g(x), \varphi_g(y)]$ is again a commutator.

Solutions of 1.5.7. (a) The existence of E is readily proved as in 1.2.4(d), taking the intersection of all the congruences in S that contain R. For a constructive procedure one can follow the same pattern as in 1.4.9(a), for congruences of lattices.

(b) In $(\mathbb{N}, +)$, the equivalence relation $m\, E\, n$ whose associated partition is $\{0\}$ and $\mathbb{N} \setminus \{0\}$ is a congruence of monoids, with $[0] = \{0\}$. The equality congruence $m = n$ has the same class $[0]$.

Solutions of 1.5.9. (a) If G has prime order p, its only subgroups are $\{1\}$ and G, by Lagrange's Theorem; therefore it is generated by any element $x \neq 1$ (if any), and isomorphic to $\mathbb{Z}/p$.

We are left with considering a group G of order 4. If G has a generator, than it is isomorphic to $\mathbb{Z}/4$. Otherwise, we write its elements as $1, x, y, z$. All elements have an order which divides 4, and cannot be 4; therefore x, y, z have order 2, and $x^2 = y^2 = z^2 = 1$.

Let us write the multiplication table of G. At the left we write what we already know, leaving the other results unprejudiced: xy, xz, etc.

$$\begin{array}{cccc} 1 & x & y & z \\ x & 1 & xy & xz \\ y & yx & 1 & yz \\ z & zx & zy & 1 \end{array} \qquad\qquad \begin{array}{cccc} 1 & x & y & z \\ x & 1 & z & y \\ y & z & 1 & x \\ z & y & x & 1 \end{array} \tag{7.12}$$

Now we observe that each row is obtained by left multiplication with a fixed element; it is thus a permutation of the set $|G|$, and contains all the four elements, without repetition. The same holds for columns, which are produced by right multiplication. Therefore xy cannot be 1, nor x, nor y, and must be z. Similarly we find $xz = y$, and all the other results are already determined, as shown in the right table above.

This table is symmetric with respect to the 'main diagonal', which means that G is commutative. It is actually isomorphic to the abelian group $\mathbb{Z}/2 \oplus \mathbb{Z}/2$ studied in Exercise 1.3.9(g), if not in a canonical way.

In fact, the group $G = \{1, x, y, z\}$, with the previous multiplication law, is a *better description* of the Klein four-group, because its three non-trivial elements are treated in the same way. (Let us note that any permutation of {x, y, z} gives an automorphism of G.)

7.1.6 Exercises of Section 1.6

Solutions of 1.6.1. (b) If the abelian group A is free with basis $(e_x)_{x\in X}$ and $a = \Sigma_{x\in X}\, \lambda_x e_x$ has $na = 0$ for some $n > 0$, we deduce that

$$\Sigma_{x\in X}\, (n\lambda_x)e_x = 0 = \Sigma_{x\in X}\, 0.e_x,$$

whence $n\lambda_x = 0$ and $\lambda_x = 0$, for all indices x.

Thus $a = 0$ is the only torsion element of A.

(c) For $x \in \mathbb{Q}$ and a positive integer n, the condition $nx = 0$ implies $x = 0$. (This is the case in any field of characteristic 0: see 2.2.1(a).)

$\mathbb{Q}$ cannot be a free abelian group, because any rational number $x = h/k \neq 0$ (with $h, k \in \mathbb{Z}^*$) generates a proper subgroup, and for any other element $y = h'/k' \neq 0$ we can find a linear combination $\lambda x + \mu y = 0$ with $\lambda, \mu \in \mathbb{Z}^*$, so that the pair x, y cannot be part of a basis. (E.g. one can take $\lambda = kh'$ and $\mu = -k'h$.)

Solutions of 1.6.6. (a) We already know that a finite cyclic group A of prime order p has no other subgroups than the trivial and the total one (Exercise 1.3.9(f)).

Conversely, let A be a simple abelian group. Then it is generated by any non-zero element, and it is finite, because $\mathbb{Z}$ has proper subgroups. If its order is a product $n = hk$, then $A \cong \mathbb{Z}/n$, whose subgroup generated by $[h]$ has order k; it follows that $k = 1$ or $k = n$, and n is a prime number.

Solutions of 1.6.7. (a) One can easily find two permutations of the set $\{1, 2, 3\}$ that do not commute. For $n > 3$ the inclusion $\{1, 2, 3\} \subset \{1, 2, ..., n\}$ gives an injective homomorphism $\underline{S}_3 \to \underline{S}_n$, where a permutation $f \in \underline{S}_3$ is extended by the identity on all higher integers.

(b) A non-trivial, proper subgroup of $\underline{S}_3$ has either 2 or 3 elements, and is thus generated by an element of order 2 or 3.

The transposition $f = (1, 2)$, which interchanges 1 and 2, is involutive (inverse to itself), and generates a subgroup $\langle f \rangle = \{\mathrm{id}\,, f\}$ which is easily seen to be non-invariant. (This also proves that our group is not commutative). The same, of course, holds for the transpositions $g = (1, 3)$ and $h = (2, 3)$.

On the other hand, the cyclic permutation $c = (1, 2, 3)$, which sends 1 to 2, 2 to 3 and 3 to 1, generates the subgroup $\langle c \rangle = \{\mathrm{id}\,, c, c^2\} = \langle c^2 \rangle$, which has order 3 and index 2: therefore it is invariant in $\underline{S}_3$ (as proved in 1.5.9).

We have thus covered the non-trivial elements of $\underline{S}_3$ (namely f, g, h, c, c^2), and all its subgroups. Only $\langle c \rangle$ is an invariant subgroup.

*(c) This proof is taken from [Bou1]. We use the number

$$w_n = \prod_{i<j} (j - i) \qquad (1 \leqslant i < j \leqslant n). \tag{7.13}$$

Now, for $f, g \in \underline{S}_n$, we prove that $\varepsilon(gf) = \varepsilon(g).\varepsilon(f)$, cancelling $w_n > 0$ in the second identity

$$w_n.\varepsilon(f) = \prod_{i<j} (f(j) - f(i)),$$

$$w_n.\varepsilon(gf) = \prod_{i<j} (gf(j) - gf(i)) = \varepsilon(f).\prod_{i<j} (g(j) - g(i))$$
$$= w_n.\varepsilon(g).\varepsilon(f).$$

Solutions of 1.6.8. *(a) Each $n \in \mathbb{N}^*$ has a unique decomposition $n = \prod_p p^{n_p}$ as a product of powers of prime numbers p; of course the natural exponents n_p are quasi null (i.e. all of them are 0, out of a finite number of prime indices p), so that the factorisation is essentially finite.

As a consequence, $\mathbb{N}_D$ can be embedded in the cartesian power $\prod_p \overline{\mathbb{N}}$ of the totally ordered set $\overline{\mathbb{N}} = \mathbb{N} \cup \{\infty\}$: we send any $n \in \mathbb{N}^*$ to the family $(n_p)_p$ indexed by all prime numbers, and $0 = \max \mathbb{N}_D$ to the constant family $(\infty)_p$.

Therefore the lattice $\mathbb{N}_D$ is distributive, and X as well.

As every cyclic group A is a quotient of $\mathbb{Z}$, the lattice Sub(A) is also distributive.

7.1.7 Exercises of Section 1.7

Solutions of 1.7.5. (a) We are assuming that $a^2/b^2 = 2$, where the positive integers a and b have no common factor except 1. Now $a^2 = 2b^2$, whence a^2 is even and a as well; then a^2 is a multiple of 4, so that b^2 is even, and b is also – a contradiction.

A similar argument works replacing 2 with any natural number which is not the square of a natural number.

(b) The interval $[0, 1]$ contains rational and irrational numbers: for instance, 0 and $\sqrt{2}/2$.

Now, any non-degenerate interval J has two elements $a < b$, and contains the interval $[a, b]$. Let $r = b - a > 0$, and take an integer $n > 2/r > 0$ (which exists, by Exercise 1.1.9(f)).

The interval $[na, nb]$ has length $n(b - a) = nr > 2$ and contains the interval $[k + 1, k + 2]$ where k is the integral part of na (in fact $na < k + 1$ and $k + 2 \leqslant na + 2 < nb$).

Translating $[0, 1]$ by adding the rational number $k + 1$, the interval $[k +$

$1, k+2]$ also contains rational and irrational numbers, and the interval $[na, nb]$ does the same; multiplying by the rational number $1/n$, also $[a, b]$ does, and J as well.

Considering the real numbers $1/n$ and $\sqrt{2}/2 - 1/n$ (for $n > 1$) we also prove that there are infinitely many rational and irrational numbers in $[0, 1]$, and therefore in J.

(c) The set $A = \{x \in \mathbb{Q} \mid x \leqslant a\}$ contains the integral part $[a]$ and is upper bounded by $[a] + 1$. Letting $s = \sup A \leqslant a$, the case $s < a$ is excluded, or the non-degenerate real interval $]s, a[$ would contain rational numbers between s and a.

7.2 Exercises of Chapter 2

7.2.1 Exercises of Section 2.1

Solutions of 2.1.1. (a) We begin with the left part of property (2.2), namely $(kx).y = k(x.y)$. If $k = 0$, this is obviously true. If this holds for some $k \geqslant 0$, then it also holds for $k+1$

$$((k+1)x).y = (kx+x).y = (kx).y + x.y = k(x.y) + x.y = (k+1)(x.y).$$

If $k = -n < 0$, our property holds as well:

$$(kx).y = -(nx).y = -n(x.y) = k(x.y).$$

The right-hand part, namely $k(x.y) = x.(ky)$, amounts to the previous one in the opposite ring R^{op}. Property (2.3) is a consequence.

(b) Let X be a boolean algebra; we write the complement of the element x as x^*.

We define the symmetric difference of two elements $x, y \in X$, extending what we have seen in a power set (Exercise 1.3.0(e))

$$x \Delta y = (x \wedge y^*) \vee (x^* \wedge y). \tag{7.14}$$

This operation is obviously commutative. It is associative, because (using distributivity and De Morgan laws, in 1.4.7(b))

$$\begin{aligned}(x\Delta y)\Delta z &= (((x \wedge y^*) \vee (x^* \wedge y)) \wedge z^*) \vee ((x^* \vee y) \wedge (x \vee y^*) \wedge z) \\ &= (x \wedge y^* \wedge z^*) \vee (x^* \wedge y \wedge z^*) \vee (((x^* \wedge y^*) \vee (y \wedge x)) \wedge z) \\ &= (x \wedge y^* \wedge z^*) \vee (x^* \wedge y \wedge z^*) \vee (x^* \wedge y^* \wedge z) \vee (x \wedge y \wedge z).\end{aligned}$$

In fact, we have obtained a symmetric result, invariant under any permutation of the variables x, y, z. Therefore $(x\Delta y)\Delta z = (z\Delta y)\Delta x$ coincides with $x\Delta(y\Delta z)$.

Moreover, $x \Delta 0 = x$ and every element x is inverse to itself: $x \Delta x = 0$. We have thus an abelian group (X, Δ).

For the meet operation, $(X, \wedge)$ is an idempotent commutative monoid, with unit 1_X. Moreover this operation distributes over the symmetric difference

$$\begin{aligned}(x \wedge z) \,\Delta\, (y \wedge z) &= (x \wedge z \wedge (y \wedge z)^*) \vee ((x \wedge z)^* \wedge y \wedge z) \\ &= (x \wedge z \wedge (y^* \vee z^*)) \vee ((x^* \vee z^*) \wedge y \wedge z) \\ &= (x \wedge z \wedge y^*) \vee (x^* \wedge y \wedge z) \\ &= ((x \wedge y^*) \vee (x^* \wedge y)) \wedge z = (x \,\Delta\, y) \wedge z,\end{aligned}$$

so that $(X, \Delta, \wedge)$ is an idempotent commutative unital ring.

Conversely, suppose we have such a ring R, with operations written in the usual way: $x + y$ and xy. The condition $xx = x$ implies that $2x = 0$ (for all x) because

$$x^2 + x^2 = x + x = (x + x)^2 = x^2 + 2xx + x^2 = x^2 + 2x + x^2.$$

Now, every element is opposite to itself. We already know that the idempotent multiplicative monoid R is a meet-semilattice, with $x \leqslant y$ when $xy = x$, and $x \wedge y = xy$. Binary joins exist, and are computed as:

$$x \vee y = x + y + xy.$$

In fact:

- $x(x + y + xy) = x + xy + xy = x, \qquad (x + y + xy)y = xy + y + xy = y,$
- if $x \leqslant z$ and $y \leqslant z$ then $(x + y + xy)z = xz + yz + xyz = x + y + xy$.

The lattice $(R, \leqslant)$ is distributive

$$(x \vee y) \wedge z = (x + y + xy)z = xz + yz + xz.yz = xz \vee yz.$$

Moreover the lattice is bounded, with $0 \vee x = x$ and $1 \wedge x = x$. It is actually a boolean algebra, with complement $x^* = 1 + x$, as

$$xx^* = x + x = 0, \qquad x \vee x^* = x + x^* + xx^* = x + 1 + x = 1.$$

Finally, the two procedures we have seen are inverse to each other. Starting from a boolean algebra X, we form the ring $(X, \Delta, \wedge)$, and we come back to the same ordered set, as we define $x \leqslant y$ precisely when $x \wedge y = x$.

Starting from an idempotent commutative unital ring $(R, +, .)$ we form the boolean algebra $(R, \leqslant)$, which gives a ring $(R, \Delta, \wedge)$ with the same original operations:

$$\begin{aligned}x \,\Delta\, y &= (x \wedge y^*) \vee (x^* \wedge y) = xy^* + x^*y + xy^*x^*y \\ &= x(1 + y) + (1 + x)y = x + xy + y + xy = x + y,\end{aligned}$$

$$x \wedge y = xy.$$

Solutions of 2.1.2. (a) The mapping f must be a homomorphism of the underlying abelian groups, of the form

$$f(k) = ka \qquad \text{(for } k \in \mathbb{Z}\text{)}, \tag{7.15}$$

where $a = f(1)$ is a well determined element of R. Using a property of multiples in R (Exercise 2.1.1(a)) we have

$$f(hk) = (hk).a, \qquad f(h).f(k) = (ha).(ka) = (hk).a^2. \tag{7.16}$$

We conclude that f is a ring homomorphism if and only if the element $a = f(1)$ is *idempotent* in R: $a^2 = a$.

(b) If R is unital, there is a unique unital homomorphism $f\colon \mathbb{Z} \to R$, namely $f(k) = k1_R$.

Solutions of 2.1.3. (a) If R is a field, each non-zero ideal contains some invertible element, and is total.

Conversely, let R be a non-trivial commutative ring; any element $x \neq 0$ generates a principal ideal $(x) \neq (0)$; if this is total, we conclude that $1 = xy$ for some $y \in R$, and x is invertible.

Solutions of 2.1.5. (a) Ideals are closed under arbitrary intersections, and we can apply Exercise 1.4.5(a).

(b) The first formula of (2.22) is already proved. For the second, the subgroup $I + J$ is plainly an ideal, and obviously the least ideal containing I and J. One can easily prove, in the same way, that an arbitrary join of ideals is a join of subgroups.

Solutions of 2.1.8. (a) Straightforward verifications.

(b) If R is a non-trivial ring and $a \in R^*$, the element $(a, 0)$ is a proper zero-divisor of $R \times R$.

(c) Property (2.26) is obvious and the rest is a consequence, by canonical factorisation of the anti-monotone mapping φ (which is monotone with values in $(\mathrm{Idl}(R))^{\mathrm{op}}$).

(d) Suppose now that R is an integral domain. If $(a) = (b) \neq (0)$, we deduce that $a = \lambda b$ and $b = \mu a$, whence $a = \lambda\mu a$ and, cancelling $a \neq 0$, we have that $\lambda\mu = 1$ and λ is invertible; otherwise $a = b = 0$ and $a = 1.b$.

Conversely, if $a = \lambda b$ with $\lambda \in \mathrm{Inv}(R)$, we have that $b \mid a$ and $a \mid b$ (in any commutative unital ring).

7.2.2 Exercises of Section 2.2

Solutions of 2.2.1. (a) (*Characteristic of fields*) For a field K, the smallest subring R_0 is an integral domain. Therefore the characteristic must be either a prime number p or zero.

In the first case the ring $R_0 \cong \mathbb{Z}/p$ is a field, whence a subfield of K and of course the smallest one.

In the second case, $R_0 \cong \mathbb{Z}$ is not a field, but we can extend the homomorphism $f\colon \mathbb{Z} \to K$, $f(k) = k1_K$ to the rational field, letting

$$g(k/n) = (k1_K).(n1_K)^{-1} \qquad (\text{for } k \in \mathbb{Z},\ n \in \mathbb{N}^*). \tag{7.17}$$

This is indeed a field homomorphism $g\colon \mathbb{Q} \to K$

$$\begin{aligned} g(h/m + k/n) &= g((hn + km)/(mn)) = ((hn + km)1_K).(mn1_K)^{-1} \\ &= (f(h).f(n) + f(k).f(m)).(f(m).f(n))^{-1} \\ &= f(h).f(m)^{-1} + f(k).f(n)^{-1} = g(h/m) + g(k/n), \end{aligned}$$

$$\begin{aligned} g((h/m).(k/n)) &= g((hk)/(mn)) = (hk1_K).(mn1_K)^{-1} \\ &= (f(h).f(k)).(f(m).f(n))^{-1} = (f(h).f(m)^{-1}).(f(k).f(n)^{-1}) \\ &= g(h/m).g(k/n). \end{aligned}$$

Since all field homomorphisms are injective, the image $K_0 = \operatorname{Im} g$ is a subfield of K isomorphic to $\mathbb{Q}$. It is the least subfield containing R_0, and therefore the least subfield of K.

(b) The set K of the real numbers of the form $x + y\sqrt{2}$, for $x, y \in \mathbb{Q}$, is easily seen to be a subfield of $\mathbb{R}$. The only non-obvious point is about inverses, which is settled remarking that

$$(x + y\sqrt{2}).(x - y\sqrt{2}) = x^2 - 2y^2,$$

and that the rational number $x^2 - 2y^2$ (with $x, y \in \mathbb{Q}$) only annihilates if $x = y = 0$ (or we would get $(x/y)^2 = 2$).

To show that K is a proper subfield (without using cardinals), it is sufficient to verify that $\sqrt{3} \notin K$. In fact the relation

$$3 = (x + y\sqrt{2})^2 = x^2 + 2y^2 + 2xy\sqrt{2} \quad (x, y \in \mathbb{Q})$$

would imply $xy = 0$, whence $y = 0$ or $x = 0$, and $3 = x^2$ or $6 = 4y^2 = (2y)^2$.

This cannot be, as we already know that $\sqrt{3}$ and $\sqrt{6}$ are both irrational.

Solutions of 2.2.3. (a) is obvious, as $x \in I$ is equivalent to $[x] = 0$ in R/I. (Note also that R/I cannot be the trivial ring, if I is a proper ideal.)

(b) We know, by (2.25), that the canonical projection $p\colon R \to R/I$ gives an isomorphism of ordered sets

$$p^*\colon \mathrm{Idl}(R/I) \longrightarrow \{J \in \mathrm{Idl}(R) \mid J \supset I\}.$$

The commutative unital ring R/I is a field if and only if it is non-trivial and has no ideals between (0) and the total one (by 2.1.3(a)), which is equivalent to saying that $I \neq R$ and there are no other ideals of R between them, or in other words that I is a maximal ideal.

(c) Obvious, because $a \mid x$ is equivalent to $x \in (a)$.

(d) If a is irreducible and $(a) \subset (b)$, then $a = bc$ and we deduce that the ideal (b) is either (a) or the ring.

Conversely, if this property holds and $a = bc$, then $(a) \subset (b)$ and we deduce that b is associated to a or to 1; in the second case, c is associated to a.

(e) If a is prime, then it is irreducible and (a) is maximal among proper (principal) ideals. Conversely, this implies that (a) is a prime ideal and a is a prime element.

(f) Consider the set X of all the proper ideals of R which contain I, ordered by inclusion.

Then $I \in X$. For a totally ordered family (J_i) of ideals belonging to X, their union J is an ideal of R: if $x, y \in J$ and $z \in R$, then x and y belong to two ideals of the family, therefore $x + y$ and zx belong to the biggest of them, and to J. Moreover the unit 1_R does not belong to J, or it would belong to some J_i.

There exists thus some maximal element in X; plainly, it is a maximal ideal of R, and contains I.

Solutions of 2.2.8. (a) Consider the mapping $f\colon \mathbb{Z} \to R$ defined by $f(k) = k1_R$, whose image is the minimal subring R_0 of R (as in 2.2.1). From $0_R < 1_R$ (by 1.1.9(c)), we have the relation $f(k+1) = f(k) + 1_R > f(k)$.

Therefore f is strictly increasing: if $h < k$ in $\mathbb{Z}$, then $f(h) < f(k)$ in R (as one can prove by induction on $k - h > 0$). Since $\mathbb{Z}$ is totally ordered, f is injective and R_0 is infinite.

(b) K has characteristic 0, and its minimal subfield Q is isomorphic to $\mathbb{Q}$. Any isomorphism $\mathbb{Q} \to Q$ must send 1 to 1_K, and coincide with (2.38).

(c) As we have seen in 1.7.5(c), every element a of K is a supremum in K of some subset $A_a \subset Q$, which is non-empty and upper bounded in Q.

Then $f(A_a)$ is non-empty and upper bounded in Q', and we can (and must) extend $f\colon Q \to Q'$ by letting $f(a) = \sup_{K'} f(A_a)$.

7.2.3 Exercises of Section 2.3

Solutions of 2.3.1. (a) The verification is easy and similar to what we have seen while studying rings.

(b) We already know that this holds for $k = 0$ and 1. If it holds for $k \geqslant 1$ then it also holds for $k' = k + 1$, as

$$k'(\lambda x) = k(\lambda x) + \lambda x = (k\lambda).x + \lambda x = (k\lambda + \lambda).x = (k'\lambda).x.$$

For negative integers, use the previous point.

(c) The axioms of $\mathbb{Z}$-module for this multiplication have already been verified, in 1.3.3(i–iii).

(d), (e) It is sufficient to prove the second point, which is straightforward.

For an abelian group M, a structure of module on the ring R/I gives a multiplication $\lambda x = [\lambda]x$ with scalars in R, which is trivial on all the scalars of I. Conversely, an R-module structure on M satisfying this property gives a well-defined multiplication $[\lambda]x = \lambda x$ with scalars in R/I. The two processes are inverse to each other.

(f) First suppose that M is a $\mathbb{Q}$-vector space. All terms kx (for $k \in \mathbb{Z}$ and $x \in M$) are determined as multiples, by the underlying structure $(M, +)$ of abelian group (as in (b)). Moreover, if $n \in \mathbb{N}^*$, the element $y = (1/n)x$ of M is also determined by $(M, +)$: in fact, it satisfies the equation $ny = x$, and is the unique solution because n is an invertible scalar of $\mathbb{Q}$.

Finally, every multiplication $(k/n)x = k((1/n)x)$ by a rational scalar is determined by $(M, +)$. We also note that, in M, all equations $ny = x$ (for $n \in \mathbb{N}^*$ and $x \in M$) have one and only one solution y. The existence says that the abelian group $(M, +)$ is divisible, while the uniqueness says that it is torsion free (Exercise 1.3.9(h)).

Conversely, it is now obvious that any abelian group which is *divisible and torsion free* can be given a (unique) structure of vector space on $\mathbb{Q}$, even though verifying every detail can be long.

Solutions of 2.3.4. (a) Take $A' = \{(a, 0) \mid a \in A\} \subset M$, and B' in the symmetric way.

(c) The meet $(a) \cap (b) = (c)$ of two subgroups of $\mathbb{Z}$ is generated by $c = \mathrm{lcm}(a, b)$; if $c = 0$ then $a = b = 0$, and $(a) + (b) = (0)$.

(d) If p is a prime number, the only subgroups of $\mathbb{Z}/p$ are the trivial and the total one. The subgroups of $A = \mathbb{Z}/4$ form a chain:

$$\{0\} \subset \langle[2]\rangle \subset A,$$

and each decomposition of A is still trivial. $\mathbb{Z}/6$ is the internal direct sum of the subgroups generated by the cosets [2] and [3].

The general pattern should now be clear: a cyclic group $\mathbb{Z}/n$ is indecomposable if and only if n is a power p^k of a prime number. Proving this requires some more work.

(f) Because of the internal decompositions $M = A \oplus B = A \oplus C$, every element $b \in B$ can be uniquely written as $b = a + c$, with $a \in A$ and $c \in C$. Sending b to c defines a homomorphism, characterised as

$$f\colon B \to C, \qquad b - f(b) \in A.$$

Symmetrically, we have a homomorphism $g\colon C \to B$, and their composites are identities: if $b = a + c$ then $c = -a + b$ and $g(f(b)) = g(c) = b$.

*As to the second point, let X be a non-trivial R-module and $A = \bigoplus_{n \in \mathbb{N}} X$. Then $A \oplus X \cong A$, but X is not trivial. In fact, in an *internal* direct sum $A = A \oplus B$, the submodule B must be trivial.*

7.2.4 Exercises of Section 2.4

Solutions of 2.4.1. (a) Any element $y \in L \setminus S$ could be written as $y = \sum_{x \in S} \lambda_x x$, and we would have a linear combination $\sum_{x \in S} \lambda_x x + (-1)y = 0$ of elements of L with at least one non-zero coefficient.

(b) Let B be a basis of M. The previous point shows that any proper subset of B cannot be a set of generators, while any proper superset cannot be linearly independent.

*(d) By hypothesis, every ideal of R is a free R-module. Two elements $a \neq b$ in R are always linearly dependent on R, as $ba - ab = 0$ and one of them, at least, is not zero. Therefore every non-trivial ideal has a singleton basis, and it is sufficient to prove that R is an integral domain. Let $a \neq 0$ and $ac = 0$; the ideal (a) has a basis-element b, which is free and cannot be a proper zero-divisor; but $cb = 0$, and therefore $c = 0$.

*(e) We assume that every R-module is free, and we already know that R is an integral domain, by (d). Take a scalar $\lambda \neq 0$ and the projection $p\colon R \to R/\lambda R = M$. Since the codomain is free over R, we can find a homomorphism $m\colon M \to R$ such that $pm = \mathrm{id}\, M$: we define it on a basis of M, by the axiom of choice, and extend it by linearity over R.

Now the submodule $N = m(M)$ (isomorphic to M) is generated by $\mu = m([1_R])$, and $\lambda\mu = m([\lambda]) = 0$. It follows that $\mu = 0$ and $N = \{0\}$, so that M is trivial as well, $\lambda R = R$ and λ is invertible.

Solutions of 2.4.7. (a) One follows the usual pattern of these topics, as for rings and R-modules.

(b) Given a homomorphism $f\colon R \to \mathrm{Cnt}(X)$ of unital rings, we define the scalar multiplication in R by means of the product in X

$$\lambda x = f(\lambda).x. \tag{7.18}$$

The axioms of R-module are plainly satisfied, as well as condition 2.4.6(ii).

Conversely, if X is a unital R-algebra, we define $f\colon R \to X$ letting $f(\lambda) = \lambda 1_X$, and we get a unital homomorphism

$$\begin{gathered} f(\lambda+\mu) = (\lambda+\mu)1_X = \lambda 1_X + \mu 1_X = f(\lambda) + f(\mu), \\ f(\lambda\mu) = (\lambda\mu)1_X = \lambda(\mu 1_X) = (\lambda 1_X).(\mu 1_X) = f(\lambda).f(\mu), \\ f(1_R) = 1_R 1_X = 1_X. \end{gathered} \tag{7.19}$$

The image of f is contained in $\mathrm{Cnt}(X)$, because

$$f(\lambda).x = \lambda x = \lambda(x.1_X) = x.(\lambda 1_X) = x.f(\lambda).$$

The two procedures are inverse to each other.

(d) We view a boolean algebra X as an idempotent commutative unital ring (Exercise 2.1.1(b)). Here $2x = 0$, for all $x \in X$, whence X is a vector space on the field $\mathbb{Z}/2$ (by Exercise 2.3.1(d)), and the structures are consistent.

(e) Let L be a finite field, and K its minimal subfield (see 2.2.1(a)). We have seen that $K \cong \mathbb{Z}/p$, for a prime number p. The K-vector space L is isomorphic to a power K^n, and its order is a power p^n, with $n \geqslant 1$.

7.2.5 Exercises of Section 2.5

Solutions of 2.5.3. (a) Letting $R = \{a_1, ..., a_n\}$, one can take the product of the polynomials $X - a_i$.

(b) We want the ring R to be idempotent and infinite. It is sufficient to take the boolean algebra $R = \mathcal{P}S$ of an infinite set, as a ring (Exercise 2.1.1(b)).

Then, the ring $R[X]$ is not idempotent, as $X \neq X^2$. But $2.1_R = 0$, and $2f = 0$ for every polynomial f.

(e) If R is the trivial ring, so is $R[X]$; otherwise the polynomial X cannot be invertible, because $\deg(Xg) = 1 + \deg g > 0$, for every $g \neq 0$ (by (d)).

If R is an integral domain, so is $R[X]$, again by point (d); the converse is obvious, because R is embedded in $R[X]$; in this case, no polynomial of degree $\geqslant 1$ can be invertible.

(f) One can take $f = \overline{2}X + 1$, where $\overline{2}$ is the coset of 2 in $\mathbb{Z}/4$.

(g) The universal property says that, given a mapping $f\colon \{X\} \to A$ with values in a commutative unital R-algebra (which simply means an element $f(X) = x \in A$), there is a unique homomorphism $g\colon R[X] \to A$ of commutative unital R-algebras such that $g(X) = f(X)$.

In fact, we must let $g(\sum_i a_i X^i) = \sum_i a_i x^i$ (computed by the R-algebra structure of A). On the other hand, defining g in this way, we do get a homomorphism as required.

(h) It is an obvious consequence, because a ring is the same as a $\mathbb{Z}$-algebra (see 2.4.6).

(i) The ideal (2) consists of all polynomials of $\mathbb{Z}[X]$ whose coefficients are even. The ideal (X) consists of all polynomials f which have constant term $f(0) = 0$ (it is the kernel of $\mathrm{ev}_0\colon \mathbb{Z}[X] \to \mathbb{Z}$).

Their join in the lattice of ideals of $\mathbb{Z}[X]$ is the ideal $I = (2, X) = (2) + (X)$, and consists of the polynomials whose constant term is even. It is not principal: taking any polynomial f, if $2 \in (f)$ then $f \mid 2$, whence either $f = \pm 1$ (and generates the total ideal) or $f = \pm 2$ (and generates the ideal (2)).

Solutions of 2.5.4. (a) Suppose that $f(a) = 0$; we prove that f is divisible by $X - a$, by complete induction on $n = \deg f \geqslant 1$.

In degree 1, $f = a_1 X + a_0$ and $a_1 a + a_0 = 0$, so that

$$f = a_1(X - a) + a_1 a + a_0 = a_1(X - a).$$

If our claim holds for all degrees $< n$, and $c(f) = b$, we let

$$f = b(X - a)^n + g,$$

where $\deg g < n$ (because f and $b(X - a)^n$ have the same degree and the same leading coefficient); by the inductive assumption $g = (X - a)h$ for some polynomial h, and

$$f = b(X - a)^n + (X - a)h = (X - a)(b(X - a)^{n-1} + h).$$

The converse is obvious. We have proved that $\operatorname{Ker} \mathrm{ev}_a = (X - a)$, so that the surjective homomorphism $\mathrm{ev}_a\colon R[X] \to R$ induces the required isomorphism.

(b) We apply Exercise 2.5.3(d). If $X - a = fg$, one of the factors has degree 1 and the other (say g) has degree zero; therefore $g = c(g)$ and $c(f).c(g) = 1$, so that g is invertible.

Let f be a polynomial of degree > 1, with a root a. Then $f = (X - a)g$, where each of the factors has degree $\geqslant 1$, and is not invertible. Therefore f is not irreducible.

(c) If $g(a) = 0$, the polynomial g is divisible by $X - a$ and $m_f(a) > m$.

Conversely, if this is the case, the polynomial $f = (X - a)^m.g$ can be factorised as $f = (X - a)^{m+1}.h$; cancelling $(X - a)^m \neq 0$ (by Exercise 2.5.3(e)), we get $g = (X - a).h$ and $g(a) = 0$.

(d) The polynomial $f = aX + b$ (with $a \neq 0$) has one root in R, namely $-ba^{-1}$, and is associated to the polynomial $a^{-1}f = X + a^{-1}b$, whence irreducible too.

(e) We know that every $a \in R$ is a root of $f = X + X^2$. Its multiplicity is always 1, because

$$(X - a)^2 = (X + a)^2 = X^2 + 2aX + a^2 = X^2 + a,$$

and $X^2 + X = (X^2 + a)g$ would give $g = 1$ (by Exercise 2.5.3(d)) and a contradiction.

(f) We already noted that the polynomial $f = X^2 + 1$ has no roots in $\mathbb{R}$. It is irreducible, because a factorisation $f = gh$ cannot have factors of degree 1 (which would have a root), whence one of them has degree 0 and is invertible. The polynomial f^2 is reducible (as f is not invertible), and has no roots: any root of f^2 would also be a root of f.

Solutions of 2.5.9. An $R[X]$-module M amounts to a pair (N, φ) where N is an R-module and $\varphi\colon N \to N$ is an R-homomorphism.

Indeed, given M, we let $N = |M|$ be the underlying R-module, by restricting the scalars to R; then we define

$$\varphi\colon N \to N, \qquad \varphi(y) = X.y \qquad\qquad (y \in N),$$

by scalar multiplication with the polynomial X, which is $R[X]$-linear and therefore R-linear.

Conversely, given a pair (N, φ) as above, we make N into an $R[X]$-module, extending the scalar multiplication to all polynomials $f = a_0 + a_1X + a_2X^2 + ... + a_nX^n$

$$f.y = a_0.y + a_1.\varphi(y) + a_2.\varphi^2(y) + \ ... \ + a_n.\varphi^n(y).$$

The two processes are inverse to each other.

Moreover, an $R[X]$-homomorphism $h\colon M \to M'$ is the same as a homomorphism of R-modules such that $h(X.y) = X.h(y)$, and this amounts to $h\varphi = \varphi'h$ (where the module M' corresponds to the pair $(|M'|, \varphi')$).

Similarly, a sub-$R[X]$-module H of M is the same as a sub-R-module stable under the homomorphism φ, in the sense that $\varphi(H) \subset H$.

7.2.6 Exercises of Section 2.6

Solutions of 2.6.2. (a) For (i), both terms are defined for matrices A, B, C which are, respectively, of type $m \times n$, $n \times p$ and $p \times q$. In both cases the (i,l)-entry of the result is given by the following sum (where $j = 1, ..., n$ and $k = 1, ..., p$)

$$\Sigma_{j,k}\ a_{ij}.b_{jk}.c_{kl}.$$

For the first equation of (ii), take $A \in \mathrm{Mat}_{mn}(R)$ and $B, C \in \mathrm{Mat}_{np}(R)$. Then both results give the same general (i,k)-entry (added for $j = 1, ..., n$):

$$\Sigma_j\ a_{ij}(b_{jk} + c_{jk}) = \Sigma_j\ a_{ij}b_{jk} + \Sigma_j\ a_{ij}.c_{jk}.$$

For the first equation of (iii), take $A \in \mathrm{Mat}_{mn}(R)$ and $I_n \in \mathrm{Mat}_{nn}(R)$. Then we get the (i,k)-entry:

$$\Sigma_j\ a_{ij}.\delta_{jk} = a_{ik}.$$

(b) Follows from (a), except the last point. If $R = \{0\}$, each ring $M_n(R)$ has one element, and is null as well. Otherwise, $0 \neq 1$ in R. In $M_2(R)$, commutativity is contradicted by simple examples, like

$$\begin{pmatrix} 0 & 1 \\ 0 & 0 \end{pmatrix} \cdot \begin{pmatrix} 0 & 0 \\ 1 & 0 \end{pmatrix} = \begin{pmatrix} 1 & 0 \\ 0 & 0 \end{pmatrix}, \qquad \begin{pmatrix} 0 & 0 \\ 1 & 0 \end{pmatrix} \cdot \begin{pmatrix} 0 & 1 \\ 0 & 0 \end{pmatrix} = \begin{pmatrix} 0 & 0 \\ 0 & 1 \end{pmatrix}.$$

For $n > 2$ one can use the same example, adding null entries at the right and below.

(c) The only non-obvious point is concerned with the product. The general term c_{ik} of AB is expressed in (2.96), and the general term of $B^{\mathrm{tr}}.A^{\mathrm{tr}}$ is

$$d_{ik} = \Sigma_j\ b_{ji}a_{kj} = \Sigma_j\ a_{kj}b_{ji} = c_{ki}.$$

(d) The condition $Ax = b$ amounts to $\Sigma_j\ a_{ij}x_j = b_i$, for $i = 1, ..., m$.

Solutions of 2.6.4. (a) Let $(e_1, ..., e_n)$ be the canonical basis of R^n, with $e_j = (\delta_{kj})_{k=1,...,n}$.

Then the matrix $A = (a_{ij})$ sends the element e_j of the basis to the column-matrix

$$\begin{aligned} f_A(e_j) = Ae_j &= (\Sigma_k\ a_{ik}\delta_{kj})_{i=1,...,m} = (a_{ij})_{i=1,...,m} \\ &= (a_{1j}, a_{2j}, ..., a_{mj}) \in R^m, \end{aligned}$$

which is indeed the j-th column of the matrix.

Now, a homomorphism $f\colon R^n \to R^m$ is determined by assigning, to any element e_j of the basis, an arbitrary element of R^m. Finally, the homomorphism ρ in (2.103) is bijective.

(b) This simply follows from the associativity of the product of matrices, when legitimate

$$(f_A.f_B)(x) = f_A(f_B(x)) = A(Bx) = (AB)x = f_{AB}(x) \qquad (x \in R^p).$$

(c) There is a unique isomorphism $\varphi\colon R^n \to N$ which takes the canonical basis $(e_1, ..., e_n)$ to the given basis $(x_1, ..., x_n)$. Similarly we have the isomorphism $\psi\colon R^m \to M$ determined by the basis $(y_1, ..., y_m)$ of M.

We can thus transform the previous isomorphism ρ, as required

$$\begin{aligned} &\rho'\colon \mathrm{Mat}_{mn}(R) \to \mathrm{Hom}_R(N, M), \\ &\rho'(A) = \psi.\rho(A).\varphi^{-1}\colon N \to R^n \to R^m \to M. \end{aligned} \tag{7.20}$$

Solutions of 2.6.6. (a) Given the affine mapping $f\colon X \to Y$, we take the pair $(f(x_0), f')$, where f' is the linear mapping defined in (2.108)

$$f'\colon V \to W, \qquad f'\langle x' - x\rangle = \langle fx' - fx\rangle. \tag{7.21}$$

Given the pair (y_0, f'), where $f'\colon V \to W$ is K-linear, we define

$$f\colon X \to Y, \qquad f(x) = y_0 + f'\langle x - x_0\rangle. \tag{7.22}$$

This is indeed an affine mapping, because formula (7.21) gives back the original f'. In fact $\langle fx' - fx\rangle = f'\langle x' - x\rangle$, because:

$$\begin{aligned} f(x') = y_0 + f'\langle x' - x_0\rangle &= y_0 + f'\langle x - x_0\rangle + f'\langle x' - x\rangle \\ &= f(x) + f'\langle x' - x\rangle. \end{aligned}$$

7.2.7 Exercises of Section 2.7

Solutions of 2.7.2. (a) The verification of the axioms of fields involving the product is straightforward and left to the reader, except two points. First, the product (obviously commutative) distributes over the sum:

$$\begin{aligned} &((a, b) + (c, d)).(x, y) = (a + c, b + d).(x, y) \\ &= (ax + cx - by - dy, bx + dx + ay + cy), \end{aligned}$$

$$(a, b).(x, y) + (c, d).(x, y) = (ax - by, bx + ay) + (cx - dy, dx + cy).$$

Secondly, the inverse of $(a, b) \neq 0$ is as stated in (2.122). Letting $c = (a^2 + b^2)^{-1} > 0$, we have:

$$(a, b).(ac, -bc) = (a^2c + b^2c,\ bac - abc) = (1, 0) = 1.$$

(b) Supposing that the field $\mathbb{C}$ has a total order $\leqslant$ consistent with the operations (see 1.1.8), we deduce that every element $z \in \mathbb{C}$ has $z^2 \geqslant 0$. But then $1 = 1^2 \geqslant 0$ and $1 = -\underline{i}^2 \leqslant 0$, a contradiction.

7.3 Exercises of Chapter 3

7.3.1 Exercises of Section 3.1

Solutions of 3.1.1. (a) We can equivalently compare the square powers of these numbers $\geqslant 0$ (by Proposition 1.7.4):

$$||x + y||^2 = \Sigma_i\,(x_i^2 + 2x_iy_i + y_i^2) = ||x||^2 + 2\,\Sigma_i\, x_iy_i + ||y||^2,$$
$$(||x|| + ||y||)^2 = ||x||^2 + 2||x||.||y|| + ||y||^2.$$

It is thus sufficient to prove the *Cauchy–Schwartz inequality*:

$$|\Sigma_i\, x_iy_i| \leqslant ||x||.||y|| \qquad (x, y \in \mathbb{R}^n). \tag{7.23}$$

If $y = 0$ this is obviously true. Otherwise, we can divide by $||y|| > 0$, and prove that

$$|\Sigma_i\, x_iy_i| \leqslant ||x|| \qquad (\text{for } x, y \in \mathbb{R}^n,\ ||y|| = 1). \tag{7.24}$$

(We are replacing y by the normalised vector $y/||y||$, renamed as y.) Letting $\lambda = \Sigma_i\, x_iy_i \in \mathbb{R}$ we have:

$$0 \leqslant ||x - \lambda y||^2 = \Sigma_i\,(x_i^2 - 2\lambda x_iy_i + \lambda^2y_i^2)$$
$$= ||x||^2 - 2\lambda.\Sigma_i\, x_iy_i + \lambda^2 = ||x||^2 - \lambda^2,$$

and $\lambda^2 \leqslant ||x||^2$, which means that $|\lambda| \leqslant ||x||$.

Solutions of 3.1.5. (b) If we take a point $b \in B_X(a, r)$, then $d(a, b) < r$. Let $r' = r - d(a, b) > 0$. Applying the triangle inequality

$$d(a, x) \leqslant d(a, b) + d(b, x),$$

any point $x \in B_X(b, r')$ has $d(a, x) < d(a, b) + r' = r$ and belongs to $B_X(a, r)$.

The rest follows from an obvious fact: a union of open subsets of X is always open.

(c) Take a point $b \in X \setminus D_X(a, r)$, so that $d(a, b) > r$. Letting $r' = d(a, b) - r > 0$, the existence of a point $x \in B_X(b, r') \cap D_X(a, r)$ would give a contradiction: $d(a, b) \leqslant d(a, x) + d(x, b) < r + r' = d(a, b)$.

The complement of $S_X(a, r)$ in X is the union of the open sets $B_X(a, r)$ and $X \setminus D_X(a, r)$.

(d) The first two properties are obvious. For (Nb.3), the intersection $N \cap N'$ of two neighbourhoods of a in X contains an intersection $B_X(a,r) \cap B_X(a,r')$ of open balls, which is the smallest of them; the empty intersection is X.

For (Nb.4), there exists some $r > 0$ such that $a \in B_X(a,r) \subset N$; by point (b), $B_X(a,r)$ is a neighbourhood of each of its points.

(e) We have already seen that a union of open subsets of X is always open. If U and V are open subsets, they are both neighbourhoods of any point $a \in U \cap V$, and $U \cap V \in \mathcal{N}(a)$ by the previous point.

(f) Taking complements in $\mathcal{P}X$ interchanges unions and intersections.

(g) Letting out the obvious case $A = \emptyset$, we fix a point $x_0 \in A$ and let $r_0 = d(0, x_0) \geqslant 0$.

If $\text{diam}(A) < r$ (a positive real number) then $d(0,x) \leqslant d(0,x_0) + d(x_0,x) < r_0 + r$ (for all $x \in A$), whence $A \subset B_n(0, r_0 + r)$, and all the properties of this exercise are satisfied.

Conversely, the weakest of them says that A is contained in some closed ball $D_n(a,r)$, and then $d(x,y) \leqslant d(x,a) + d(a,y) \leqslant 2r$ (for all $x, y \in A$).

Solutions of 3.1.7. (a) The properties (i)–(iii) are obvious. As to (iv), it is sufficient to note that the complement-operator is an anti-automorphism of the boolean algebra $\mathcal{P}X$, which interchanges open and closed subsets.

As to (v), if $a \in X$, one and only one of these conditions is met:

- there is a neighbourhood of a in X contained in A (i.e. $x \in A^\circ$),
- there is a neighbourhood of a in X contained in $C_X A$ (i.e. $x \in (C_X A)^\circ$),
- every neighbourhood of a in X meets A and $C_X A$ (i.e. $x \in \partial A$).

Property (vi) is a consequence: $\overline{A}$ is the complement of $(C_X A)^\circ$, and coincides with $A^\circ \cup \partial A$; moreover $A^\circ \cup \partial A \subset A \cup \partial A \subset \overline{A}$.

Now (viii) follows from (vi), and (vii) is a consequence: A is open if and only if $C_X A$ is closed, which means that $\partial_X A \subset C_X A$, or equivalently that $\partial_X A$ does not meet A.

(b) We already know that $B_X(a,r)$ is open, and coincides with its interior, while $D_X(a,r)$ is closed, and coincides with its closure. Taking $X = \mathbb{R}^n$ and letting

$$U = X \setminus D_n(a,r) = \{x \in X \mid d(a,x) > r\},$$

we prove now that:

(*) every point $x \in S_n(a,r)$ is adherent to $B_n(a,r)$ and to U.

In fact, considering the half-line L from a to x, we see that

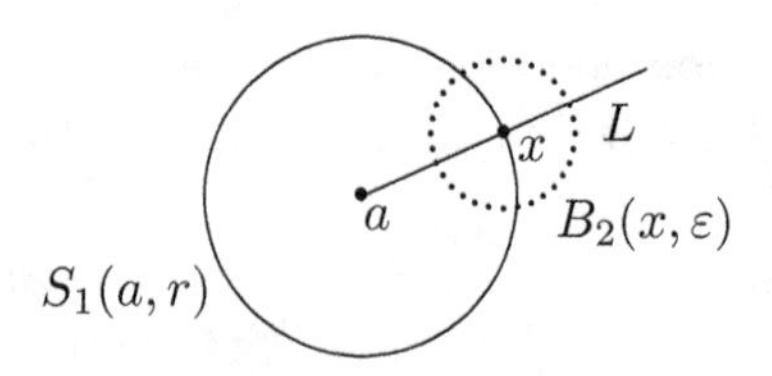

every neighbourhood of x meets both $L \cap B_n(a, r)$ and $L \cap U$. (The figure is drawn in dimension 2, or in a plane containing L for higher dimension.)

This is geometrically obvious. Analytically, the half-line L of $\mathbb{R}^n$ can be parametrised as

$$x_t = (1 - t)a + tx \qquad (t \geqslant 0). \tag{7.25}$$

The distance of the point x_t from a and x is

$$d(x_t, a) = ||x_t - a|| = ||tx - ta|| = t.r,$$
$$d(x_t, x) = ||x_t - x|| = ||(1 - t)a - (1 - t)x|| = |1 - t|.r,$$

so that, letting $0 < \varepsilon < r$:

- x_t lies in the open ball $B_n(a, r)$ for $0 \leqslant t < 1$, in the open set U for $t > 1$,
- x_t lies in the open ball $B_n(x, \varepsilon)$ for $|t - 1| < \varepsilon/r$.

Finally, the open ball $B_n(x, \varepsilon)$ contains (infinitely many) points of $B_n(a, r)$ (for $1 - \varepsilon/r < t < 1$) and of U (for $1 < t < 1 + \varepsilon/r$).

Now $D_n(a, r)$ is a closed subset containing $B_n(a, r)$; it is actually the closure of the latter, because every point $x \in D_n(a, r) \setminus B_n(a, r) = S_n(a, r)$ is adherent to $B_n(a, r)$.

Similarly, $B_n(a, r)$ is an open subset contained in $D_n(a, r)$; it is actually the interior of the latter, because a point $x \in S_n(a, r)$ is adherent to U and cannot have a neighbourhood contained in $D_n(a, r)$.

Also because of (*), the interior of $S_n(a, r)$ is empty.

We conclude that:

$$B_n(a, r) \text{ is open}, \quad (B_n(a, r))^- = D_n(a, r), \ \partial(B_n(a, r)) = S_n(a, r),$$
$$D_n(a, r) \text{ is closed}, \ (D_n(a, r))^\circ = B_n(a, r), \ \partial(D_n(a, r)) = S_n(a, r),$$
$$S_n(a, r) \text{ is closed}, \qquad (S_n(a, r))^\circ = \emptyset, \qquad \partial(S_n(a, r)) = S_n(a, r).$$

(c) If we take the euclidean space $X = \mathbb{Z} \subset \mathbb{R}$, and $a \in \mathbb{Z}$, we have:

$$B_X(a, 1) = \{a\},$$
$$D_X(a, 1) = \{a - 1, a, a + 1\} = B_X(a, 3/2),$$
$$S_X(a, 1) = \{a - 1, a + 1\}, \qquad S_X(a, 3/2) = \emptyset.$$

Here every point is open, and therefore every subset is open (and closed).

The closure of any subset is the subset itself, the closure of $B_X(a,1)$ is not $D_X(a,1)$, and the boundary of any subset is empty.

(d) The subset A has no interior points in $\mathbb{R}^n$. Its closure is:

$$\{x \in \mathbb{R}^n \mid 0 \leqslant x_1 \leqslant 1,\ x_2 = ... = x_n = 0\}.$$

The boundary in $\mathbb{R}^n$ is the same.

(e) The intervals which are (topologically) open in $\mathbb{R}$ are those of the following kinds, for $a < b$:

$$\emptyset, \qquad]a,b[, \qquad]a,+\infty[, \qquad]-\infty,b[, \qquad \mathbb{R}.$$

The intervals which are (topologically) closed in $\mathbb{R}$ are those of the following kinds, for $a < b$:

$$\emptyset, \qquad \{a\}, \qquad [a,b], \qquad [a,+\infty[, \qquad]-\infty,b], \qquad \mathbb{R}.$$

The remaining intervals are of the following types, for $a < b$:

$$[a,b[, \qquad\qquad]a,b],$$

and they are neither open nor closed in $\mathbb{R}$.

(f) The property (cl.1) is obvious. For the second point, $\overline{A}$ is plainly the least closed subset of X containing itself. For the last, $\overline{A} \cup \overline{B}$ is a closed subset containing $A \cup B$, and any closed subset which contains both must contain $\overline{A}$ and $\overline{B}$.

Solutions of 3.1.9. (a) The subset A is the f-preimage of the open interval $]-\infty, \lambda[$ of the real line, and is thus an open subset of X. Similarly, B and C are closed.

(b) The previous facts hold for every continuous f. But A is empty for the constant function $f(x) = \lambda + 1$ and total in X for the constant function $f(x) = \lambda - 1$. Similarly, B and C are open and closed for some continuous function f.

(c) For any $i = 1, ..., n$, the subset $A_i = \{x \in \mathbb{R}^n \mid 0 \leqslant x_i \leqslant 1\}$ is closed in $\mathbb{R}^n$, by point (a). Their intersection is $\mathbb{I}^n$.

7.3.2 Exercises of Section 3.2

Solutions of 3.2.1. (a) The discrete and the indiscrete topology coincide on the empty set and on each singleton. They are different on each set having at least two elements, including the set $S = \{0,1\}$. The *Sierpinski* topology on the set S has three open sets: $\emptyset$, $\{1\}$, S. Interchanging the roles of 0 and 1 we have the fourth (and last) topology, obviously homeomorphic to the previous one.

One can note that the characteristic function $\chi_A : X \to \{0, 1\}$ of a subset A in the space X is continuous with respect to the Sierpinski topology on $\{0, 1\}$ if and only if A is open in X.

(b) The first two cases are obvious. If f is constant at $b \in Y$, and V is open in Y, either b belongs to V or it does not; in the first case $f^{-1}(V)$ is X, in the second is empty. In each case the preimage of V is open in X.

(d) If f is bijective, the preimages of the inverse mapping f^{-1} are the same as the direct images of f.

(e) The mapping t_v is bijective, with inverse t_{-v}. It preserves distances (it is an *isometry*), and takes open balls to open balls

$$t_v(B_n(a, r)) = B_n(a + v, r).$$

Taking into account that images preserve arbitrary unions, this proves that t_v is an open mapping; but its inverse is also, whence it is a homeomorphism.

The second part is similarly proved. The mapping m_λ is bijective, with inverse $m_{\lambda^{-1}}$. It multiplies distances by the modulus of λ: $d(\lambda x, \lambda y) = |\lambda|.d(x, y)$, and takes open balls to open balls

$$m_\lambda(B_n(a, r)) = B_n(\lambda a, |\lambda| r).$$

Solutions of 3.2.2. (a) In $\mathcal{PS}$ binary intersections distribute over arbitrary unions:

$$A \cap (\cup_i B_i) = \cup_i (A \cap B_i). \tag{7.26}$$

Therefore the set $\mathcal{O}$ is also stable under finite intersections, and gives the least topology containing $\mathcal{O}_0$.

(d) The proposed closed subsets are plainly stable under arbitrary intersections and finite unions. It is thus sufficient to prove that any singleton $\{a\}$ is closed in the euclidean topology of $\mathbb{R}^n$.

This can be proved directly, by the same arguments used in Exercise 3.1.7(b); or we can remark that $\{a\}$ is the intersection of two (closed) tangent spheres, for instance $S_n(a - e_1, 1)$ and $S_n(a + e_1, 1)$, where $e_1 = (1, 0, ..., 0) \in \mathbb{R}^n$.

Solutions of 3.2.3. (b) Suppose we are given the family $(\mathcal{N}(a))_{a \in X}$, satisfying the axioms (Nb.1–4). We define the open subsets as required, and we obtain a set $\mathcal{O}(X)$ which is a topology; in fact it is plainly stable under arbitrary unions, and is also stable under finite intersections, because the neighbourhoods of any point are stable in this sense.

This procedure is inverse to the previous one described in 3.2.3, from the open subsets to the neighbourhoods.

Solutions of 3.2.4. (a) The proof is the same as for Exercise 3.1.7(a).

(b) If X is a topological space and we take $\mathrm{cl}(A) = \overline{A}$, the Kuratowski axioms are satisfied (by the same easy arguments of Exercise 3.1.7(f)). The property (cl.1) is obvious. For (cl.2), $\overline{A}$ is the least closed subset of X containing itself. For (cl.3), $\overline{A} \cup \overline{B}$ is a closed subset containing $A \cup B$, and any closed subset which contains both must contain $\overline{A}$ and $\overline{B}$.

On the other hand, let us start from a given closure operator. Defining the closed subsets as suggested, we get a subset $\mathcal{C}l(X)$. It is stable under finite unions, because of (cl.3). To verify that it is stable under arbitrary intersections, take a family A_i $(i \in I)$ of subsets such that $\mathrm{cl}(A_i) = A_i$, for all indices i. Then, using formula (3.22), $\mathrm{cl}(\bigcap A_i) \subset \bigcap \mathrm{cl}(A_i) = \bigcap A_i$.

Also here the two processes are inverse to each other.

(c) All the three properties say that every point $a \in X$ is adherent to A.

(d) Every neighbourhood $]a - r, a + r[$ of $a = \sup A$ contains points of A (because a is an upper bound of A and $a - r/2$ is not) and points of the complement $\mathbb{R} \setminus A$ (because a is an upper bound of A and $a + r/2$ is not).

(e) In the euclidean line $\mathbb{R}$ every non-empty open set contains some non-trivial interval $]a, b[$ (which is an open ball of $\mathbb{R}$). Any such interval contains (infinitely many) rational and irrational numbers, as we have seen in Exercise 1.7.5(b).

Solutions of 3.2.6. (a) We know that the topology generated by any set $\mathcal{B} \subset \mathcal{P}S$ consists of the unions of finite intersections of elements of $\mathcal{B}$.

If $\mathcal{B}$ satisfies (BO.1, 2), this simply amounts to all the unions of elements of $\mathcal{B}$. Conversely, if $\mathcal{B}$ is the basis of the topology that it generates, each finite intersection of elements of $\mathcal{B}$ is an open subset and therefore a union of elements of $\mathcal{B}$.

(e) Obviously $\mathcal{B} \subset \mathcal{O}(X)$. Conversely, for any open subset U and any $a \in U$ there exists a neighbourhood $N_a \in \mathcal{B}(a)$ which is contained in U, and their union is U.

(g) The f-preimage of an open interval $]r, s[$ is the set

$$U = \{x \in \mathbb{R}^n \mid r < ||x|| < s\}.$$

This is empty if $s \leqslant 0$, and coincides with:

- the open ball $B_n(0, s)$, if $r < 0 < s$,
- the open set $B_n(0, s) \setminus \{0\}$, if $r = 0 < s$,
- the open set $B_n(0, s) \setminus D_n(0, r)$, if $0 < r < s$.

Solutions of 3.2.7. (a) We start from a preordered set X. The downward closed subsets are easily seen to be stable in $\mathcal{P}X$ under arbitrary unions and arbitrary intersections, forming an Alexandrov topology.

An open subset $U = X \setminus C$ is upward closed: if $x \in U$ and $x \prec x'$ in X, also $x' \in U$ or we would get $x \in C$. In the same way the complement of an upward closed subset is downward closed, and therefore closed in our topology. The rest is obvious.

(b) If $x \prec x'$ in $S(X)$, we have $x \in \overline{x'}$ and $f(x) \in f(\overline{x'}) \subset \{f(x')\}^-$, which means that $f(x) \prec f(x')$ in $S(Y)$.

(c) Taking a downward closed subset D of Y, its preimage $C = f^{-1}(D)$ is downward closed in X, because $x \prec c \in C$ implies $f(x) \prec f(c) \in D$, whence $f(x) \in D$ and $x \in C$.

(d) Starting from a preordered set X, the Alexandrov space $A(X)$ gives a preorder relation xPx' defined by $x \in \overline{x'} = \downarrow(x')$, which coincides with the original one.

Starting from an Alexandrov topological space X, the preordered set $S(X)$ gives an Alexandrov topology $X' = A(S(X))$. By the previous argument, $S(X') = S(X)$; therefore the Alexandrov spaces X and X' have the same underlying set and the same specialisation preorder, and coincide.

(e) The closed subsets of $\mathbb{R}_A$ comprise the empty subset and every lower unbounded interval

$$]-\infty, a[, \qquad]-\infty, a], \qquad \mathbb{R}.$$

There are no other closed subsets: let C be closed and non empty. If it is not upper bounded, then it is the total subset: for every $x \in \mathbb{R}$ there is some $c \in C$ with $x < c$, and therefore $x \in C$. If C is upper bounded, and $a = \sup C$, we prove in the same way that $]-\infty, a[\subset C \subset]-\infty, a]$, and C must coincide with one of them.

The non-empty open sets are the upper-unbounded intervals.

7.3.3 Exercises of Section 3.3

Solutions of 3.3.2. (a) Take two distinct real numbers λ_i, for $i = 1, 2$. We can find two disjoint neighbourhoods N_i of them, for instance $N_i = B_1(\lambda_i, \varepsilon)$, where $\varepsilon = d(\lambda_1, \lambda_2)/2 > 0$.

But then no sequence can lie eventually in both. (As we will see in Section 3.6, we have used the separation axiom T_2, which holds in all the euclidean spaces.)

(b) For a real number $\varepsilon > 0$, let $h = [\varepsilon^{-1}]$, the integral part of ε^{-1} (defined in Exercise 1.1.9(f)), so that $\varepsilon^{-1} < h+1$. If $k \geqslant h+1$, then $0 < 1/k \leqslant (h+1)^{-1} < \varepsilon$.

(c) The (eventually constant) sequence $-1, 1, ..., 1, ...$ converges to 1. The sequence $1, -1, ..., -1, ...$ converges to -1. The sequence $x_k = (-1)^k$ has no limit.

Note that, for a sequence (x_k) with image $\{-1, 1\}$ there can be no other outcome: if $\lambda \notin \{-1, 1\}$, the sequence cannot fall in a neighbourhood of λ which does not contain these two points (and such a neighbourhood exists).

(d) Let $\lambda = \lim x_k$ and $\mu = \lim y_k$. If $\mu < \lambda$ and $\varepsilon = (\lambda - \mu)/2$, then $x_k \in B_1(\lambda, \varepsilon)$ and $y_k \in B_1(\mu, \varepsilon)$, for k sufficiently large. Then $y_k < \mu + \varepsilon = (\lambda + \mu)/2 = \lambda - \varepsilon < x_k$, a contradiction.

For the last remark, we have $1/k > 0$ (for $k \geqslant 1$) and $\lim_k 1/k = 0$.

(e) For every $\varepsilon > 0$ there exists some $h \in \mathbb{N}$ such that, for $k \geqslant h$:

$$\lambda - \varepsilon < x_k \leqslant z_k \leqslant y_k < \lambda + \varepsilon.$$

(f) If a sequence is not bounded, it cannot lie eventually in any interval $]\lambda - 1, \lambda + 1[$, and cannot have a limit. Secondly, let (x_k) be bounded and $\mathrm{Im}\, x \subset]-r, r[$, for some $r > 0$. Assuming that $\lim y_k = 0$:

- for every $\varepsilon > 0$ there is some $h \in \mathbb{N}$ such that, for every $k \geqslant h$, $|y_k| < \varepsilon/r$, and then $|x_k y_k| < r.\varepsilon/r = \varepsilon$ (for all $k \geqslant h$).

(g) For the sum, the assertion follows easily from the following inequality (see 1.1.9(d))

$$|x_k + y_k - (\lambda + \mu)| \leqslant |x_k - \lambda| + |y_k - \mu|.$$

For the product, we use the fact that our sequences are bounded, and there is some $r > 0$ such that $|x_k| < r$ (for all k). Then

$$\begin{aligned} 0 &\leqslant |x_k y_k - \lambda\mu| = |x_k y_k - x_k\mu + x_k\mu - \lambda\mu| \\ &\leqslant |x_k y_k - x_k\mu| + |x_k\mu - \lambda\mu| = |x_k|.|y_k - \mu| + |x_k - \lambda|.|\mu| \\ &\leqslant r.|y_k - \mu| + |x_k - \lambda|.|\mu|. \end{aligned}$$

Now both sequences $(r.|y_k - \mu|)_k$ and $(|x_k - \lambda|.|\mu|)_k$ converge to 0, by the previous point. It follows that their sum converges to 0, and

$$\lim |x_k y_k - \lambda\mu| = 0$$

(by the Squeeze Theorem). This means that $\lim(x_k y_k - \lambda\mu) = 0$ and $\lim x_k y_k = \lambda\mu$.

(h) The first point is obvious: there is some $K \geqslant 0$ such that, for every $k \geqslant K$, $|x_k - \lambda| < \lambda/2$ and therefore $x_k > \lambda/2$. We now use the relation

$$|x_k^{-1} - \lambda^{-1}| = |x_k - \lambda|.x_k^{-1}.\lambda^{-1} \qquad (\text{for } k \geqslant K).$$

The sequence $(|x_k - \lambda|)_k$ is infinitesimal, while the sequence $(x_k^{-1})_{k \geqslant K}$ is bounded: $0 < x_k^{-1} < 2/\lambda$. Therefore, by (f), the previous sequence converges to 0, and $(x_k^{-1})_{k \geqslant K}$ converges to λ^{-1}.

The last assertion is an obvious consequence.

(i) We already know that a convergent sequence is bounded.

Suppose now that the sequence (x_k) is increasing and upper bounded in $\mathbb{R}$, with $a = \sup\{x_k\}$. It is obviously lower bounded, by x_0.

For every $\varepsilon > 0$ there is some term $x_h \in]a - \varepsilon, a]$, because $a - \varepsilon$ is not an upper bound of the image. But then, for all $k \geqslant h$, $x_k \geqslant x_h$ and therefore $x_k \in]a - \varepsilon, a]$.

Solutions of 3.3.4. (a) For every $a \in \mathbb{R}^n$ the open balls $B_n(a, 1/k)$ form a countable local basis (for $k \in \mathbb{N}^*$).

(b) We already know that the countable set $\mathbb{Q}$ is dense in $\mathbb{R}$ (see Exercise 3.2.4(e)).

Consider now the countable set $\mathcal{B}$ of all open intervals $]a, b[$ of the real line, where $a < b$ in $\mathbb{Q}$. These are open sets of the euclidean topology, and every open interval $]\alpha, \beta[$ of the real line is the union of all the intervals $]a, b[$, where $\alpha < a < b < \beta$ and $a, b \in \mathbb{Q}$.

One can easily extend these results to the euclidean space $\mathbb{R}^n$. For second-countability, all open *n-dimensional cubes* $]a, b[^n$, with $a < b$ in $\mathbb{Q}$, are a countable basis of the euclidean topology of $\mathbb{R}^n$.

(c) If the space X has a countable basis $\mathcal{B}$, every point a has a local basis $\mathcal{B}(a)$ formed by all the open sets $U \in \mathcal{B}$ which contain a. Moreover, if we choose a point in each non-empty element of $\mathcal{B}$, we form a countable subset $A \subset X$ which is obviously dense in X: every non-empty open set U is a union of elements of $\mathcal{B}$, and therefore contains some point of A.

The reader can note that we have not used this fact in point (a): the local basis at $a \in \mathbb{R}$ (for instance) used there is much simpler and more useful than that formed by all the open intervals with rational endpoints, containing a.

Solutions of 3.3.6. (a) To prove that the mappings $s, m \colon \mathbb{R}^2 \to \mathbb{R}$ are continuous at $(a, b) \in \mathbb{R}^2$, we apply the characterisation of continuity by convergent sequences, in 3.3.5(g).

We take a sequence (x_k, y_k) which converges to (a, b) in $\mathbb{R}^2$. Then $x_k \to a$ and $y_k \to b$ in $\mathbb{R}$, because the projections $\mathbb{R}^2 \to \mathbb{R}$ are continuous. Applying 3.3.2(g), it follows that the sequence $(x_k + y_k)$ converges to $a+b$ in $\mathbb{R}$, and the sequence $(x_k y_k)$ converges to ab.

(b) A consequence of (a), because the functions $f + g$ and h are the composites

$$s(f,g)\colon X \to \mathbb{R}^2 \to \mathbb{R}, \qquad m(f,g)\colon X \to \mathbb{R}^2 \to \mathbb{R}.$$

(c) A consequence of (b), multiplying the constant mapping $f(x) = \lambda$ by $\mathrm{id}\,\mathbb{R}$.

(d) The euclidean space $\mathbb{R}^*$ is also first-countable. We take a sequence (x_k) which converges to a in $\mathbb{R}^*$; applying 3.3.2(h), the sequence (x_k^{-1}) is defined for k sufficiently large, and converges to a^{-1} in $\mathbb{R}^*$.

For the second statement we note that, with the previous notation for m and i

$$f/g = m(f, ig)\colon X \to \mathbb{R}{\times}\mathbb{R}^* \to \mathbb{R}.$$

(e) Follows from the previous results: a polynomial function is a linear combination of monomial functions, which are products of projections.

(f) From (d) and (e).

7.3.4 Exercises of Section 3.4

Solutions of 3.4.2. (d) First, $\mathrm{cl}_X(S) \cap A$ is closed in A and contains S, whence it contains $\mathrm{cl}_A(S)$. Conversely, if $x \in \mathrm{cl}_X(S) \cap A$, every neighbourhood of x in A is the trace $N' = N \cap A$ of some neighbourhood N of x in X; therefore N meets $S \subset A$, and also N' meets S.

The second point follows from $f(X) \subset (f(A))^-$ (in 3.2.5).

(e) In both cases, the open sets of B are the subsets $(U \cap A) \cap B = U \cap B$, for U open in X.

(f) If $h\colon X' \to A$ is continuous, the composed mapping $jh\colon X' \to A \to X$ is also. Conversely, let $g = jh$ be continuous; any open subset V of A is the pre-image $j^{-1}(U)$ of an open subset of X, so that $h^{-1}(V) = h^{-1}j^{-1}(U) = g^{-1}(U)$ is open in X'.

(g) The mapping $f_A\colon A \to Y$ is given by composing f with the inclusion $A \to X$; then we restrict f_A on the codomain, using the previous exercise. The last point follows from restricting f and its inverse.

Solutions of 3.4.4. (a) First we draw, in the space $[0,1]^2$, two closed subsets, on the left, and two open subsets, on the right

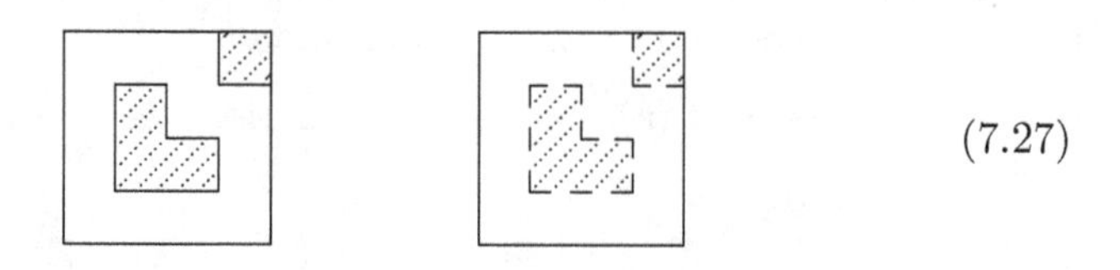

(7.27)

We are following the convention that a part of the boundary of a subset S of the plane is dashed when its points are not supposed to belong to S. Note that an open subset of $[0,1]^2$ need not be open in the plane

Similarly, we draw two closed and two open subsets of the space $]0,1[^2$

(7.28)

Now, a closed subset of $]0,1[^2$ need not be closed in the plane.

Drawing similar closed and open subsets in the space $]0,1[\times[0,1]$ is left to the reader.

(c) The intersection of any open ball of $\mathbb{R}^2$ with $\mathbb{S}^1$ is an open arc of the latter (possibly total or empty). This fact is geometrically obvious; an analytic proof may need a time which can be better used for more interesting things.

(d) First let us note that, for every $a \in \mathbb{R}$ and $\lambda \in \mathbb{R}^*$, the affine transformation

$$f\colon \mathbb{R} \to \mathbb{R}, \qquad f(x) = \lambda x + a, \tag{7.29}$$

is a homeomorphism, as composed of a homothety and a translation (Exercise 3.2.1(e)); or just because the inverse mapping is also affine. It is now simpler to let $\lambda > 0$.

Restricting this homeomorphism (as in 3.4.2(g)), the open interval $]0,1[$ is homeomorphic to its image $]a, a+\lambda[$, and therefore to any interval $]a,b[$ (for $a < b$). Similarly, all the closed intervals $[a,b]$ are homeomorphic to each other (for $a < b$).

In the same way, all the intervals $[a,b[$ are homeomorphic to each other, and all the intervals $]a,b]$ are also. But the former are homeomorphic to the latter, as $[0,1[\,\cong\,]0,1]$, via $f(x) = 1 - x$.

Similarly, all the intervals $[a, +\infty[$ are homeomorphic to $[0, +\infty[$, all the intervals $]-\infty, b[$ are homeomorphic to $]-\infty, 0[$ and $[0, +\infty[\ \cong\ ]-\infty, 0[$.

We use now some rational homeomorphisms (see Exercise 3.3.6(d)) that link bounded and non-bounded intervals

$$\begin{aligned} &f\colon]0,1] \rightleftarrows]-\infty,-1] : g, \quad f(x) = -1/x, \quad g(y) = -1/y, \\ &f\colon [-1,0[\rightleftarrows [1,+\infty[: g, \quad f(x) = -1/x, \quad g(y) = -1/y. \end{aligned} \tag{7.30}$$

Restrictions of the first pair prove that $]0,1[\ \cong\ ]-\infty,-1[$.

Finally, a homeomorphism

$$f\colon]-1,1[\to \mathbb{R}, \tag{7.31}$$

can be obtained on two finite closed covers of these spaces, pasting at 0 the rational homeomorphisms

$$\begin{aligned} &f_1\colon]-1,0] \rightleftarrows]-\infty,0] : g_1, \quad f_1(x) = x/(1+x), \quad g_1(y) = y/(1-y), \\ &f_2\colon [0,1[\rightleftarrows [0,+\infty[: g_2, \quad f_2(x) = x/(1-x), \quad g_2(y) = y/(1+y), \end{aligned}$$

according to Lemma 3.4.3(c).

(Using transcendental functions, as in 3.1.3(a), is quicker.)

Solutions of 3.4.6. (a) If $h\colon X/R \to Y$ is continuous, the composed mapping $hp\colon X \to X/R \to Y$ is also.

Conversely, let $g = hp$ be continuous; any open subset V of Y has an open preimage $g^{-1}(V) = p^{-1}h^{-1}(U)$ in X, which means that $h^{-1}(U)$ is open in X/R.

(b) The mapping $f'\colon X \to Y/S$ is continuous, by composing f with the projection $Y \to Y/S$.

Then we consider the mapping $f''\colon X/R \to Y/S$ induced by f (legitimate because of the hypothesis (3.35)); $f''p = f'\colon X \to Y/S$ is continuous, whence f'' is also, by (a). For the last point, one applies the previous result to f and f^{-1}.

Solutions of 3.4.8. (a) A consequence of Exercise 3.4.6(b).

(b) The equivalence relation R_f associated to f is the congruence modulo 2π

$$\begin{aligned} f(t) = f(t') &\Leftrightarrow t - t' = 2k\pi, \text{ for some } k \in \mathbb{Z} \\ &\Leftrightarrow t - t' \in 2\pi\mathbb{Z}. \end{aligned} \tag{7.32}$$

The canonical factorisation of f gives a continuous map $g\colon \mathbb{R}/R_f \to \mathbb{S}^1$. Moreover the restriction $f'\colon \mathbb{R} \to \mathbb{S}^1$ is open, because every open interval

$]a,b[$ of the line is transformed by f into an open arc of the circle (and actually the total one whenever $b-a \geqslant 2\pi$); this is an open subset of $\mathbb{S}^1 \subset \mathbb{R}^2$ (Exercise 3.4.4(c)). It follows that g is also open, and therefore a homeomorphism.

(c) The restriction $[0,2\pi[\, \to \mathbb{S}^1$ is not open. For instance, the interval $[0,\pi[$ is open in $[0,2\pi[$, but its image in the circle is not: the point $f(0)=(1,0)$ is not interior in the half-circle $f([0,\pi[)$ (as a subset of the circle).

Solutions of 3.4.9. (a) Plainly, j is injective and p is surjective.

To prove that j is a topological embedding, let V be open in A; then the set $U=p^{-1}(V)$ is open in X, and $V=j^{-1}p^{-1}(V)=j^{-1}(U)$ is the preimage of an open set of X. To prove that p is a topological projection, suppose that $V\subset A$ has an open preimage $U=p^{-1}(V)$ in X; then $V=j^{-1}(U)$ is open in A.

(c) Use the cover of $\mathbb{R}$ formed of three closed intervals

$$]-\infty,a], \qquad [a,b], \qquad [b,+\infty[,$$

and Lemma 3.4.3(b), to define a continuous mapping $p\colon \mathbb{R}\to[a,b]$ which is the identity on the interval $[a,b]$.

As any constant mapping is continuous, one can simply take $p(x)=a$ for $x\leqslant a$ and $p(x)=b$ for $x\geqslant b$.

7.3.5 Exercises of Section 3.5

Solutions of 3.5.1. (a) Obvious.

(b) We deduce this point from the previous one.

Take a point $a=(a_i)\in X$. For each index i, the basis $\mathcal{B}_i$ gives a local basis $\mathcal{B}_i(a_i)$, formed of all the open sets of $\mathcal{B}_i$ which contain a_i, and we get a local basis $\mathcal{B}(a)$ at $a\in X$, as in the previous point. The union $\mathcal{B}'$ of all these local basis $\mathcal{B}(a)$, for $a\in X$, is a basis of open sets for X. Now, $\mathcal{B}'\subset\mathcal{B}$, because every element of $\mathcal{B}'$ is a product of open sets in $\mathcal{B}_i$.

On the other hand, $\mathcal{B}\subset\mathcal{B}'\cup\{\emptyset\}$, because a non-empty element $U=\prod_i U_i$ of $\mathcal{B}$ has some point $a=(a_i)$, and belongs to the local basis $\mathcal{B}(a)$.

Finally, $\mathcal{B}$ and $\mathcal{B}'$ are bases of the same topology on X.

(c) If $\mathbb{R}^n$ is endowed with the cartesian power topology, each point a has a local basis formed by the products $\prod_i\,]a_i-r_i,\, a_i+r_i[$. Then we also have a local basis $\mathcal{B}(a)$ formed by all the open cubes $N_r(a)=\prod_i\,]a_i-r,\, a_i+r[$, centred at a. Each of them contains an open ball, and is contained in another

$$B(a, r) \subset N_r(a) \subset B(a, r\sqrt{n}),$$

so that the cartesian power topology of $\mathbb{R}^n$ coincides with the euclidean one.

(An n-dimensional cube of edge-length $2r$ has a semi-diagonal of length $r\sqrt{n}$, by repeatedly applying Pythagoras' Theorem.)

Solutions of 3.5.3. (a), (b) As in 3.5.1.

(c) For every index i, the preimage $p_i^{-1}(C_i)$ is closed in X_i, and $C = \bigcap_i C_i$.

(d) One implication follows from the continuity of projections.

Conversely, suppose that every sequence $(p_i x_k)_k$ converges to a_i. Let $U = \prod_i U_i$ be an open neighbourhood of a in X, where $U_i = X_i$ except for a finite set of indices $J \subset I$.

For every $i \in J$ there exists some $k_i \in \mathbb{N}$ such that the sequence $p_i x_k \in U_i$, for $k \geqslant k_i$. Then all this holds for $k \geqslant \overline{k} = \max_{i \in J} k_i$, and $x \in U = \prod_i U_i$ for $k \geqslant \overline{k}$ (because the indices $i \notin J$ give no condition).

(f) Every element of the canonical basis is infinite, and every non-empty open set is also; in particular, no singleton is open.

(g) The product topology $Y = \prod_i A(X_i)$ is less fine than the Alexandrov topology $A(X)$: a canonical open set of Y is a product of upward-stable subsets of the factors, and therefore upward-stable in X.

The other way round, suppose first that the set I of indices is finite: then these two topologies on the product set coincide. In fact, take a point $a = (a_i) \in X$; the single open set $U = \uparrow a$ is a local basis at a for $A(X)$ (Exercise 3.2.7(f)). But $U = \prod_i \uparrow a_i$ is a canonical open set of the product topology Y. It follows that every open set of $A(X)$ is open in Y.

If the set I is infinite, and each factor X_i is not chaotic (and therefore has at least two points), the two topologies do not coincide. In fact we can choose in each X_i a point a_i such that $\uparrow a_i \neq X_i$. We have thus a point $a = (a_i) \in X$ such that $\uparrow a$ is open in $A(X)$, but is not open in the product topology: each image $p_i(\uparrow a) = \uparrow a_i$ is non-total (and non-empty).

(h) A canonical open subset $U = \prod_i U_i$ of the product $\prod_i X_i$ is sent to the subset $f(U) = \prod_i f_i(U_i)$, which is a canonical open set of $\prod_i Y_i$: every $f_i(U_i)$ is open in X_i, and is the total subset except for a finite subset of indices.

Solutions of 3.5.4. (a) The product of the inclusion mappings $m_i\colon A_i \to X_i$ is the inclusion $m\colon A \to X$, which is thus continuous.

Conversely, if $V = \prod_i V_i$ is a canonical open set of A for the product topology, let $V_i = U_i \cap A$, where each U_i is open in X_i, and the total one for nearly all indices i. Then $U = \prod_i U_i$ is open in X, and $U = U \cap A$ is open in the topology of A as a subspace of X.

If each inclusion $m_i \colon A_i \to X_i$ has a retraction $p_i \colon X_i \to A_i$, then their product $m \colon A \to X$ has a retraction, the product $p = \prod p_i \colon X \to A$.

(b) The mapping m_i is continuous, because all its components $p_j m_i$ are (for $j \in I$); in fact $p_i m_i = \mathrm{id}\, X_i$, and for $j \neq i$ the mapping $p_j m_i \colon X_i \to X_j$ is constant at a_j. Since $p_i m_i : X_i \to X_i$ is an identity, X_i is a retract of X, and m_i is a topological embedding (see 3.4.9(a)).

(d) The first point is obvious: every $y \in Y$, if any, gives an embedding $m_y \colon X \to X \times Y$, and the mapping $f_y = f m_y$ is a restriction of f. Conversely, the preimage $f^{-1}(W)$ of an open subset W of Z is the union of all subsets $f_y^{-1}(W) \times \{y\}$ of $X \times Y$ (for $y \in Y$), and each of them is open, in our hypotheses.

Solutions of 3.5.5. (a) Letting $p \colon X \to X/R$ and $p_i \colon X_i \to X_i/R_i$ be the canonical projections, the mapping $fp = \prod_i p_i$ is continuous

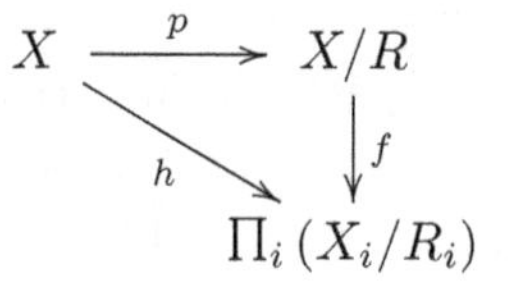

$$\begin{array}{ccc} X & \xrightarrow{p} & X/R \\ & {\scriptstyle h} \searrow & \downarrow {\scriptstyle f} \\ & & \prod_i (X_i/R_i) \end{array} \qquad h = \textstyle\prod_i p_i. \tag{7.33}$$

Then f is also, by the universal property of a quotient (in 3.4.6(a)).

(b) First, the map $h = \prod_i p_i$ is open, by Exercise 3.5.3(h). Therefore R, which is the equivalence relation of X associated to the open surjective map h, is open.

Finally, if V is open in X/R, then $f(V) = fp(p^{-1}(V)) = h(p^{-1}(V))$ is open in $\prod_i (X_i/R_i)$, which proves that f is open.

Solutions of 3.5.8. (a) The saturation of an open set U of X is the union of all sets $g_X(U)$, for $g \in G$; these sets are open, because each $g_X : X \to X$ is a homeomorphism.

7.3.6 Exercises of Section 3.6

Solutions of 3.6.2. (a) In both cases, the property of limit is a transcription of the definition of continuity at a, of f or its extension g.

(b) Let us fix a homeomorphism $\varphi \colon [0, r[\to [0, +\infty[$ with inverse ψ, and

$\varphi(0) = 0$ (the last point is a consequence, as we will see later, but here we need not bother).

Using the normalisation mapping $N\colon \mathbb{R}^n \setminus \{0\} \to \mathbb{S}^{n-1}$ (defined in (3.42)), the following mappings are inverse homeomorphisms

$$f\colon B_n(0,r) \setminus \{0\} \to \mathbb{R}^n \setminus \{0\}, \qquad f(x) = \varphi(||x||).N(x), \tag{7.34}$$

$$g\colon \mathbb{R}^n \setminus \{0\} \to B_n(0,r) \setminus \{0\}, \qquad g(x) = \psi(||x||).N(x). \tag{7.35}$$

(Each point x moves on its half-line $\{\lambda x \mid \lambda > 0\}$ from the origin, according to φ and ψ.)

Moreover $\lim_{x\to 0} f(x) = 0$, as an easy consequence of the fact that $||f(x)|| = \varphi(||x||)$ tends to 0 when x tends to 0. By the previous point, we can extend f to a continuous mapping $B_n(0,r) \to \mathbb{R}^n$, letting $f(0) = 0$ (with some abuse of notation); the similar extension of g is its inverse.

(c) It is sufficient to work on the cube $]-r,r[^n$. One can proceed as in the previous case, replacing the euclidean norm with the l_∞-norm

$$||-||_\infty\colon \mathbb{R}^n \to \mathbb{R}, \qquad ||x||_\infty = \max|x_i|, \tag{7.36}$$

whose open balls at the origin are the cubes $]-r,r[^n$. The replacement includes a new normalisation mapping, $N_\infty(x) = x/||x||_\infty$ (for $x \neq 0$).

All this will become clearer when we deal with the l_p-distances and norms, in Section 4.5.

Solutions of 3.6.3. (a) Let us suppose that f admits two limits b_i at a (for $i = 1,2$). For every $N_i \in \mathcal{N}(b_i)$ there is some $M_i \in \mathcal{N}(a)$ such that $f(M_i \cap A) \subset N_i$. The neighbourhood $M = M_1 \cap M_2$ of a contains some $x \in A$ and then $f(x) \in N_1 \cap N_2$.

Therefore b_1 and b_2 have no disjoint pair of neighbourhoods in Y, and coincide.

Solutions of 3.6.5. (a) If $x \neq y$ in $\mathbb{R}^n$, let $2r = d(x,y) > 0$. Then the open balls $B_n(x,r)$ and $B_n(y,r)$ are disjoint.

(c) Take $x, y \in X$ with $Gx \neq Gy$ in X/G. We want to prove the existence of two disjoint neighbourhoods of x and y in X, which are saturated for the action of G.

For all $g \in G$ we can find disjoint open neighbourhoods U_g and V_g of x and gy; then the finite intersection $U = \bigcap_g U_g$ is disjoint from each V_g.

Now, each set $g^{-1}V_g$ is an open neighbourhood of y, and $V = \bigcap_g g^{-1}V_g$ is also. Moreover, for every $g \in G$

$$U \cap gV \subset U \cap g(g^{-1}V_g) = U \cap V_g = \emptyset,$$

whence, for all $g, h \in G$, $hU \cap gV = h(U \cap h^{-1}gV) = \emptyset$.

We have proved that the saturated sets $\mathrm{sat}_G(U)$ and $\mathrm{sat}_G(V)$ are disjoint. They are open neighbourhoods of x and y (as already remarked in Exercise 3.5.8(a)).

(d) Trivially, the quotient X/R of any space by the chaotic equivalence relation is a singleton, and T_2. More interestingly, we will see that every space has a universal Hausdorff quotient, in 6.5.1.

On the other hand, the quotient $\mathbb{R}/\mathbb{Q}$ of the real line given by the congruence modulo $\mathbb{Q}$ (where $x \equiv y$ if $x - y \in \mathbb{Q}$), has the chaotic topology: any neighbourhood of x in $\mathbb{R}$ contains an open interval $]x-r, x+r[$, whose saturated set modulo $\mathbb{Q}$ is the whole line.

(e) If X is Hausdorff, all its subspaces X_i are. Conversely, suppose that all X_i are Hausdorff and take two distinct points x, y in X. Either they belong to the same space X_i, and have disjoint neighbourhoods there (preserved in X), or $x \in X_i$ and $y \in X_j$, with $i \neq j$, and we have got the (open) neighbourhoods we are looking for.

Solutions of 3.6.7. (a) Consider the continuous mapping

$$h = (f, g) \colon X \to Y \times Y.$$

Then $\mathrm{Eq}(f, g) = h^{-1}(\Delta_Y)$ is closed in X, by Proposition 3.6.7. The rest is obvious.

(b) Suppose that $A \subset X$ is a retract of the T_2-space X, with retraction $p \colon X \to A$. Then A is the equaliser of $jp \colon X \to X$ and $\mathrm{id}\, X$.

7.4 Exercises of Chapter 4

7.4.1 Exercises of Section 4.1

Solutions of 4.1.6. (a) First, our relation is reflexive, via the constant path 0_x at any point x. Second, it is symmetric, as proved by reversing a path a

$$-a \colon \mathbb{I} \to X, \quad (-a)(t) = a(1-t). \tag{7.37}$$

Third, it is transitive, as proved by *concatenating* a path a from x to y, with a *consecutive* path b from y to z

$$(a+b)(t) = \begin{cases} a(2t), & \text{for } 0 \leqslant t \leqslant 1/2, \\ b(2t-1), & \text{for } 1/2 \leqslant t \leqslant 1. \end{cases}$$

The mapping $a + b$ is well defined and continuous: we are pasting on a finite closed cover of $\mathbb{I}$ two continuos functions which take the same value at 1/2, namely $a(1) = y = b(0)$ (see Lemma 3.4.3(c)).

This 'partial' operation is not associative (when legitimate): it is only so up to homotopy with fixed endpoints. See the construction of the homotopy groupoid of the space X, in [Bro1] or [Bro2].

(b) Given two points x, y in $\mathbb{R}^n$, we can use the $\mathbb{R}$-linear structure of the latter to construct an affine linear mapping

$$a\colon \mathbb{R} \to \mathbb{R}^n, \qquad a(t) = x + t(y - x), \tag{7.38}$$

whose restriction to the interval $\mathbb{I}$ gives a path from x to y.

(c) The first part is as above: our path $a\colon \mathbb{I} \to \mathbb{R}^n$ describes the line segment from x to y. For the second part, use Exercise (a).

(d) The image of any path $a\colon \mathbb{I} \to X$ is a connected subset of X, whence the path-component of x is contained in $\mathrm{Cc}(x)$.

*To construct a connected space which is not path-connected, we start from the graph of a real function

$$A = \{(x, y) \in \mathbb{R}^2 \mid x > 0,\ y = \sin(1/x)\}. \tag{7.39}$$

This subspace $A \subset \mathbb{R}^2$ is path-connected, being the image of the map

$$a\colon]0, +\infty[\to \mathbb{R}^2, \qquad a(x) = (x, \sin(1/x)),$$

defined on a real interval.

Therefore also its closure B in $\mathbb{R}^2$ is connected; one can see that we must add to A a segment of adherent points, in the cartesian axis y

$$B = \overline{A} = A \cup (\{0\} \times [-1, 1]). \tag{7.40}$$

Intuition tells us that B is not path-connected: the new points cannot be linked to the old ones by a path in B. This can be proved. Finally, B has two path-components, of which A is not closed in B.*

(e) Obvious: if $f\colon X \to Y$ is a surjective continuous mapping, any pair of points $y = f(x)$ and $y' = f(x')$ of Y is linked by a path $fa\colon \mathbb{I} \to Y$, where $a\colon \mathbb{I} \to X$ is a path from x to x' in X.

(f) A path $a\colon \mathbb{I} \to X$ with values in product $X = \prod_i X_i$, from $x = (x_i)$ to $y = (y_i)$, amounts to a family of paths $a_i\colon \mathbb{I} \to X_i$ from x_i to y_i.

Solutions of 4.1.7. (b) The idea is to distinguish the presence of the end-points by properties of connectedness.

Let X be a connected space. We say that a point $x \in X$ *disconnects* the space if $X \setminus \{x\}$ is no longer connected. Plainly, given a homeomorphism $f\colon X \to Y$, the point x disconnects X if and only if its image $f(x)$ disconnects Y.

We can say something more: f restricts to a bijection $f'\colon A \to B$ between the subsets of the points which disconnect their space; and of course it also restricts to a bijection between their complements, the subsets of points which do not disconnect the space. (If useful, we can also remark that we have two homeomorphisms between the corresponding subspaces.)

Now, in the interval $]0, 1[$ every point disconnects the space. The interval $[0, 1[$ has precisely one point which does not disconnect it, namely 0. The interval $[0, 1]$ has two such points. We have also shown that the spaces $[0, 1[$ and $[0, 1]$ are not topologically homogeneous (see 3.2.8(c)).

(The property of compactness, or the Extreme Value Theorem, can also be used to show that the homeomorphism type of $[0, 1]$ is distinct from the other two: see the next section.)

(c) We can take $n \geqslant 1$, because the singleton A_0 is already set-theoretically distinguished from all the other infinite sets.

The space A_n is connected; taking out the origin, we get a space with $2n$ components, the half-axes (each of them homeomorphic to the line). Taking out any other point, we get a space with 2 components. If $A_n \cong A_1$, we deduce that $n = 1$ (or A_n would have a point which can disconnect it in more than 2 components). If $A_n \cong A_m$ and $n, m > 1$, we deduce that $2n = 2m$.

Note that we are using 'disconnection' in a finer way than in the previous point, counting the connected components of the disconnected space $X \setminus \{x\}$.

(d) It is sufficient to prove that a space $A(X)$ with Alexandrov topology is connected if and only if the equivalence relation $x \approx y$ in X is chaotic.

If this is the case, the connected component of x, being open and closed (by Theorem 4.1.3(b)), must be upwards and downwards closed in the preordered set, and therefore total.

Conversely, if the relation $x \approx y$ is not chaotic, we can decompose the preordered set X in a disjoint union $A \cup B$ of non-empty subsets where no point of A is comparable to any point of B. But then A and B are closed in the Alexandrov topology.

Solutions of 4.1.8. (a) It is sufficient to note that the open balls $B_n(a, r)$ are (obviously!) starred with respect to their centre a: see Exercise 4.1.6(c).

(It is also true that the open balls $B_n(a, r)$ are convex subsets of $\mathbb{R}^n$; but this would require some computation, even though geometrically obvious.)

(b) By connected induction, in the sense of 4.1.1(g). Let X be locally path-connected, and A the path-component of a point $x \in X$.

The (non-empty) subset A is open: if $y \in A$ and U is a path-connected open neighbourhood of y in X, then U is contained in A (by the transitivity of the relation $\sim_P$). But all the other path-components are open as well, and so is their union: therefore A is also closed. As X is connected, $X = A$.

*(c) The subspace $B \subset \mathbb{R}^2$ is connected, but all the 'small' open neighbourhoods of a point $(0, x) \in B \setminus A$ in $\mathbb{R}^2$ intersect B in disconnected sets. Drawing A (the graph of a C^∞-function) this is graphically obvious.

Solutions of 4.1.9. (a) The euclidean space A (a triangle, if the interval I is bounded) is easily seen to be convex, and therefore connected. We define a function

$$g\colon A \to \mathbb{R}, \qquad g(x_1, x_2) = (f(x_2) - f(x_1))/(x_2 - x_1),$$

which is continuous, as a quotient of continuous functions, of which $x_2 - x_1 > 0$, in A.

If f is injective, $g(x_1, x_2) \neq 0$ on A. But $g(A)$ is a connected subset of $\mathbb{R}$, whence it is an interval non containing 0: either $g(A) \subset]0, +\infty[$ or $g(A) \subset]-\infty, 0[$. In the first case f is strictly increasing, in the second strictly decreasing.

(b) We know that $J = f(I)$ is an interval, by 4.1.4(b) and 4.1.5(a). We also know that f is strictly increasing or strictly decreasing, by the previous point; therefore f is an isomorphism or an anti-isomorphism of totally ordered sets (see 1.4.8(e)).

In the first case, as a consequence of the characterisation of real intervals (Proposition 1.7.3), an isomorphism $f\colon I \to J$ of ordered sets takes intervals of I to intervals of J.

Moreover, it takes intervals open in I to intervals open in J. In fact, an interval of I is open in I if and only if it has no minimum (except possibly the minimum of I) nor maximum (except possibly the maximum of I); these conditions are preserved by an order isomorphism.

Thus f and its inverse function are both open mappings. The same holds for an anti-isomorphism.

(c) The function f is continuous, by Exercise 3.3.6(e). It is easily proved to be strictly increasing on I, and therefore injective. Its image is an interval contained in I, which contains 0 and is upper unbounded, because $x^n \geqslant x$ when $x \geqslant 1$. Finally $f(I) = I$, and we can apply the previous point.

7.4.2 *Exercises of Section 4.2*

Solutions of 4.2.1. (b) The open intervals $]-r, r[$, for $r > 0$, form a totally ordered open cover of $\mathbb{R}$. Any finite number of them is contained in the largest, and cannot cover $\mathbb{R}$.

Similarly, $]a, b[$ is covered by the open intervals $]a+\varepsilon, b-\varepsilon[$ (for $0 < 2\varepsilon < b-a$), and $[a, b[$ by its open subsets $[a, b-\varepsilon[$ (for $0 < \varepsilon < b-a$).

(d) The condition on neighbourhoods is obviously necessary. Suppose that it is satisfied, and take an arbitrary open cover $\mathcal{U}$ of X. For every $x \in X$ we can choose an open set N_x that contains x and belongs to $\mathcal{U}$; there is thus a finite subcover of $(N_x)_{x \in X}$, and a finite subcover of $\mathcal{U}$.

(f) Take an open cover $\mathcal{U}$ of the space X, and let U be a non-empty set belonging to this family (if any). Then its complement in X is finite, and covered by a finite subset of $\mathcal{U}$; we only have to add U to the latter.

Solutions of 4.2.8. (a) For any point $a \in \mathbb{R}^n$, the compact balls $D_n(a, r)$, with $r > 0$, form a fundamental system of compact neighbourhoods.

(b) Take a point $a \in \mathbb{Q}$ and suppose it has a compact neighbourhood K in $\mathbb{Q}$. Then K contains the rational trace of an interval $]a-r, a+r[$ of the line; being closed in $\mathbb{R}$ (as any compact subset) it contains the whole interval, which is a contradiction.

7.4.3 *Exercises of Section 4.3*

Solutions of 4.3.4. (c) We use the stereographic projection of the pierced sphere

$$f\colon \mathbb{S}^n \setminus \{N\} \to Y, \qquad Y = \mathbb{R}^n \times \{-1\} \subset \mathbb{R}^{n+1}, \tag{7.41}$$

from the 'North pole' $N = \{0, ..., 0, 1\} \in \mathbb{R}^{n+1}$ to the hyperplane Y, tangent at the sphere, at the 'South pole' $S = \{0, ..., 0, -1\}$

This mapping is obtained taking, for any point $x \in \mathbb{S}^n \setminus \{N\}$, the half-line from N to x, and letting $f(x)$ be the intersection of this half-line with the hyperplane Y. The reader can accept that f is a homeomorphism, which is geometrically evident in low dimension. (Working out an analytic computation of f and its inverse may be too long to be really interesting.)

By Theorem 4.3.3, $\mathbb{S}^n$ is the Alexandrov compactification of $\mathbb{R}^n$.

7.4.4 Exercises of Section 4.4

Solutions of 4.4.4. (a) Viewing $\mathbb{S}^1$ in the complex field, we consider the surjective continuous mapping

$$f \colon \mathbb{S}^1 \to \mathbb{S}^1, \qquad f(z) = z^2, \tag{7.42}$$

which wraps the circle $\mathbb{S}^1$ on itself, twice. The associated equivalence relation is the congruence modulo the group $T = \{1, -1\}$.

Therefore f induces a bijective continuous mapping $\mathbb{S}^1/T \to \mathbb{S}^1$, which is a homeomorphism – by the usual argument on compact Hausdorff spaces.

(b) We replace f with the mapping $f_n(z) = z^n$, which wraps $\mathbb{S}^1$ on itself n times, and T with the multiplicative group U_n of n-th complex roots of 1 (isomorphic to the additive group $\mathbb{Z}/n$): see 2.7.6.

We get again an induced homeomorphism $\mathbb{S}^1/U_n \to \mathbb{S}^1$.

7.4.5 Exercises of Section 4.5

Solutions of 4.5.0. (a) The first part is as in Exercise 3.1.5(b), the second is obvious.

(c) The Hausdorff property is proved as in Exercise 3.6.5(a). Every point has a countable local basis, e.g. $B_X(a, 1/n)$ (for $n > 0$). The rest is already known: see 3.6.3 and 3.3.5(d), (e).

(f) Suppose that $d(a, b) \geqslant d(a', b')$, so that

$$d(a, b) \leqslant d(a, a') + d(a', b') + d(b', b),$$
$$|d(a, b) - d(a', b')| = d(a, b) - d(a', b') \leqslant d(a, a') + d(b, b').$$

(g) Let X be a semimetric space. If there are two distinct points a, b with $d(a, b) = 0$, then $d(a, x) = d(b, x)$ (for every $x \in X$); therefore the points a, b have the same open balls, and the same neighbourhoods: X cannot even be T_0.

Solutions of 4.5.2. (a) The mapping $f \colon (\mathbb{Z}, d) \to (\mathbb{Z}, D)$ which is the identity on the set $\mathbb{Z}$ is a weak contraction, because $D(m, n) \leqslant d(m, n)$.

Its inverse mapping $g \colon (\mathbb{Z}, D) \to (\mathbb{Z}, d)$ is also uniformly continuous: for every $\varepsilon > 0$, it is sufficient to take $\delta = 1$, so that $D(m, n) < 1$ implies $m = n$ and $d(m, n) = 0 < \varepsilon$.

Solutions of 4.5.3. (a) We only have to check the triangle inequality:

$$\begin{aligned} d_\infty(x,y)+d_\infty(y,z) &= \max{}_i\, d_i(x_i,y_i)+\max{}_i\, d_i(y_i,z_i)\\ &= \max{}_{i,j}\,(d_i(x_i,y_i)+d_j(y_j,z_j))\\ &\geqslant \max{}_i\,(d_i(x_i,y_i)+d_i(y_i,z_i))\\ &\geqslant \max{}_i\,(d_i(x_i,z_i)) = d_\infty(x,z). \end{aligned}$$

(b) The projections $p_i\colon X\to X_i$ are weak contractions for the l_∞-metric on X. Now we have a (finite) family of mappings $f_i\colon Y\to X_i$ (defined on a metric space) and consider the mapping $f\colon Y\to X$ of these components.

- If the f_i are continuous, also f is (easily proved to be).
- If the f_i are uniformly continuous, also f is (easily proved to be).
- If each f_i has a Lipschitz constant L_i, then f admits $L=\max_i L_i$ as a Lipschitz constant.
- If the f_i are weak contractions, so is f

$$d_\infty(f(y),f(y')) = \max{}_i\, d_i(f_i(y),f_i(y')) \leqslant d_Y(y,y').$$

(c) The first point of the previous exercise also holds when Y is a general topological space, and proves that the topology of X associated to the l_∞-metric is the product topology of the spaces X_i.

As to $\mathbb{R}^n$, the euclidean distance is obtained as follows, from the euclidean distance of $\mathbb{R}$, $d(x,y)=|x-y|$

$$d_2(x,y)=\sqrt{\Sigma_i\,(x_i-y_i)^2}=\sqrt{\Sigma_i\, d(x_i,y_i)^2},$$

and this makes evident the inequalities (4.47).

Solutions of 4.5.4. (a) Suppose that the sequence (x_k) converges to x in X. Then for every $\varepsilon>0$ there is some $h\in\mathbb{N}$ such that, for every $n\geqslant h$, $d(x_n,x)<\varepsilon/2$.

Thus, if $m,n\geqslant h$, we have:

$$d(x_m,x_n)\leqslant d(x_m,x)+d(x,x_n)<\varepsilon.$$

(b) Let (x_k) be a Cauchy sequence in X. Taking $\varepsilon=1$, we can choose an $h\in\mathbb{N}$ such that, for every $m,n\geqslant h$, $d(x_m,x_n)<1$ and compare all points x_n to x_h.

Letting $L=\max\{d(x_h,x_n)\mid n\leqslant h\}+1$, we have $d(x_h,x_n)<L$ for all $n\in\mathbb{N}$, and $d(x_m,x_n)\leqslant d(x_m,x_h)+d(x_h,x_n)<2L$ (for all m,n).

(c) Suppose that $f\colon X\to Y$ is uniformly continuous and (x_k) a Cauchy sequence in X. Starting from a real number $\varepsilon>0$

- there is some $\delta>0$ such that, if $d(x,x')<\delta$ then $d(f(x),f(x'))<\varepsilon$,

- there is some $h \in \mathbb{N}$ such that, for every $m, n \geqslant h$, $d(x_m, x_n) < \delta$.

Now, if $m, n \geqslant h$, we conclude that $d(f(x_m), f(x_n)) < \varepsilon$.

(d) Let $f\colon X \to Y$ be a uniform isomorphism, and suppose that the metric space X is complete.

By the previous point, a Cauchy sequence (y_k) in Y is taken by f^{-1} to a Cauchy sequence (x_k) in X. The latter converges to a point x, and is taken back, by f, to the sequence (y_k); therefore (y_k) converges to $f(x)$ in Y.

(e) It is easy to see that, in a finite product of metric spaces, a sequence is Cauchy if and only if all its projections are (in the factor spaces). We already know that a sequence in the product converges if and only if all its projections do.

7.4.6 Exercises of Section 4.6

Solutions of 4.6.2. (a) By 3.5.4(a), the topological product $H \times H$ is a subspace of $G \times G$. The multiplication $m\colon H \times H \to H$ is continuous, as a restriction of the multiplication of G. Similarly, the inversion of H is continuous, as a restriction of the inversion of G.

(b) The multiplication m of G forms a commutative diagram (where m_i is the multiplication of G_i)

$$\begin{array}{ccc} G \times G & \xrightarrow{m} & G \\ {\scriptstyle p_i \times p_i}\downarrow & & \downarrow{\scriptstyle p_i} \\ G_i \times G_i & \xrightarrow[m_i]{} & G_i \end{array} \qquad m((g_i), (h_i)) = (g_i h_i). \tag{7.43}$$

Therefore each component $p_i m$ is continuous, and m is also. The continuity of the inversion of G is proved in a similar way.

(c) First, the quotient space G/H is the same as the orbit space of G modulo the left (or right) action of the group H, whence the projection $p\colon G \to G/H$ is open (by 3.5.8(a)).

The multiplication m' of G' forms a commutative diagram

$$\begin{array}{ccc} G \times G & \xrightarrow{m} & G \\ {\scriptstyle p \times p}\downarrow & & \downarrow{\scriptstyle p} \\ G' \times G' & \xrightarrow[m']{} & G' \end{array} \qquad m'([g], [h]) = [gh]. \tag{7.44}$$

Now the surjective continuous mapping $p \times p$ is open (Exercise 3.5.3(h)), and is thus a topological projection (equivalent to the canonical projection $(G \times G)/(H \times H)$). Therefore the continuity of the composite $m'(p \times p)$

implies the continuity of m'. The continuity of the inversion of G is proved in a similar, even easier way.

The last point follows from 4.6.1(e).

Solutions of 4.6.3. (a) Follows from 3.3.6.

(b) The additive group $\mathbb{C}$ is the topological group $\mathbb{R}^2$. The operations of the multiplicative group $\mathbb{C}^*$ are computed on the topological space $\mathbb{R}^2$ as in 2.7.2

$$\begin{aligned}(a,b).(c,d) &= (ac-bd, bc+ad),\\ (a,b)^{-1} &= (a/(a^2+b^2), -b/(a^2+b^2)).\end{aligned} \tag{7.45}$$

Identifying $\mathbb{R}^2 \times \mathbb{R}^2$ with $\mathbb{R}^4$, the two components $m_k = p_k m\colon \mathbb{R}^2 \times \mathbb{R}^2 \to \mathbb{R}$ of the complex multiplication are polynomials in four real variables

$$m_1(a,b,c,d) = ac - bd, \qquad m_2(a,b,c,d) = bc + ad,$$

therefore they are continuous (by 3.3.6(e)), and so is m.

Similarly the components $i_k = p_k i\colon (\mathbb{R}^2 \setminus \{0,0\}) \to \mathbb{R}$ of the complex inversion are quotients of continuous functions, with a denominator that takes values in $\mathbb{R}^*$

$$i_1(a,b) = a/(a^2+b^2), \qquad i_2(a,b) = -b/(a^2+b^2),$$

therefore they are continuous (by 3.3.6(d)), and so is i.

(c) An obvious consequence of the previous points.

Solutions of 4.6.5. (a) The axioms (N.1, 2) obviously imply the axioms (M.0–2) of 4.5.0, for the distance function $d(x,y) = ||x-y||$. The continuity of the sum and scalar multiplication follows from (N.1) and (N.2), respectively.

(b) Axiom (N.3) is equivalent to (M.3), and we have already seen that the latter is equivalent to the Hausdorff property (in Exercise 4.5.0(g)).

7.5 Exercises of Chapter 5

7.5.1 Exercises of Section 5.2

Solutions of 5.2.5. (a) Reflexivity and symmetry of the homotopy relation are obvious. For transitivity, one pastes two homotopies, from f to f' and from f' to f'', using the closed cover of $X \times \mathbb{I}$ formed by the half-cylinders $X \times [0,1/2]$ and $X \times [1/2,1]$.

For consistency with composition, one composes two homotopies φ, ψ 'along the diagonal': $\vartheta(x,t) = \psi(\varphi(x,t),t)$.

Solutions of 5.2.7. (a) Every category C has a least congruence, namely $f = g$, whose quotient can be identified with C. It also has a coarsest congruence $f \sim g$, saying that f and g are parallel morphisms. The quotient category $\mathsf{C}/\!\sim$ is the (possibly large) set $\mathrm{Ob}\,\mathsf{C}$, preordered by the relation described in the statement.

(c) The cardinal $2 = \{0, 1\}$ is viewed as a discrete category on two objects.

The additive group $(\mathbb{Z}/2, +)$ 'is' a groupoid with one object $*$ and a non-trivial automorphism $i\colon * \to *$, which is involutive: $ii = \mathrm{id}\,(*)$.

The multiplicative monoid $(\mathbb{Z}/2, .)$ 'is' a category with one object $*$ and a non-trivial endomorphism $e\colon * \to *$, which is idempotent: $ee = e$.

7.5.2 *Exercises of Section 5.3*

Solutions of 5.3.5. (a) The homomorphism $\varphi_R\colon R \to R$ of left R-modules is the left multiplication by the scalar $\lambda = \varphi(1)$, which commutes with any other $\mu \in R$, because $\lambda\mu = \varphi(\mu) = \mu\varphi(1) = \mu\lambda$.

For every left R-module A, the naturality of φ on the homomorphism $f\colon R \to A$ that sends 1 to an element $a \in A$ gives the relation:

$$\varphi_A(a) = \varphi_A(f(1)) = f(\varphi_R(1)) = f(\lambda) = \lambda a.$$

Solutions of 5.3.7. For (a), (b), (f) and (g) one can use the characterisation of equivalences, by property 5.3.6(iii).

(c) Use the category T of 5.2.8(a).

(d) The construction is simpler if the functor U is essentially surjective on objects, as in the previous case: then we construct a category C' with the same objects as A and hom-sets $\mathsf{C}'(X,Y) = \mathsf{C}(UX, UY)$. Otherwise one can add to C' the objects of C, with suitable morphisms.

(h) It is sufficient to prove that an equivalence F between two skeletal categories is an isomorphism.

In fact F reflects isomorphisms, because it is full and faithful (see Exercise 5.3.3(c)), and is essentially surjective on objects: it follows that it is bijective on objects and morphisms.

7.5.3 *Exercises of Section 5.4*

Solutions of 5.4.2. (b) The forgetful functor $U\colon G\mathsf{Set} \to \mathsf{Set}$ simply forgets the action. For a set X, we form the G-set $|G| \times X$, with the obvious left action of the group G

$$g.(\gamma, x) = (g\gamma, x) \qquad (g, \gamma \in G,\ x \in X). \tag{7.46}$$

The embedding $\eta\colon X \to U(|G|{\times}X)$ sends x to $(1, x)$. For every mapping $h\colon X \to U(A)$ with values in a G-set, the mapping

$$k\colon |G|{\times}X \to A, \qquad k(\gamma, x) = \gamma.f(x),$$

is a morphism of G-sets, because:

$$k(g(\gamma, x)) = k(g\gamma, x) = g\gamma.h(x) = g.k(\gamma, x).$$

Moreover it extends h, and is the unique G-morphism with this property, as $(\gamma, x) = \gamma.(1, x)$.

We note now that the action of G on $|G| \times X$ is fixed-point free: if $g.(\gamma, x) = (\gamma, x)$ then $g\gamma = \gamma$ and $g = 1$. Conversely, if G acts freely on a set A, we can *choose* a subset $X \subset |A|$ containing a single point in each orbit of A; then the mapping

$$|G|{\times}X \to A, \qquad (\gamma, x) \mapsto \gamma x, \tag{7.47}$$

is obviously bijective, and an isomorphism of G-sets.

We conclude thus that a G-set is free (with respect to the forgetful functor U) if and only if its action is fixed-point free. (We have used the axiom of choice.)

(d) We extend the construction of the polynomial algebra $R[X]$, in 2.5.2, to any set $\mathcal{X}$ of indeterminates.

First we form the free commutative monoid $\mathbb{N}\mathcal{X} = \bigoplus_{X\in\mathcal{X}} \mathbb{N}$ (see the solution of 1.6.2(b)), written *in multiplicative notation*: an element is an essentially finite product $\prod_{X\in\mathcal{X}} X^{n_X}$ of powers of indeterminates, with a quasi null family of natural exponents (n_X); this means a monomial in the indeterminates $X, Y, Z, ... \in \mathcal{X}$, like X^2YZ^3; of course, $XYZXZZ$ is the same monomial.

Then we let $R[\mathcal{X}]$ be the free R-module generated by the set $\mathbb{N}\mathcal{X}$ of these monomials. An element, called a *polynomial* in the indeterminates of $\mathcal{X}$, is a formal linear combination of monomials

$$f = \textstyle\sum_{M\in\mathbb{N}\mathcal{X}} a_M M, \tag{7.48}$$

with quasi null coefficients $a_M \in R$.

The R-module $R[\mathcal{X}]$ becomes a (commutative, unital) R-algebra, by an R-linear extension of the product of monomials

$$fg = \Sigma_{M,N\in\mathbb{N}\mathcal{X}}\, a_M b_N (MN), \tag{7.49}$$

where $g = \Sigma_{N\in\mathbb{N}\mathcal{X}}\, b_N N$.

The composed embedding $\eta\colon \mathcal{X} \to \mathbb{N}\mathcal{X} \to R[\mathcal{X}]$ satisfies the universal property, by combining the universal properties of the two constructions.

(e) As in the previous case, we proceed in two steps, with respect to the forgetful functors $R\mathsf{Alg} \to \mathsf{Mon} \to \mathsf{Set}$.

First we form the free monoid $M(\mathcal{X})$ on the set $\mathcal{X}$, by finite words on this alphabet, or 'non-commutative monomials', where $XYX \neq X^2Y$. Then we get the free unital R-algebra $RM(\mathcal{X})$ on this monoid, by R-linear combinations of these monomials; the result is the unital R-algebra of 'non-commutative polynomials', with variables in $\mathcal{X}$ and coefficients in R.

In particular, for the ring $R = \mathbb{Z}$ of integers, we get the free unital ring on the set $\mathcal{X}$. Its elements are 'non-commutative polynomials', with variables in $\mathcal{X}$ and integral coefficients; here the element $2 + XYX - X^2Y$ cannot be 'simplified'.

(f) Independently of the functor U, there is a unique arrow $\eta\colon \emptyset \to U(\perp)$, and its universal property amounts to saying that $\perp$ is an initial object.

(g) The category Fld cannot have an initial object, because it is disconnected: a field-homomorphism $K \to K'$ can only exist when K and K' have the same characteristic (see 2.2.1(a)).

The same fact shows that the free field on an arbitrary set X cannot exist: if $(K, \eta\colon X \to U(K))$ is a candidate universal arrow, take a field K' of a characteristic different from that of K. Then there exists a mapping $X \to U(K') \neq \emptyset$, and it cannot be extended to a homomorphism $K \to K'$.

(h) We know that the category Fld_0 has an initial object, namely $\mathbb{Q}$, the minimal field of characteristic 0.

The argument for a non-empty set X essentially works as follows. If $x \in X$, take a field K of characteristic 0, containing the generator x; then x is either transcendental or algebraic over the minimal subfield, and these properties are preserved by any homomorphism $K \to K'$. But then x cannot be sent to an arbitrary element of K'.

Solutions of 5.4.3. (b) An object of the free category $\mathsf{C}(\Gamma)$ on the graph Γ is a vertex of the graph. A morphism is a finite family of consecutive arrows in the graph, also viewed as a 'path' in Γ

$$\begin{gathered} f = (f_0, f_1, ..., f_n)\colon X \to Y, \\ \mathrm{Dom}\, f_0 = X, \quad \mathrm{Cod}\, f_{i-1} = \mathrm{Dom}\, f_i, \quad \mathrm{Cod}\, f_n = Y \quad (1 \leqslant i \leqslant n). \end{gathered} \tag{7.50}$$

For every object X we are also adding the 'empty path' 1_X at X. The new morphisms compose by concatenation, as in the free monoid generated by a set.

The embedding $\eta\colon \Gamma \to U\mathsf{C}(\Gamma)$ of graphs is obvious (by unary paths).

Every graph-morphism $F\colon \Gamma \to U\mathsf{C}$ with values in a category can be uniquely extended to a functor $G\colon \mathsf{C}(\Gamma) \to \mathsf{C}$, which acts as F on the objects and takes the morphism (7.50) to the composed map

$$F(f_n).\dots.F(f_0)\colon F(X) \to F(Y).$$

(c) We take the congruence E of semigroups (see 1.5.7) on S generated by relating xy to yx, for all $x, y \in S$. The quotient semigroup S/E is commutative, and the canonical projection $\eta\colon S \to S/E$ is the universal arrow we are looking for.

Congruences of monoids and groups work in the same way (see also 1.5.8), so that the same construction also applies to monoids and groups. However, if S is a group, the quotient S/E is usually (and more effectively) described as below.

(d) The universal arrow from the group G to the embedding $\mathsf{Ab} \to \mathsf{Gp}$ is the canonical projection on the abelianised quotient group G^{ab}

$$\eta\colon G \to G^{\mathrm{ab}} = G/[G,G]. \tag{7.51}$$

The universal property says that every homomorphism $f\colon G \to A$ with values in an abelian group induces a (unique) homomorphism $g\colon G^{\mathrm{ab}} \to A$ (as it takes all commutators of G to 0_A).

If G is a non-commutative *simple* group, then $[G,G]$ contains a non-trivial element and must be G.

Solutions of 5.4.8. (a) A natural transformation $\varphi\colon F \to G\colon \mathsf{C} \to \mathsf{D}$ gives a functor

$$\begin{aligned} &\Phi\colon \mathsf{C}\times\mathbf{2} \to \mathsf{D}, \qquad \Phi(-,0) = F, \quad \Phi(-,1) = G,\\ &\Phi(f\colon X \to X', \iota\colon 0 \to 1) = \varphi(f)\colon F(X) \to G(X'), \end{aligned} \tag{7.52}$$

where $\varphi(f)$ is defined in (5.22). Similarly, we have a functor

$$\begin{aligned} &\Phi'\colon \mathsf{C} \to \mathsf{D}^{\mathbf{2}}, \qquad \Phi'(X) = \varphi_X\colon F(X) \to G(X),\\ &\Phi'(f\colon X \to X') = (Ff, Gf)\colon \varphi_X \to \varphi_{X'}. \end{aligned} \tag{7.53}$$

Both processes are invertible.

(b) A functor $H\colon G \to \mathsf{Set}$ amounts to a set $X = H(*)$ equipped with an action of G, where $gx = H(g)(x)$.

Similarly, a natural transformation $\varphi\colon H \to H'\colon G \to \mathsf{Set}$ amounts to a mapping $\varphi(*)\colon X \to X'$ consistent with the actions of G.

7.5.4 Exercises of Section 5.5

Solutions of 5.5.4. (a) In pOrd and Ord every mono is injective, with the same argument as in Set, based on the ordered singleton.

To prove that epis are surjective in pOrd one can use a two-point set with chaotic preorder (as in (5.48)). In Ord the conclusion is the same, but the proof is more complicated: for a monotone mapping $f\colon X \to Y$ and a point $y_0 \in Y \setminus f(X)$, one can construct an ordered set Y' where y_0 is duplicated.

(b) In Hsd monomorphisms coincide with the injective continuous mappings, with the same proof as in Top, based on the topological singleton.

The cancellation property of maps with a dense image is proved in Exercise 3.6.7(a).

(Our argument for the surjectivity of epimorphisms in Top is based on the indiscrete two-point space, and does not apply here.)

(d) We take two natural transformations $\varphi, \psi\colon H \to K\colon \mathsf{C}' \to \mathsf{C}$ whose codomain is the domain of F. If F is faithful, one can easily see that $F\phi = F\psi$ implies $\phi = \psi$.

To prove the converse it is sufficient to take the singleton category $\mathsf{C}' = \mathbf{1}$, so that the functors H and K are arbitrary objects of C and $\varphi, \psi\colon H \to K$ are arbitrary parallel morphisms of C.

Solutions of 5.5.8. (a), (b) The first point is a rewriting of the definition of a monomorphism; the second is a consequence.

7.6 Exercises of Chapter 6

7.6.1 Exercises of Section 6.1

Solutions of 6.1.4. (d) If $m\colon A \to X$ and $p\colon X \to A$ are such that $pm = \mathrm{id}\,A$, it is easy to see that m is the equaliser of mp and $\mathrm{id}\,X$.

7.6.2 Exercises of Section 6.2

Solutions of 6.2.5. (d) The product in $\mathcal{S}$ of a family (X_i) of sets is easily constructed 'along the equivalence' $R\colon \mathsf{Set}_\bullet \rightleftarrows \mathcal{S} : S$

$$R(\Pi_i\, S(X_i)) = (\Pi_i\,(X_i \cup \{\overline{x}_i\})) \setminus \{(\overline{x}_i)\}, \tag{7.54}$$

first adding a base point $\overline{x}_i \notin X_i$ to each factor, then taking their cartesian product in $\mathsf{Set}_\bullet$ and discarding the base point of the product.

Thus the cardinal of the product in $\mathcal{S}$ of two finite sets of m and n elements is $(m+1)(n+1)-1 = mn+m+n$.

*(g) The proof (also written in [M2], Section V.2, Proposition 3) is simpler than one might expect.

Assuming that C is not a preordered set, there is some hom-set $\mathsf{C}(A,B)$ with at least two distinct arrows $f \neq g$. Take the (small) cartesian power $C = B^J$, where J is the cardinal of $\mathrm{Mor}\,\mathsf{C}$; then the cardinal of the set $\mathsf{C}(A,C) \cong \mathsf{C}(A,B)^J$ is at least 2^J, a contradiction.

Solutions of 6.2.9 (Pullbacks and pushouts). (a) The pullback A is constructed as the equaliser of the maps $fp_0, gp_1 \colon X_0 \times X_1 \to X$. The second claim is plain.

Dually, the pushout in (6.18) is the coequaliser of two obvious maps $X \to X_i \to X_0 + X_1$.

(d) We have seen in (b) that the pullback of a symmetric pair (f,f) can only be a symmetric pair (h,h) when f is mono.

In particular, if $f \colon X \to \top$ takes values in the terminal object, the pullback will be the pair of projections $X \times X \to X$, which only coincide when $X \to \top$ is mono: in Set this means that X has at most one element.

(g) In Set these facts are obvious. In Top and pOrd they are a consequence, because in both categories equalisers and coequalisers are computed as in Set, and equipped with the induced structure.

(i) In Set, the property is easy to prove. In Top our property holds in two main, well-known situations: when the subspaces X_i are both open in X, or both closed: see Lemma 3.4.3.

*(j) See Exercise 3.6.7(a) and its solution.

7.6.3 Exercises of Section 6.3

Solutions of 6.3.4. (d) DX is the set X with the discrete order $x = x'$, while CX is the set X with the indiscrete, or chaotic, preorder: $x, x' \in X$.

Here D has a left adjoint $\pi_0 \colon \mathsf{pOrd} \to \mathsf{Set}$, where $\pi_0(X)$ is the quotient of the preordered set X modulo the equivalence relation spanned by the relation: $x \prec x'$.

(e) DX is now the *discrete* category on the set X (only having one identity arrow 1_x for each $x \in X$) and CX is the *indiscrete* category on X (having precisely one arrow $x \to x'$ for each pair of elements of X).

The functor π_0 takes a category C to its set of connected components, namely the quotient of the set $\mathrm{Ob}\,\mathsf{C}$ modulo the equivalence relation generated by the existence of a morphism $x \to x'$. This quotient is the singleton if C is not empty, and moreover any pair of objects is linked by a chain of morphisms $x_0 \to x_1 \leftarrow x_2 \ldots \to x_n$.

(f) It is the functor $\mathrm{po}\colon \mathsf{Cat} \to \mathsf{pOrd}$, where $\mathrm{po}(\mathsf{S})$ is defined in 5.2.7(a). The unit of the adjunction is the canonical projection $\eta\colon \mathsf{S} \to \mathrm{po}(\mathsf{S})$.

(g) The reader can easily give an analytic proof, proving that a universal cocone $(L, (u_i\colon X_i \to L))$ of the functor $X\colon \mathsf{I} \to \mathsf{C}$ is transformed by a left adjoint F into a cocone $(FL, (Fu_i\colon FX_i \to FL))$ of the functor $FX\colon \mathsf{I} \to \mathsf{D}$, which is also universal.

In fact, if $(Y, (g_i\colon FX_i \to Y))$ is a cocone of FX in D, the adjunction gives a cocone $(GY, (f_i\colon X_i \to GY))$ of X in C; then we have a unique map $f\colon L \to GY$ such that $fu_i = f_i$ (for all $i \in \mathrm{Ob}\,\mathsf{I}$), which corresponds to a unique map $g\colon FL \to Y$ such that $g.Fu_i = g_i$ (for all i).

*As a more formal alternative, one can use the characterisation of colimits as representative objects, in 6.2.2.

We have an adjunction $F \dashv G$, with $F\colon \mathsf{C} \to \mathsf{D}$, and a functor $X\colon \mathsf{I} \to \mathsf{C}$. By 6.2.2 its colimit L represents the covariant functor

$$H = \mathsf{C}^{\mathsf{I}}(X, \Delta(-))\colon \mathsf{C} \to \mathsf{Set},$$

via the morphism $u\colon X \to \Delta L$ in C^{I}. Then FL represents the covariant functor

$$K = \mathsf{D}^{\mathsf{I}}(FX, \Delta(-))\colon \mathsf{D} \to \mathsf{Set},$$

via the morphism $Fu\colon FX \to \Delta FL$, by composing three functorial isomorphisms in the variable Y

$$\begin{aligned}&\mathsf{D}(FL, Y) \to \mathsf{C}(L, GY) \to \mathsf{C}^{\mathsf{I}}(X, \Delta GY)\\ &= \mathsf{C}^{\mathsf{I}}(X, G^{\mathsf{I}}(\Delta Y)) \to \mathsf{D}^{\mathsf{I}}(F^{\mathsf{I}}(X), \Delta Y) = K(Y).\end{aligned}$$

The first comes from the adjunction $F \dashv G$, the second from L representing the functor $H\colon \mathsf{C} \to \mathsf{Set}$, the third from the extended adjunction $F^{\mathsf{I}} \dashv G^{\mathsf{I}}$ (see 6.3.3(d)).*

(h) The functor $D\colon \mathsf{Set} \to \mathsf{Top}$ does not preserve infinite products and cannot have a left adjoint, while C does not preserve (even binary) sums and cannot have a right adjoint.

(i) In each of these cases, the initial object is not preserved. The second assertion follows from Exercises 5.4.4(a), (b).

(j) We have a universal arrow $(R^+, \eta R\colon R \to R^+)$ from the ring R to the embedding $\mathsf{Rng} \to \mathsf{Rng}'$. The ring R^+ is obtained by adding a unit (without considering whether R has a unit or not): it has additive part $\mathbb{Z} \oplus R$, and the following multiplication and unit

$$(h, r)(k, s) = (hk, hs + kr + rs), \qquad \underline{1} = (1, 0).$$

The injection $\eta R \colon R \to R^+$ is obvious: $\eta(r) = (0, r)$; every homomorphism $f \colon R \to S$ with values in a unital ring is uniquely extended to a unital homomorphism $g \colon R^+ \to S$, letting $g(h, r) = f(r) + h1_S$.

The embedding $\mathsf{Rng} \to \mathsf{Rng}'$ cannot have a right adjoint, because it does not preserve the initial object, which is $\mathbb{Z}$ for unital rings and the trivial ring for the general ones.

(In Algebra it is often preferred to define R^+ by 'adding a unit when it is not already there'; this process is idempotent, but is not a functor $\mathsf{Rng}' \to \mathsf{Rng}$.)

Solutions of 6.3.9. (a) The set of cones $\mathsf{Set}^{\mathsf{I}}(\Delta\{*\}, X)$ can be rewritten as $\mathsf{Set}(\{*\}, \mathrm{Lim}X) = \mathrm{Lim}X$, by the adjunction $\Delta \dashv \mathrm{Lim}$.

(b) By 6.3.5(c), Δ has a left adjoint S (resp. a right adjoint P) if and only if C has binary sums (resp. products)

$$\begin{array}{lll} S \colon \mathsf{C}^2 \to \mathsf{C}, & S(X, Y) = X + Y & (S \dashv \Delta), \\ P \colon \mathsf{C}^2 \to \mathsf{C}, & P(X, Y) = X \times Y & (\Delta \dashv P). \end{array} \qquad (7.55)$$

(c) If C is a category of modules, the functors S and P coincide: the direct sum $B(X, Y) = X \oplus Y$ is at the same time

- left adjoint to Δ, with a unit $\eta \colon (X, Y) \to (X \oplus Y, X \oplus Y)$ giving the injections of the sum,
- right adjoint to Δ, with a counit $\varepsilon \colon (X \oplus Y, X \oplus Y) \to (X, Y)$ giving the projections of the product.

(d) In any category C there is a natural isomorphism

$$\mathsf{C}(\Sigma_{i \in I} X, Y) \to \mathsf{C}(X, \Pi_{i \in I} Y),$$

whenever sums and products indexed by the set I exist.

(e) Recall the functor $(-)^+ \colon \mathsf{Rng}' \to \mathsf{Rng}$, left adjoint to the embedding of unital rings (in 6.4.4(j)).

The equivalence we want is realised by the following functors

$$\begin{array}{c} H \colon \mathsf{Rng}' \rightleftarrows \mathsf{Rng}/\mathbb{Z} : K, \\ H(R) = (R^+, p_1 \colon R^+ \to \mathbb{Z}), \qquad K(R, p \colon R \to \mathbb{Z}) = \mathrm{Ker}\, p. \end{array} \qquad (7.56)$$

where $p_1 \colon R^+ \to \mathbb{Z}$ is the first projection.

Starting from a ring R, we have $KH(R) = \mathrm{Ker}\, p_1 \cong R$. The other way round, if we start from a copointed unital ring $(R, p\colon R \to \mathbb{Z})$, we get an isomorphic copointed ring $(\mathrm{Ker}\, p)^+$, with a natural isomorphism

$$R \to (\mathrm{Ker}\, p)^+, \qquad r \mapsto (p(r), r - p(r).1_R).$$

7.6.4 Exercises of Section 6.4

Solutions of 6.4.1. (a) We let x vary in X, and y in Y.

Assuming (i) we prove (ii). The relation $f(x) \leqslant f(x)$ gives $x \leqslant g(f(x))$; moreover, if $x \leqslant g(y)$ then $f(x) \leqslant y$, and $f(x)$ is the required minimum.

Assuming (ii) we prove (iii). The mapping f is easily seen to be increasing. As $f(x)$ belongs to $\{y \in Y \mid x \leqslant g(y)\}$, we get $x \leqslant g(f(x))$; moreover $f(g(y)) = \min\{t \in Y \mid g(y) \leqslant g(t)\} \leqslant y$.

Assuming (iii) we prove (i). From $f(x) \leqslant y$ we get $x \leqslant g(f(x)) \leqslant g(y)$; from $x \leqslant g(y)$ we get $f(x) \leqslant f(g(y)) \leqslant y$.

(b) The right adjoint to the embedding $i\colon \mathbb{Z} \to \mathbb{R}$ is the integral-part function, or *floor* function

$$[-]\colon \mathbb{R} \to \mathbb{Z}, \qquad [x] = \max\{k \in \mathbb{Z} \mid k \leqslant x\}. \tag{7.57}$$

The left adjoint is the *ceiling* function

$$\min\{k \in \mathbb{Z} \mid k \geqslant x\} = -[-x], \tag{7.58}$$

related (here) to the right adjoint by the anti-isomorphism $x \mapsto (-x)$ of the real and integral lines.

(c) An irrational number has no 'best' rational approximation, lower or upper. For instance, the subset $\{x \in \mathbb{Q} \mid x \leqslant \sqrt{2}\}$ has no maximum, while $\{x \in \mathbb{Q} \mid x \geqslant \sqrt{2}\}$ has no minimum.

7.6.5 Exercises of Section 6.5

Solutions of 6.5.4. (a) Let X be a topological space and $x, y \in X$; if $\mathcal{N}(x) = \mathcal{N}(y)$ then $\overline{x} = \overline{y}$ and $x \prec y \prec x$ in the specialisation preorder.

Thus X is T_0 if and only if this preorder is an order.

(c) Let X be a topological space. If $x \prec y$ in the specialisation preorder, then $x \in \overline{y}$ and every neighbourhood of y contains x.

Therefore, X is T_1 if and only if the specialisation preorder is discrete.

(d) The least open neighbourhood of x in $A(\mathbb{R})$ is the interval $\uparrow x = [x, +\infty[$. The open subsets are of the following form

$$[x, +\infty[, \qquad]x, +\infty[, \qquad \emptyset, \qquad \mathbb{R}.$$

References

[AHS] J. Adámek, H. Herrlich, G.E. Strecker, Abstract and concrete categories. The joy of cats, J. Wiley and Sons, 1990.

[BBP] M.A. Bednarczyk, A.M. Borzyszkowski, W. Pawlowski, Generalized congruences–epimorphisms in *Cat*, Theory Appl. Categ. 5 (1999), No. 11, 266–280.

[Bc] J.M. Beck, Triples, algebras and cohomology, Repr. Theory Appl. Categ. 2 (2003).

[Be] M.K. Bennett, Affine and projective geometry, J. Wiley and Sons, 1995.

[Bi] G. Birkhoff, Lattice theory, 3rd ed., Amer. Math. Soc. Coll. Publ. 25, 1973.

[Bo1] F. Borceux, Handbook of categorical algebra 1, Cambridge University Press, 1994.

[Bo2] F. Borceux, Handbook of categorical algebra 2, Cambridge University Press, 1994.

[Bo3] F. Borceux, Handbook of categorical algebra 3, Cambridge University Press, 1994.

[Bou1] N. Bourbaki, Algebra I, Chapters 1–3, Hermann and Addison–Wesley, 1974.

[Bou2] N. Bourbaki, General Topology I, Chapters 1–4, Springer, 1989.

[Bou3] N. Bourbaki, General Topology II, Chapters 5–10, Springer, 1989.

[Bra] H. Brandt, Über eine Verallgemeinerung des Gruppenbegriffes, Math. Ann. 96 (1927), 360–366.

[Bro1] R. Brown, Elements of modern topology, McGraw–Hill, 1968.

[Bro2] R. Brown, Topology and groupoids, Third edition of Elements of modern topology, BookSurge, 2006.

[Bu] D.A. Buchsbaum, Appendix, in: H. Cartan, S. Eilenberg, Homological algebra, Princeton University Press, 1956.

[Car] H. Cartan, Théorie des filtres, C. R. Acad. Sci. Paris, 205 (1937), 595–598.

[ClP] A.H. Clifford, G.B. Preston, The algebraic theory of semigroups, Vol. 1, Math. Surveys of the Amer. Math. Soc. 7, 1961.

[Coh] P.M. Cohn, Universal algebra, Harper and Row, 1965.

[Cor] J.F. Cornwell, Group theory in physics, An introduction, Academic Press, 1997.

[Da] M.M Day, Normed linear spaces, Third edition, Springer, 1973.

[Dg] J. Dugundij, Topology, Allyn and Bacon, 1966.

[Du] J. Duskin, Variations on Beck's tripleability criterion, Lecture Notes in Math. 106, Springer, 1969, pp. 74–129.
[EiM] S. Eilenberg, S. Mac Lane, General theory of natural equivalences, Trans. Amer. Math. Soc. 58 (1945), 231–294.
[Fk] A.A. Fraenkel, Abstract set theory, Third revised edition, North-Holland Publishing Co., 1966.
[Fr1] P. Freyd, Abelian categories, An introduction to the theory of functors, Harper and Row, 1964.
Republished in: Reprints Theory Appl. Categ. 3 (2003).
[Fr2] P. Freyd, On the concreteness of certain categories, in: Symposia Mathematica, Vol. IV, INDAM, 1968/69. Academic Press, 1970, pp. 431–456.
[Fr3] P. Freyd, Homotopy is not concrete, in: The Steenrod algebra and its applications, Battelle Memorial Inst., 1970, Lecture Notes in Math. Vol. 168, Springer, 1970, pp. 25–34.
[Fu] L. Fuchs, Abelian groups, Springer, 2015.
[GaZ] P. Gabriel, M. Zisman, Calculus of fractions and homotopy theory, Springer, 1967.
[G1] M. Grandis, Directed Algebraic Topology, Models of non-reversible worlds, Cambridge University Press, 2009.
Available at: http://www.dima.unige.it/~grandis/BkDAT_page.html
[G2] M. Grandis, Homological Algebra, The interplay of homology with distributive lattices and orthodox semigroups, World Scientific Publishing Co., 2012.
[G3] M. Grandis, Homological Algebra in strongly non-abelian settings, World Scientific Publishing Co., 2013.
[G4] M. Grandis, Category Theory and Applications: A textbook for beginners, World Scientific Publishing Co., 2018.
[GT] M. Grandis, W. Tholen, Natural weak factorisation systems, Archivum Mathematicum (Brno) 42 (2006), 397–408.
[Gr1] G. Grätzer, Universal algebra, Van Nostrand Co., 1968.
[Gr2] G. Grätzer, General lattice theory, Academic Press, 1978.
[Gt] A. Grothendieck, Sur quelques points d'algèbre homologique, Tôhoku Math. J. 9 (1957), 119–221.
[Ha] P.R. Halmos, Naive set theory, Van Nostrand, 1960.
[Hat] A. Hatcher, Algebraic Topology, Cambridge University Press, 2002.
Available at: https://www.math.cornell.edu/~hatcher/AT/ATpage.html
[Hd] F. Hausdorff, Grundzüge der Mengenlehre, Veit, 1914.
[HiW] P.J. Hilton, S. Wylie, Homology theory: An introduction to algebraic topology, Cambridge University Press, 1960.
[Ho] J.M. Howie, An introduction to semigroup theory, Academic Press, 1976.
[Hu] J.F. Humphreys, A course in group theory, Oxford Science Publications, 1996.
[Je] T. Jech, Set theory, Academic Press, 1978.
[Ka] D.M. Kan, Adjoint functors, Trans. Amer. Math. Soc. 87 (1958), 294–329.
[Kap] I. Kaplansky, Set theory and metric spaces, Allyn and Bacon, 1972.
[Ke] J.L. Kelley, General topology, Van Nostrand, 1955.
[Ko1] T.W. Körner, Vectors, pure and applied. A general introduction to linear algebra, Cambridge University Press, 2013.
[Ko2] T.W. Körner, Where Do Numbers Come From? Cambridge University Press, 2019.
[La1] S. Lang, Linear Algebra, Third Edition, Springer, 1987.
[La2] S. Lang, Algebra, Revised third edition, Springer, 2002.

[Law] M.V. Lawson, Inverse semigroups, The theory of partial symmetries, World Scientific Publishing Co., 1998.

[Lw] F.W. Lawvere, Metric spaces, generalized logic and closed categories, Rend. Sem. Mat. Fis. Univ. Milano 43 (1974), 135–166. Republished in: Reprints Theory Appl. Categ. 1 (2002).

[M1] S. Mac Lane, Natural associativity and commutativity, Rice Univ. Studies 49 (1963), 28–46.

[M2] S. Mac Lane, Categories for the working mathematician, Springer, 1971.

[MaM] S. Mac Lane, I. Moerdijk, Sheaves in geometry and logic. A first introduction to topos theory, Springer, 1994.

[Mas1] W.S. Massey, Algebraic topology: An introduction, Harcourt, Brace and World, 1967.

[Mas2] W.S. Massey, Singular homology theory, Springer, 1980.

[Mat] M. Mather, Pull-backs in homotopy theory, Can. J. Math. 28 (1976), 225–263.

[May] J.P. May, Simplicial objects in algebraic topology, Van Nostrand Co., 1967.

[Mit] B. Mitchell, Theory of categories, Academic Press, 1965.

[Mo] K. Morita, Duality for modules and its applications to the theory of rings with minimum condition, Sci. Rep. Tokyo Kyoiku Daigaku, Sect. A 6 (1958), 83–142.

[Mu] J.R. Munkres, Topology: a first course, Prentice Hall, 1975.

[No1] E. Noether, Invariante Variationsprobleme, Nachrichten von der Gesellschaft der Wissenschaften zu Göttingen, Mathematisch-Physikalische Klasse, 1918, pp. 235–257.

[No2] E. Noether, Abstrakter Aufbau der Idealtheorie in algebraischen Zahl- und Funktionenkörpern, Math. Ann. 96 (1927), 26–61.

[Nr] D.G. Northcott, An introduction to homological algebra, Cambridge University Press, 1960.

[Pe] G. Peano, Calcolo geometrico secondo l'Ausdehnungslehre di H. Grassmann, preceduto dalle operazioni della logica deduttiva, Fratelli Bocca Editori, Torino, 1888.

[Pu] D. Puppe, Korrespondenzen in abelschen Kategorien, Math. Ann. 148 (1962), 1–30.

[Ro] J.J. Rotman, An introduction to the theory of groups, Fourth edition, Springer, 1995.

[Se] Z. Semadeni, Banach spaces of continuous functions, Polish Sci. Publ., Warszawa, 1971.

[Sp] E.H. Spanier, Algebraic topology, Mc Graw–Hill, 1966.

[SW] H.H. Schaefer, M.P. Wolff, Topological vector spaces, Second edition, Springer, 1999.

[Ty1] A. Tychonoff, Über die topologische Erweiterung von Räumen, Math. Ann. 44 (1930), 514–561.

[Ty2] A. Tychonoff, Über die Abbildungen bikompakter Räume in Euklidische Räume, Math. Ann. 111 (1935), 760–761.

[Vi] J.W. Vick, Homology theory. An introduction to algebraic topology, Second edition, Springer, 1994.

[Wa] B.L. van der Waerden, Algebra, Vol. I–II, Based in part on lectures by E. Artin and E. Noether, Springer, 1991.

[Wo] W.A. Wooster, Tensors and group theory for the physical properties of crystals, Clarendon Press, 1973.

[ZS] O. Zariski, P. Samuel, Commutative Algebra, Volume I, Van Nostrand, 1958.

Index

$\in$, membership, 17
$\subset$, weak inclusion, 6, 17, 21
$\emptyset$, empty set, 18
$\cup, \cap$, union & intersection, 6, 21
$\setminus$, difference of sets, 6, 21
$\Leftrightarrow, \Rightarrow$, implication, 21
$\sharp$, cardinal of a set, 19, 80
$\downarrow, \uparrow$, down- & up-closed, 47
$\vee, \wedge$, join & meet, 33, 49, 281
$\bot, \top$, minimum & maximum, 45
  initial & terminal object, 268
$\triangleleft$, normality relation, 60, 85

$\mathbf{0}, \mathbf{1}, \mathbf{2}, ...$, ordinal categories, 239
$\cong$, isomorphism relation, 11, 236
$\simeq$, homotopy relation, 237
$\simeq$, equivalence of categories, 248
$\dashv$, relation of adjunction, 286
$\rightarrowtail, \twoheadrightarrow$, mono & epi, 260

(A.1)–(A.9), axioms of fields, 6
(A.10)–(A.16), 15
$\aleph_0$, a cardinal, 80
$\underline{A}_n$, the alternating group, 75
abelian category, 303
abelian group, 31
  Ab, their category, 233
  as a $\mathbb{Z}$-module, 98
abelianised group, 254
absorbing element, 58
(AC), axiom of choice, 79
action
  of a group on a set, 72
  of a group on a space, 176
  of a topological group, 222
adherent point, 145, 152
adjoint functor, 286
  AdjCat, their category, 288
  and comma categories, 293
  between ordered sets, 294
  composition, 287
  existence and uniqueness, 290
  faithful, full, 290
  preserving limits, 289
Adjoint Functor Theorem, 290
affine combination, 127
affine space on a field, 127
algebra on a ring R, 109
  RAlg, their category, 233
  commutative –, 110
    RCAlg, their category, 233
algebraic
  character of a category, 304
  element over a ring, 118
  equation, 116
algebraically closed field, 116
atom in a preordered set, 45
Aut, group of automorphisms
  for a semigroup, 62
  for a topological space, 158

$\mathcal{B}_X(a)$, a set of open balls, 143
Banach space, 224
basis of a module, 104
basis of a topology, 154
bijective mapping, bijection, 19
biproduct, 274, 294
(BO.1–2), 155
boolean algebra, 52
  as a ring, 83
boundary of a subset, 145, 153
bounded lattice, 50
bounded subset, 15, 141, 210
  totally –, 218

C_X, complement in a set, 6
$\mathbb{C}$, the complex field, 131
  as a field of real matrices, 136
  as an algebraic extension, 135
χ, characteristic function, 22
cancellable element, 88
cancellation laws, 8, 31
canonical, 22
  versus natural, 246
canonical factorisation
  of a continuous mapping, 168
  of a homomorphism, 40
  of a mapping, 26
cardinal sets, 19, 80
cartesian product
  in a category, 268
  of abelian groups, 35
  of categories, 237
  of modules, 101
  of ordered structures, 54
  of pointed sets and spaces, 269
  of rings, 85
  of sets, 22
  of topological spaces, 171
casting out nines, 38
Cat, a category of categories, 243
  as a 2-category, 247

category, 234
  additive, 274
  balanced, 260
  comma –, 291
  complete & cocomplete, 277
  concrete, 244
  discrete, 235, 255
  indiscrete, 249, 255
  slice –, 292
  small, large, 235
  well powered, 266
  with small hom-sets, 235
Cauchy sequence, 214
Cauchy–Schwartz inequality, 336
chain in an ordered set, 79
chaotic, *see* indiscrete
characteristic of a ring, 90
$\mathcal{Cl}(X)$, set of closed sets, 144, 149
(cl.1–3), axioms of closure, 146
(Cl.1–2), axioms of closed sets, 145
closed ball, *see* disc
closed mapping, 150
closed subset, 144, 149
closure of a subset, 145, 152
cluster point, 153
Cnt, centre, 63, 84
Cod, codomain of a map, 234
coequaliser, 272
cofinite subset or topology, 151
coherent system of isomorphisms, 245
Cok, cokernel, 302
cokernel pair, 282
commutative diagram, 19, 245
commutators, 64
compact Hausdorff space, 193
  CmpHsd, their category, 305
compact space, 189
  euclidean –, 192
  metric –, 218
  sequentially –, 218

compactification
Alexandrov one-point –, 195, 197
Stone–Čech universal –, 306
compactifications of the plane, 206
complete
category, 277
lattice, 50, 279
metric space, 214, 300
totally ordered field, 15
uniform space, 220
completion of a metric space, 215
concatenation, 70, 352
cone & cocone of a functor, 275
in comma categories, 293
congruence, *see* quotient
conjugate in the complex field, 132
connected component, 184
connected space, 182
continuous mapping, 140, 150
at a point, 152
induced on quotients, 167
restricted on subspaces, 164
convergence, *see* limit
convex subset, 187
coprime numbers, 50
coset, in a quotient, 37
counit, *see* unit
countable set, 80
cover of a space, 164, 189
cyclic groups, 35, 43, 135

$\mathbb{D}^n$, a disc, or closed ball, 148
Δ, symmetric difference, 32, 83
δ_{ij}, Kronecker delta, 122
dense subset, 153
determinant, 124
diagram in a category, 244
difference, 9, 31
dimension of a vector space, 106
direct sum
of abelian groups, 41
of modules, 101, 263
disc, or closed ball, 144
disjoint union of sets, 24
distance, *see* metric
divisibility relation, 46, 50, 89
divisible abelian group, 99
Dom , domain of a map, 234
dual, *see* opposite

empty mapping, empty family, 19
empty set, 18
End, endomorphisms
for a set, 58
for an abelian group, 34
epimorphism, epi, 260
normal –, 303
regular –, 272
equaliser, 269
for topological spaces, 180
equational algebra, 29
equipotent sets, 19, 80
equivalence of categories, 248
equivalence relation, 24
open or closed, 167
essentially finite, 32
euclidean division, 38
of polynomials, 118
euclidean space, 139, 150
its metric and norm, 139
eventually, 159
exponential (complex), 137
extension of a field, 110
Extreme Value Theorem, 192

f_*, f^*, image & preimage, 21, 43, 283, 296
$\mathbb{F}_2$, the 2-element field, 9
$\mathbb{F}_q$, a Galois field, 111
field, 7
Fld, their category, 253
field of fractions, 92, 253

filter, 200
first-countable space, 160
frame, 301
free
  abelian group, 68
  abelian monoid, 71
  category on a graph, 254
  commutative unital algebra, 253
  G-set, 253
  group, 71
  module, 104
  monoid, 70
  object as a left adjoint, 289
  object, on a set, 252
  object, on a weaker object, 253
  unital algebra, 253
    on a monoid, 254
free product of groups, 272
functor, 242
  -category, 258
  additive, 274
  and limits, 277, 280, 289
  contravariant, 242
  diagonal –, 277
  exact, 303
  faithful, full, 244, 263
  forgetful, 243
  representable, 255, 280

G-set, 72, 259
G-space, 176, 259
Galois connection, 294
  AdjOrd, their category, 296
GL, general linear group, 128
graph, 235
group, 58
  Gp, their category, 233
groupoid, 236

Hausdorff space, 178, 298
  Hsd, their category, 233, 298

Heine–Borel Theorem, 192
Heine–Cantor Theorem, 218
Hom
  for abelian groups, 33
  for modules, 101
hom-functor, 255
homeomorphism, 140, 150
homogeneous space, 158
homomorphism
  of abelian groups, 32
  of an algebraic structure, 29
  of fields and rings, 11, 83, 84
  of lattices, 49
  of modules, 99
  of semigroups and groups, 59
hoTop, a category, 237

$\mathbb{I}$, standard interval, 148
$\mathbb{I}^n$, standard cube, 148
ideal in a ring, 84
Idl(R), lattice of ideals, 87
Im, image, 21, 33
increasing, *see* monotone
index of a subgroup, 67
indexed family, 19
induction, 27
  connected –, 183
inf, greatest lower bound, 16, 47
initial object, 253, 271
injective mapping, 19
inner automorphism, 60
integral domain, 88
integral part $[x]$, 17, 369
interior of a subset, 145, 152
Intermediate Value Theorem, 186
interval, 76, 146, 166, 187
Inv, 8, 61
invariant subgroup, 60
inverse mapping, 19
invertible element, 8, 61
invertible matrix, 124

invertible real function, 188
irreducible element, 91
isolated point, 153
isometry, 211
isomorphism
  in a category, 236
  of an algebraic structure, *see* homomorphism
  of categories, 242
  of functors, 247
  of ordered sets, 45
  theorems of –, 40

jointly mono & epi, 260

Ker, kernel, 33, 84, 302
kernel pair, 282
Klein bottle, 208
Klein four-group, 43, 321

(L.1–4), axioms of lattices, 51
l_p-metric and norm, 213, 224
Lagrange's Theorem, 67
lattice, 49
  Lth, their category, 301
  as an equational algebra, 51
  complete, 50, 279
  distributive, modular, 52
  of subgroups, 33
Lebesgue Number Lemma, 219
limit
  of a sequence, 158, 160
  of a mapping, 177, 178
  at infinity, 195
  at a filter, 200
limit & colimit of a functor, 275
  and comma categories, 293
  as adjoints to the diagonal, 290
  by representable functors, 277
  by universal arrows, 277
  in functor categories, 290
  weak –, 275
linear
  combination, 14, 35, 99
  dependence, 106
  differential equation, 129
  group, 128
  mapping, 99
    represented by a matrix, 125
  space, *see* vector space
  system, 121, 123
    homogeneous –, 126
Lipschitz
  -equivalent metrics, 212
  isomorphism, 212
  mapping, 211
locale, 301
locally compact space, 194
locally connected space, 188

(M.0–3), axioms of metrics, 209
mapping, 18
$\mathrm{Mat}_{mn}(R)$, a set of matrices, 122
matrix
  of functions, 129
  on a ring, 121
max & min, 15, 45
maximal element, 79
maximal ideal, 91
metric space, 209
  Mtr, their category, 233
metric, or distance, 209
  topologically equivalent –, 210
minimal subring or subfield, 90
Möbius band, 208
modular arithmetic, 37, 87
module on a ring R, 97
  R Mod, their category, 233
module on the ring $\mathbb{Z}/n$, 98
modulus
  of a complex number, 132
  of a real number, 16
monad, 304

monoid, 57
  Mon, their category, 233
  as a category, 238
monomorphism, mono, 260, 283
  normal –, 302
  regular –, 269
monotone mapping, 45
Mor , set of morphisms, 235
  as a functor, 243
Morita equivalence of rings, 249
multiple, in an abelian group, 34

$\mathbb{N}$, the set of natural numbers, 6
$\mathbb{N}_D$, divisibility lattice, 46, 50, 75
$\mathbb{N}^\bullet$, a compact space, 195
$\mathbb{N}X$, free abelian monoid, 71
$\mathcal{N}_X(a)$, 144, 152
natural transformation, 246
  $\mathrm{Nat}(F, G)$, a set of –, 256
  their compositions, 247
(Nb.1–4), 145, 152
nbd, *see* neighbourhood
neighbourhood, 144, 152
  fundamental system of –, 154
    nested case, 155
  local basis of –, 155
  of a subset, 180
normal subgroup, 60

$\mathcal{O}(X)$, set of open sets, 144, 149
Ob , set of objects, 234
(Op.1–2), axioms of open sets, 145
open ball, 143, 210
open mapping, 150
open subset, 144, 149
opposite
  category, 235, 248
    and adjunctions, 288
  group, 59
  preordered set, 45
  ring, 83

orbit of an action, 72
ordered algebraic structures, 94
ordered set, 44
  Ord, their category, 233
  as an Alexandrov T_0-space, 299
orthogonal group, 128

$\mathcal{P}X$, power set, 21
  as a boolean algebra, 53
  as a complete lattice, 50
  as a ring, 32
$\mathrm{P}^n\mathbb{R}$, projective space, 202
pair of elements, 23
partial mapping, 250
partition of a set, 25
path-connected space, 187
pid, *see* principal ideal domain
pointed set, 31
  $\mathsf{Set}_\bullet$, their category, 233
    as a slice category, 293
pointed topological space, 234
  $\mathsf{Top}_\bullet$, their category, 234
    as a slice category, 293
polar coordinates, 133
polynomial, 111
  function, 113, 114
    as a continuous mapping, 163
  on a field, 119
preordered set, 44
  pOrd, their category, 233
  as a category, 238
  as an Alexandrov space, 156, 299
presheaf, 259
prime element, 91
prime factor decomposition, 27
prime ideal, 91
principal ideal, 85
principal ideal domain, 89
projective real space, 202
pullback & pushout, 281

$\mathbb{Q}$, the rational field, 9
quasi null family, 32
quaternions, 132
Quo(A), 265
quotient
  in a category, 265
  of a category, 237
  of a group, 66
  of a lattice, 56
  of a module, 103
  of a ring, 86
  of a semigroup, 65
  of a set, 25
  of a topological space, 166
  of an abelian group, 36

$\mathbb{R}$, the field of real numbers, 6
  as a complete metric space, 214
  as a totally ordered field, 15
$\overline{\mathbb{R}}$, extended line, 76
  as a compact space, 196
$\mathbb{R}^{\bullet}$, a compact space, 196
$\mathbb{R}^n$, an ordered set, 48, 55
  as a topological space, 150
  as a vector space, 105
RX, free R-module, 104
rational function, 120, 163
retract, 170, 260
ring, 8, 82
  Rng′, their category, 233
    as a slice category, 294
  unital, 8, 82
    Rng, their category, 233
root of a complex number, 134
root of a polynomial, 116
root of a real number, 77, 189

$\underline{S}_n$, finite symmetric group, 63
$\mathbb{S}^1$, standard circle
  as a group, 135, 137
  as a space, 141
$\mathbb{S}^n$, standard sphere, 148, 201, 264
$\mathcal{S}$, category of partial maps, 250
  its limits & colimits, 279
$\sum_i x_i$, essentially finite sum, 32
saturated subset, 166
second-countable space, 161
semigroup, 30, 57
semiring, 96
separable space, 161
sequence, 158, 160
  bounded, 159
  infinitesimal, 158
Set, category of sets, 232
Sierpinski space, 150
simple group, 74
simplicial set, 259
singleton, 18
skeleton of a category, 249
specialisation preorder, 156
sphere, 144
Sub(A), 33, 50, 265
subcategory, 236
  full, wide, 237
  reflective & coreflective, 291
subfield, 10
subgroup, 32, 60
  in $\mathbb{Z}$, 38
sublattice, 49
submodule, 99
submonoid, 60
subobject in a category, 265
subring, 10, 84
subsemigroup, 60
subset, 17
subspace, 163
sum
  in a category, 271
  of monoids and groups, 272
  of pointed sets and spaces, 271
  of sets, 24
  of topological spaces, 175

sup, least upper bound, 16, 47
surface, 208
surjective mapping, 19
Sym(X), symmetric group, 63

T_2-space, *see* Hausdorff
T_0- to T_4-space, 181, 298
terminal object, 268
topological
  dimension, 142, 264
  embedding, projection, 168
  group, 221
  ring, field, 223
  shape, 141
  vector space, 223
topological space, 149
  Top, their category, 233
topology, 149
  Alexandrov –, 156
  box –, 174
  coarser, finer, 151
  cofinite, 151
  discrete, indiscrete, 150, 255
  euclidean, 150, 163
  generated by subsets, 151
  initial, final, 157
  pointless –, 301
torsion abelian group, 43
  tAb, their category, 291
torsion-free abelian group, 43
  tfAb, their category, 291
torus, 142, 207
totally ordered field, 15
transcendental element, 118
transpose matrix, 123
triangle inequality
  of a metric, 209
  of the euclidean distance, 139
  of the modulus, 17

U_n, cyclic subgroup of $\mathbb{C}^*$, 135
uniform isomorphism, 212
uniform space, 219
uniformly continuous map, 211
uniqueness and existence, 252
unit of a monad, 304
unit of an adjunction, 287
universal arrow, 251
  and representability, 258
universal property of
  a categorical (co)limit, 275
  a free object, *see* free
  a product, *see* product
  a quotient space, 167
  a subspace, 164
  a sum, *see* (direct) sum
  making Hausdorff a space, 298
  metric completion, 215
  the abelianised group, 254
  the field of fractions, 92
universe (Grothendieck), 235, 241

variety of algebras, 29
vector space on a field K, 13, 98
  K Vct, their category, 233
vector space on the field $\mathbb{Q}$, 99
vector subspace, 100
(VS.1–8), 13

weak contraction, 211
weak limit & colimit, 275

Yoneda embedding, 259
Yoneda Lemma, 256

$\mathbb{Z}$, the ring of integers, 8
$\mathbb{Z}/n$, additive cyclic group, 38
  as a ring, 88
$\mathbb{Z}X$, free abelian group, 68
zero object, 274
zero-divisor, 89
(ZF), (ZFC), 79
Zorn's Lemma, 79

CPSIA information can be obtained
at www.ICGtesting.com
Printed in the USA
BVHW072109080820
585873BV00007B/21